Palgrave Studies in the History of Science and Technology

Designed to bridge the gap between the history of science and the history of technology, this series publishes the best new work by promising and accomplished authors in both areas. In particular, it offers historical perspectives on issues of current and ongoing concern, provides international and global perspectives on scientific issues, and encourages productive communication between historians and practicing scientists.

More information about this series at
http://www.palgrave.com/gp/series/14581

Bruce B. Campbell

The Radio Hobby, Private Associations, and the Challenge of Modernity in Germany

palgrave
macmillan

Bruce B. Campbell
William & Mary
Williamsburg, VA, USA

Palgrave Studies in the History of Science and Technology
ISBN 978-3-030-26533-5 ISBN 978-3-030-26534-2 (eBook)
https://doi.org/10.1007/978-3-030-26534-2

Cover illustration: imageBROKER / Alamy Stock Photo

This Palgrave Macmillan imprint is published by the registered company Springer Nature Switzerland AG.
The registered company address is: Gewerbestrasse 11, 6330 Cham, Switzerland

Contents

Abbreviations

ARB	*Arbeiter Radio Bund* Worker's Radio Union
BArch	*Bundesarchiv* German Federal Archives
BAMA	*Bundesarchiv-Militärarchiv* Military Branch of the German Federal Archives
BZ	*Britische Zone* British Zone of Occupation
DARC	*Deutscher Amateur-Radio-Club* German Amateur Radio Club
DASD	*Deutscher Amateur- Sende- und Empfangs-Dienst* German Amateur Transmission and Monitoring Service
DDR	*Deutsche Demokratische Republik* German Democratic Republic (East Germany)
DED	*Deutscher Empfangsdienst* German Reception (Monitoring) Service
DFTV	*Deutscher Funktechnischer Verband* German Radio-Technical Union
DSD	*Deutscher Sende Dienst* German Transmission Service
e.V.	*Eingeschriebener Verein* Registered Association (club)
FDGB	*Freier Deutscher Gewerkschaftsbund* Free German Association of Unions
FM	Frequency Modulation
FRB	*Freier Radio-Bund* Free Radio Union

FTV	*Funktechnischer Verein* Radio Technical Association
Gestapo	*Geheime Staatspolizei* Secret State Police
GST	*Gesellschaft für Sport und Technik* Association for Sport and Technology
IARU	International Amateur Radio Union
ITU	International Telecommunications Union
KdT	*Kammer der Technik* Chamber of Technology
MfS	*Ministerium für Staatssicherheit (Stasi)* Ministry for State Security
NSDAP	*Nationalsozialistische Deutsche Arbeiterpartei* National Socialist German Labor Party (the Nazi Party)
NVA	*Nationale Volksarmee* National People's Army
OFV	*Oberdeutscher Funkverband* Upper-German Radio Association
RGBl	*Reichsgesetzblatt* Imperial Legal Gazette
RSHA	*Reichssicherheitshauptamt* Reich Security Main Office
RSV	*Radiosport Verband der DDR* Radio Sports Association of the German Democratic Republic
SA	*Sturmabteilung* The Nazi Brownshirts or SA
SAC	*Samstag Abend Club* Saturday Evening Club
SMAD	*Sowjetische Militäradministration in Deutschland* Soviet Military Administration in Germany
SS	*Schutzstaffel* The Nazi SS
UKW	*Ultra- Kurz- Welle* (VHF or Very High Frequency)
VHF	Very High Frequency
WBRC	*Württemburgischer-Badischer Radioklub* Radio Club of Württemberg and Baden

CHAPTER 1

Introduction

Radio is magic. Voices, sounds from across the globe magically appear in our living rooms and kitchens. In the early twentieth century, this magic was new, revolutionary, and poorly understood. It was the object of scientific investigation, but, more importantly, it was also the domain of tinkerers, "hackers", citizen scientists, and hobbyists. Radio was not only a symbol of modernity, but it was also a site where individuals wrestled with and ultimately came to terms with the new and often frightening wave of inventions, which washed over popular culture and individual lives like a great technological tsunami. Today, radio is so commonplace in industrialized societies that most of us don't even realize that it is there. When have you last seen an automobile without a radio? Our phones, computers, stereos, and even often our toasters and refrigerators are all today radios as well. What in the 1900s was cutting-edge science and by the 1920s and 1930s was the dominant form of mass media is today just one component of a web-like media system which reaches from our ear buds to the internet. We need to look back at the time when radio was new to understand how we got to our current reality, where radio is ubiquitous. It wasn't always this way.

This book shows how radio, then a wild and mysterious technology, was appropriated and mediated by ordinary individuals in Germany in the first half of the twentieth century as a hobby or leisure-time activity. In particular, I look at how private associations, clubs, became the locus of

B. B. Campbell, *The Radio Hobby, Private Associations, and the Challenge of Modernity in Germany*, Palgrave Studies in the History of Science and Technology, https://doi.org/10.1007/978-3-030-26534-2_1

this process. Within these clubs, technology, in the form of radio, intersected with the public sphere. This is a much wider process and really concerns all industrialized societies at the time, but I use Germany as a case study, for reasons I explain below.

If there is a single main point, it is precisely that "it takes a village" of social forms and institutions for humans to interact with technology in a meaningful way. Our engagement with technology is always rooted in social structures, institutions, and frameworks.[1] The radio clubs in Germany provided much of the social context within which individuals could come to grips with radio. Without all of this human "packaging" around the basic technology, there could have been no human interaction with it.

What the Book Is, and Isn't

This book is a history of radio *as a hobby* in Germany, and the private associations within which that hobby was pursued, in the first half (roughly) of the twentieth century. This is not a history of broadcast radio, though broadcast radio plays an important part in the story. It is also not a history of radio technology or equipment or the science of radio or even a history of radio scientists, though these, too, are important elements of the larger context. And while it does tell the story from a German point of view, the history of radio, and particularly of the radio hobby, can't be told without periodic reference to other countries and to the international community of radio hobbyists, which began to develop even before the First World War. There are certainly going to be references back to the US or to Britain, simply because the US was the main point of reference for the rest of the world when it came to radio, while Great Britain was a major point of reference for radio enthusiasts in Europe.

When we study the new mass media of the early twentieth century, we often neglect looking at the social structures (both formal and informal) that grew up around them. The informal social structures that developed to help people cope with/use/assimilate the new technology are particularly important, not least because they are very understudied, certainly com-

[1] My thinking is fundamentally influenced by scholars within the History of Technology community who follow the Social Construction of Technology (SCOT) paradigm, such as Wiebe E. Bijker and Thomas P. Hughes. Hughes' notion of the technical and the social imbedded in a "seamless web" is particularly important. See the end of this introduction for more on the theoretical underpinnings of my analysis.

pared to the official structures of the media or the governmental/legal frameworks within which radio was organized. The story of radio is not simply one of governments, universities, and businesses. Hobbyists and private organizations created to further their interests are a central part of the story. Moreover, individuals engaged with technology in their own ways, and for their own reasons. To be sure, they did so within a specific political/cultural/economic/institutional context, which set certain limits and tended to direct engagement in particular ways. Yet within these limits, individuals (users) engaging with the new technology did not restrict themselves to what politicians, scientists, and industry had imagined, but followed their own paths, making the technology truly their own. While this is true of many, if not all, mass technologies, it was particularly important at the dawn of the radio age, for reasons both scientific and social.

Technological Revolution in the Twentieth Century

The history of the twentieth century is deeply marked by rapid technological progress. It affected then, and still affects today nearly every aspect of society, from sexuality and personal relationships to economics and warfare. No history of modern times is complete without an understanding of the technological context, which has been characterized not least by constant and sometimes violent change. This is certainly true of communications media, which have gone through a series of "revolutions" since the invention of printing with movable type. But technology doesn't exist in a vacuum, independent of human agency. Technology is a human product and, as such, is inextricably embedded in multiple webs of human agency. Communications technology conforms to this paradigm particularly closely, for its fundamental purpose is to allow humans to communicate with one another.[2] Radio is a crucial media to study in this regard. The adoption of radio was extremely rapid, and the medium is also highly strategic.

At the same time, since the seventeenth century, at least in the Western world, the role of independent or spontaneous private associations (clubs) has become central to modern life. Together with new media technologies,

[2] Or, as McLuhan puts it, media (such as communications technology) extend the reach of humans. See: Marshall McLuhan, *Understanding Media: the Extensions of Man (New York: McGraw Hill, 1964).*

private associations have brought about a "public sphere", independent from government and organized religion, which characterizes modern life.[3] This remains the case today even in the most oppressive dictatorship, for even there, huge effort must be expanded to control and suppress the development of an independent public sphere, a task made increasingly difficult as technology empowers individuals to better communicate with one another. When (mass) technology and the public sphere collide, things happen.

The importance of the topic is huge. Though radio was not the first of our modern technological revolutions, and though it might seem somewhat antiquated today in our era of internet streaming and high-definition images coming at us from all directions, it is still ubiquitous—remember, your smartphone is nothing more than a radio grafted onto a computer. Moreover, it was a quintessentially modern technology, not least because almost immediately after it was discovered, a web of hobbyists and enthusiasts grew up around it—hobbyists who often made central contributions to the science and technology itself.[4] Radio is so very modern because it is an object of play (a hobby) as well as a tool for work.[5]

A great wave of new technologies and mass medias rolled across the world in the first half of the twentieth century, and invaded nearly every facet of life, in the industrialized countries, and then, increasingly, beyond their borders. Some will argue that the sheer amount of new technology with which people were confronted was greater at the beginning of the Industrial Revolution, while others will argue that it is greater—and more

[3] See, for example, Jürgen Habermas, *The Structural Transformation of the Public Sphere* (Cambridge, MA: MIT Press, 1989).

[4] The contribution of amateurs and hobbyists to the development of radio as both science and technology has been widely touted not least by the hobbyists themselves, who in this way participate and shape directly the trajectory of the technology. See, for example, Clinton B. Desoto, *200 Meters and Down: The Story of Amateur Radio* (Newington CT: American Radio Relay League, 1985). I am also arguing that hobbies or leisure-time activities are a particular component of modernity itself.

[5] I understand that this is a rather idiosyncratic and certainly incomplete definition of modernity. I see it linked with more conventionally understood notions of the modern autonomous self. It is also linked to "modern" economic conditions, which at least in developed countries produce a surplus of resources and free time which allows hobbies or free-time activities not only to take place but to become part of the definition of what being modern means. Benjamin's *flâneur* needs free time and a certain economic well-being in order to *flâner*. See Walter Benjamin, *Das Passagenwerk*, in: *Gesammelte Schriften*, vols. V, 1–2 (Frankfurt: Suhrkamp Verlag, 1991). For a critical view of Benjamin, see: Martina Lauster, "Walter Benjamin's Myth of the 'Flâneur' ", *The Modern Language Review* 102, no. 1 (January 2007):139–156.

complex—today. Yet the first half of the twentieth century saw not only (or not so much) the introduction of new technologies but, more importantly, their mass adoption. Technology, technological change, and concomitant social change became not only a central characteristic of modern life but also something which, on the whole, generated real enthusiasm.

It also generated a nearly equal amount of fear, or at least discomfort and concern. The pace of life seemed to quicken, certainty seemed destroyed (even in the solidity of the physical world), and people came to feel that they were the pawns of huge forces they could not control. They also feared that the change brought by new technology would cause them to be left behind. But this book argues that many sought to address their fear by embracing the technology. To understand the new technologies, even to control them, many people sought to know them. Knowledge, not least of complex technical systems, really is power. By the early twentieth century, people no longer sought this knowledge within the bounds of existing institutions such as universities or churches—at least not exclusively so. Instead, they very literally brought the mystifying and potentially scary science of radio into their own homes, domesticating it by putting it on the kitchen table and tinkering with it until it revealed all its secrets.[6]

[6]This process is beautifully described by Steve Jobs, in an interview with Daniel Morrow of the Smithsonian Institution in 1995: "There was a man who moved in down the street, maybe about six or seven houses down the block who was new in the neighborhood with his wife, and it turned out that he was an engineer at Hewlett-Packard and a ham radio operator and really into electronics. (…) I got to know this man, whose name was Larry Lang, and he taught me a lot of electronics. He was great. He used to build Heathkits. Heathkits were really great. Heathkits were these products that you would buy in kit form. You actually paid more money for them than if you just went and bought the finished product if it was available. These Heathkits would come with these detailed manuals about how to put this thing together and all the parts would be laid out in a certain way and color coded. You'd actually build this thing yourself. I would say that this gave one several things. It gave one a understanding of what was inside a finished product and how it worked because it would include a theory of operation but maybe even more importantly it gave one the sense that one could build the things that one saw around oneself in the universe. These things were not mysteries anymore. I mean you looked at a television set you would think that 'I haven't built one of those but I could. There's one of those in the Heathkit catalog and I've built two other Heathkits so I could build that'. Things became much more clear that they were the results of human creation not these magical things that just appeared in one's environment that one had no knowledge of their interiors. It gave a tremendous level of self-confidence that through exploration and learning one could understand seemingly very complex things in one's environment. My childhood was very fortunate in that way". Oral History interview with Steve Jobs, 1995. Accessed July 24, 2017. http://americanhistory.si.edu/comphist/sj1.html

Moreover, radio enthusiasts formed new institutions—or, more properly, adapted existing ones—to be able to pursue their new hobby in the company of like-minded others. Radio and other technological hobbies thus quickly joined gardening, sewing, singing, and sports as the focus of private clubs or associations, even as it simultaneously became the purpose of countless new businesses, and found another major social form in radio broadcasting.

The "Radio Hobby"

The focus of this book is what I call the "radio hobby" community in Germany. That requires some definition because we conceive radio differently today. By the term "radio hobby" I mean all those who occupied themselves with radio in an active manner as a free-time activity. Those who simply listened passively to broadcast radio—surely the largest group in German society at the time—are not included, though all those who are included did also listen to radio broadcasts. But to be a hobby, a person's engagement with radio had to have an active component.

What I call "the radio hobby" spanned a diverse but very large group of radio enthusiasts, ranging from do-it-yourselfers (who simply wished to build a usable receiver, since commercially manufactured radios in Germany remained very expensive until 1933–1934), to those whom we would today call "makers" (seeking to improve on existing technology and design) to "hi-fi enthusiasts", all the way to the small but vibrant network of "ham" radio operators who transmitted their own messages as part of an international hobby community. What this diverse group has in common is an active, hands-on, and constructive yet playful (hobby) engagement with the technology and medium of radio. Some of these people certainly hoped to turn their hobby into a profession. For many others, it was both a profession and a hobby; they worked in the radio or electronics industry, but they were also active hobbyists in their free time. Some few were even inventors or researchers at the forefront of radio science. The distinction between these categories is, and was, very fluid, not least because, in Germany, all of these interests were accommodated within the framework of the same radio hobby clubs.

Today, if we can imagine radio as a hobby at all, what springs to mind is limited to only amateur or "ham" radio, then, as now, a fairly small group. Few others, today, actively tinker or play with the technical components of radios for fun. We need to remember that things in the first half of the twentieth century were different. The radio hobby was new, and quite diverse. It was also quite large: added together, the radio hobby in Germany encompassed a significant number of people: I estimate that there were well over 300,000 radio hobbyists in Germany in 1930, organized into local, regional, national, and even international associations and clubs.[7] When we think of the richness of the radio hobby in the 1920s and 1930s and 1940s, we should keep in mind the more recent history of another technological hobby—personal computers. We all know the story of how three hobbyists founded Apple Computer in their garage.[8] This teaches us that the borders between hobbyists, makers, inventors, and science are still very fluid, indeed. We need to keep this sense of wonder and possibility when we think about early radio.

The "Radio Revolution" of the Early Twentieth Century

In 1895, exactly one person in the entire world, Guglielmo Marconi, owned a functioning radio transmitter and receiver. He (together with his assistants) was the only listener, and the only transmitter. He could only hear himself (or his assistants), and only in Morse code.[9] By 1930, even in the depths of the Great Depression, close to two-thirds of all US households owned a radio. In Germany, the development was a bit slower (not least because the standard of living was a bit lower), but even so, roughly one-third of all households owned a radio by then, and radio ownership was growing rapidly, so that even in Germany, by 1938, the figure was two-thirds of all households. By 1930, nearly all sea-going vessels worldwide were equipped with radio transceivers, and most coastal vessels had at least a receiver for weather broadcasts. The world's navies were thoroughly equipped with radios by then, and the world's armies were on their

[7] In 1930, there were between 62 million and 65 million inhabitants in Germany.

[8] See Michael Swaine and Paul Freiberger, *Fire in the Valley: the Birth and Death of the Personal Computer* (Raleigh, NC: Pragmatic Bookshelf, Third Edition 2014), chapter 7 and Andy Herzfeld, *Revolution in the Valley: the Insanely Great Story of How the Mac was Made* (Sebastopol, CA: O'Reilly Media, 2004).

[9] This is, of course, not the full story, even if it corresponds to our understanding of the history of both radio technology and radio pioneers, and to our romantic image of the Promethean inventor. The history of radio is more complex, and Marconi was certainly not alone. See particularly Marc Raboy *Marconi: The Man Who Networked the World* (Oxford and New York: Oxford University Press, 2016).

way. Radio was an immensely useful technology, yet the sheer utility of radio does not alone explain its explosive growth in the first half of the twentieth century. To go from one radio to millions in just 40 years, hundreds of millions of people had to learn how to use the new technology. The new medium was so interesting, so loaded with promise (and hype), that many even developed what at the time was called "radio fever" or "radio madness", a burning interest and enthusiasm for the new technology bordering on an obsession. But remember that mass enthusiasm for technology (of all kinds) is as characteristic of the twentieth century as are world wars and genocide. For ordinary people, radio was the future. It was at the cutting edge of science and modernity, and it created a massive social movement the likes of which had never before been seen; a movement which was the direct ancestor of our modern enthusiasm for computers and the internet.[10]

Nothing about the new tool predestined it for success, and the social forms of its use were in no way pre-destined by the technology itself. No one could have predicted that an obscure and ill-understood natural phenomenon (the ability of waves of energy to carry information over long distances without physical connection) would become the object of mass popular enthusiasm and create profound social change. Yet in the 1920s and 1930s, millions of individuals—probably several hundred thousand in Germany alone—put radio at the center of their *private* interests. Some made it a business, others made it a career, and many more made it into a *hobby*, a new sort of mass free-time activity characteristic of industrial societies. For many, the frightening, exciting, and complex technology of radio became an obsession. For them and countless others, it also was fun.

If the technology was new, the social forms it took at first were mostly old, borrowed from existing social models and simply used in new ways. Businesses already in the electrical industry soon came to build radios, as well,[11] and even the countless small radio start-ups all took over the existing

[10] There were, of course, other iconic technologies at the time, such as the airplane, the luxury liner, the automobile, and the zeppelin. See, for example, Peter Fritzsche, *A Nation of Fliers: German Aviation and the Popular Imagination* (Cambridge: Harvard University Press, 1992).

[11] For example, AEG and Siemens formed a joint venture, Telefunken, to produce radios for the German government. See: Michael Friedewald, *Die "Tönenden Funken". Geschichte eines frühen drahtlosen Kommunikationssystems, 1905–1914* (Aachener Beitrage zur Wissenschafts- und Technikgeschichte des 20. Jahrhunderts 2) (Berlin: Diepholz/GNT Verlag, 1999).

company model within a thoroughly capitalist economy (except in the USSR!). Broadcast radio was quickly taken over by other companies, even if their original purpose was to sell furniture or run a hotel, or, in the case of many European countries, by the state. In Germany, it was introduced and jealously watched over by the state Postal Ministry, which argued that the Imperial Telegraph Law of 1892[12] gave it the sole legal authority to supervise the new technology. Radio became a state monopoly, just as telegraph before it, pressing the new media into the well-known form of bureaucratic state monopoly. And the hundreds of thousands of radio hobbyists in Germany and around the world (France, Great Britain, Austria, the US, Japan, etc.) formed clubs structured just like existing ones for other hobbies and interests. People who were busy inventing radio didn't have time to re-create the (social) wheel. That happened more gradually.

With radio, even hobbyists were also inventors and scientists. Society already had models for this, as well. If, by the early twentieth century, the locus of invention had shifted from the basement or cowshed to the laboratory and university, earlier, more individualistic models still held sway over the popular imagination, such as the dilettante scientist of the eighteenth century and the entrepreneur-inventor of the nineteenth century. Hadn't Marconi himself combined all of these models? Equally as important for the popular absorption of radio, by the early twentieth century, basic education was universal in developed societies, and increasingly available in others. This meant that when radio (just like the automobile or earlier, the camera) became accessible as mass hobbies, it would be approached as science, rather than religion or the supernatural, and even individuals without deep scientific education would seek to learn and advance the technology systematically through experimentation and observation within the mental framework of existing scientific information and models. This phenomenon has continued to play a major role in the popular imagination and in our social appropriation of technology until today, when the "citizen scientist" and the "maker" are more celebrated than ever, and public libraries and universities are rushing to establish public "maker spaces".[13] All of this represents coping with new technology

[12] "Gesetz über das Telegrafenwesen des Deutschen Reiches" of April 6, 1892, RGBl p. 467 and the "Gesetz zur Abänderung des Telegrafengesetzes" of March 7, 1908, RGBL, p. 79.

[13] Henry Sauermann and Chiara Franzoni, "Crowd Science User Contribution Patterns and Their Implications", *Proceedings of the National Academy of Sciences* (PNAS) 112, no. 3 (January 20, 2015): 679–684. My own university, the College of William and Mary,

and with technological change. It also represents the adaptation of society and social forms to use new technology.

In the case of radio, the amateurs, hobbyists, and entrepreneurs in the early twentieth century had a real chance to contribute to both the basic science of the technology and the specific design of radio components for several reasons. First, the science of radio was not fully understood at the beginning; Marconi could not completely explain why his radios worked as they did. Moreover, in grappling with the new science of radio, the professional scientists early on made a key error in assuming that longer wavelengths would travel (propagate) over longer distances than shorter wavelengths. In fact, due to environmental factors that reflect and bounce radio waves around, the opposite is true (at least until one descends to extremely small wavelengths, which again tend to travel only in straight lines).[14] The result was that when the first mix of practice and legislation began to divide the radio spectrum into blocks of frequencies to be set aside for particular uses, business and government in the US and elsewhere were willing to abandon certain frequencies to the hobbyists. They did so in the mistaken belief that they were not good for much in commercial or strategic terms because they simply did not understand how radio waves traveled or propagated. This gave hobbyists a chance to contribute empirical observations to extend and elaborate the basic science.

Since the propagation of radio waves turns out to be rather complicated, and to involve natural phenomena which were also either unknown or ill-understood when radio was first invented, figuring it all out required very broad-based and detailed observations on a mass scale.[15] The scientific community was simply too small to carry out this sort of mass observation alone, but the radio hobby community was willing to contribute

now has several maker spaces. Carly Martin, "Building the BioMaker Space", *The Flat Hat*, October 12, 2016. Accessed May 4, 2018. (http://flathatnews.com/2016/10/12/building-the-biomakerspace/

[14] On the science of radio propagation, see: Henry L. Bertoni, *Radio propagation for Modern Wireless Systems* (Upper Saddle River, NJ: Prentice Hall, 2000) or Eric P. Nichols, *Propagation and Radio Science* (Newington, CT: American Radio Relay League, 2015).

[15] The travel or propagation of radio waves is influenced by (among other factors), the frequency of the wave, the height and shape of the antenna, the quality of the electrical ground of both the antenna and transmitter, the immediate surrounding geography (hills, buildings), meteorological conditions, external electromagnetic radiation (from the sun, etc.), and the configuration of one or more of the earth's atmospheric layers. See footnote 14 above.

enthusiastically. The sheer complexity of radio wave propagation opened the door for the collaboration and contribution of hobbyists. And as it turned out, the "useless" frequencies set aside for hobbyists included that part of the radio spectrum we today call "short waves", and which turned out to be the frequencies most useful for long distance communications. By the time government and industry caught up, amateurs represented the single largest group with experience actually using the shortwave portion of the spectrum. This left the hobbyists in a relatively powerful position.

Characteristically, this empirical observation of propagation not only made the hobbyists real partners of the scientists, but quickly led to the development of a social practice, the sending of reception reports or "QSL cards". In the early days of radio, not only hobbyists who might also transmit, but all sorts of simple radio listeners sent in reports to transmitters to document the fact that they had heard their transmission, in what quality, and under which conditions. This practice is still at the heart of "ham radio" and thus much of the remaining radio hobby today. Part of the enthusiasm of radio for hobbyists was the possibility of making a real discovery, possibly even a financially lucrative one, and if not, at least being able to feel that they were helping to define the technology and thus control it, as opposed to being simply its passive object. Sending in random reception reports allowed any listener to feel that they were an active contributor and was the gateway to more organized, systematic, and productive observations. This feeling of being an inside participant instead of a passive consumer was also at the root of the early enthusiasm of radio hobbyists for the many highly mediated scientific expeditions characteristic of the 1920s through the 1950s. In these cases, the thrill of having privileged access to events by being able to hear transmissions directly from the explorers combined with the scientific aspect.[16] Radio brought its listeners a feeling of having access to both the wider world and the practice of science, and a sense of being an active participant in both. This was both exciting and empowering.

[16] For a good example of the enthusiasm for scientific expeditions in the 1920s and 1930s, think of the iconic movie "King Kong" (RKO Pictures, 1933), or the vogue for ancient Egypt after the discovery of the tomb of "King Tut" in 1922 by the archaeologist Howard Carter. See Chap. 3 below.

Why Germany?

Germany is a good site for this investigation. Germany was and is a major industrial power, whose history has deviated slightly but in important ways from the paradigmatic histories of the liberal democracies—Britain, France, and the US.[17] This is true even when it comes to the history of radio. No technology, and certainly not one as strategic and as potentially lucrative as radio, can ever exist in a vacuum independent of other parts of the culture, and certainly never divorced from political or economic realities. Radio science and technology developed in Germany in much the same ways as in other contemporary industrial powers, but the social and legal frameworks into which the medium of radio was pressed were often quite different. Moreover, Germany has a very strong and highly developed network of private associations that has historically been a central part of the public sphere, and which were immediately drawn into the new media by both popular interest and law. Because of this, to study the hobby culture of radio in Germany is also to examine the more general history of private associations and clubs during the first half of the twentieth century, and thus to investigate a very important and vibrant part of German society. In German, such clubs and associations are called *Vereine* and are a cornerstone of social life, as well as being an object of study in their own right. Surprisingly, with a few exceptions, this history of private associations in Germany in the twentieth century has been largely left up to the associations themselves and thus rarely meets academic standards. One purpose of this book is therefore to shed light on these associations, as a major reflection of the public sphere. Moreover, as an aside to any fellow historians who may be reading this book, these private associations left organized bodies of records, which thus allow them and their activities to be studied. It would be much more difficult to track the individual engagement with radio in a society which has less formal organization to the public sphere.

Germany is an important location for an investigation of the intersection of technology and everyday life for another reason. Most existing studies of this intersection look at the history of technology in the US or Great Britain. These were and are world-leading industrial powers, but

[17] On the *Sonderweg*, see: Helga Grebing, *Der "deutsche Sonderweg" in Europa 1806–1945: Eine Kritik* (Stuttgart: W. Kohlhammer, 1986) and David Blackbourn and Geoff Eley, *The Peculiarities of German History: Bourgeois Society and Politics in Nineteenth Century Germany* (Oxford: Oxford University Press, 1984).

they are also politically two of the most stable societies in the world, whose forms of government did not change much at all during the entire process of industrialization. Rather than being the model, this makes the US or Britain outliers or exceptions in many ways. Germany is different, quite different.[18] Industrialization began in Germany before it was a united nation, at a time when there were many independent and competing principalities, kingdoms, and city-states. Germany became a united empire only in 1871, albeit one dominated by Prussia, and in a way that deliberately excluded Austria. In 1918, after losing the First World War, the Empire gave way to the Weimar Republic. It, in turn, was captured in 1933 by the Nazis, and Germany was a fascist dictatorship until 1945. After the end of the Second World War, Germany as a country ceased to exist, and it was replaced by four zones of occupation, each administered by one of the victorious Allies: Britain, France, the US, and the Soviet Union. When independence was regained in 1949, there were two German states—one socialist, the other capitalist—a seemingly immutable state of affairs—until it all changed once again with the end of the Cold War and German (re-)Unification in 1990. Over the time period covered by this book, Germans experienced four different governments, plus foreign occupation, and lost two world wars. And yet over the same period, Germany, certainly with ups and downs, remained one of the world's leading industrial powers and scientific innovators. This fact alone makes German history particularly interesting and allows this book to investigate how hobby culture and technological change interacted with changing and wildly different political contexts.

The time period covered by this book, roughly 1920–1955, corresponds to the birth and adolescence of radio as a technology, a medium of culture and form of communications, and as a hobby. It is also precisely the time during which the influx of new technologies, medias, and "modernity", in general became most pressing—and often oppressing for ordinary citizens in the industrialized world, just as it also opened up great advantages and provoked impressive social change. It is simply the most exciting period of social and technological change ever seen in the West.

[18] For basic works on German history in the nineteenth, twentieth, and twenty-first centuries, see: Gordon A. Craig, *Germany, 1866–1945* (The Oxford History of Modern Europe) (Oxford: Oxford University Press, 1978) or Dietrich Orlow, *A History of Modern Germany, 1871 to the Present* (7th. ed.) (New York: Routledge, 2016). See also, with a slightly different focus, Hans Ulrich Wehler, *Deutsche Gesellschaftsgeschichte, 1700–1990* (C.H. Beck, 2008).

As the writer of this book, I bring a particular skill set to my research and interpretation. I am an historian of modern Germany, who specializes in the first half of the twentieth century. I have always mixed political, cultural, and social history. I have previously worked on right-wing paramilitary groups,[19] and particularly the National Socialist *Sturmabteilung* (SA).[20] As an historian, I have also done research into the political use of culture in Weimar and the German Youth Movement.[21] However, as much as I self-identify as an historian, for the last 20 years, I have held an academic position in German Studies within a Department of Modern Languages and Literatures. As such, I have become a teacher and practitioner of Cultural Studies, and I have published on German detective fiction[22] and radio,[23]

[19] Bruce Campbell and Arthur Brenner, eds., *Death Squads in Global Perspective: Murder with Deniability* (New York: St. Martin's Press, hardcover, 2000, 2002).

[20] Bruce B. Campbell, *The SA Generals and the Rise of Nazism* (Lexington: University Press of Kentucky, 1998, 2004). See also: Bruce B. Campbell, "The SA in the *Gleichschaltung*: Stress, Organization and Violence", in *From Weimar to Hitler: Studies on the Dissolution of Weimar Democracy and the Establishment of the Third Reich, 1932–34*, ed. Larry Eugene Jones and Hermann Beck (New York: Berghahn, 2018) 194–221; Bruce B. Campbell, "Autobiographies of Violence: The SA in its own Words", in: *Central European History*, special issue on the SA, 46, No. 2 (June 2013), 217–237; Bruce B. Campbell, "Gewalt bis in die obersten Ränge. Die Höheren SA-Führer der SA-Gruppe Berlin-Brandenburg", in: *Der SA-Terror als Herrschaftssicherung. "Köpinecker Blutwoche" und Öffentliche Gewalt im frühen Nationalsozialismus* ed. Stefan Hördler (Berlin: Metropol Verlag, 2013), 62–82; Bruce B. Campbell, "The SA After the Röhm Purge and the Institutionalized State of Emergency in the Third Reich", in: *Journal of Contemporary History* 28, No. 4 (October 1993), 659–674.

[21] Bruce B. Campbell, " '*Kein schöner Land*': The Spielschar Ekkehard and the Struggle to Define German National identity in the Weimar Republic", in *Music and German National Identity*, ed. Pamela Potter and Celia Applegate (Chicago: University of Chicago Press, 2002), 128–139; Bruce B. Campbell, "Gerhard Roßbach, the Spielschar Ekkehard, and the Cultural Attack on the Weimar Republic", in *Weimar 1930. Politik und Kultur im Vorfeld der NS-Diktatur*, ed. Lothar Ehrlich and Juergen John (Weimar, Böhlau, 1997), 243–259; Bruce B. Campbell, "The Schilljugend from Wehrjugend to Luftschutz: Bündisch Youth in the Weimar Republic in Search of a Political Voice", in *Politische Jugend in der Weimarer Republik*, ed. Wolfgang R. Krabbe (Bochum, Universitätsverlag Dr. Brockmeyer, 1993), 183–201.

[22] Bruce B. Campbell, Alison Guenther-Pal and Vibeke Petersen (eds.) *Detectives, Dystopias and Poplit: Studies in Modern German Genre Fiction* (Elizabethtown, NY: Camden House, 2013). See especially my essay in this volume: "The Detective Was the Killer: the Memory of the Nazi Past in Modern German Detective Fiction" pp. 133–151.

[23] "The Radio Hobby, Government, and the Discourse of Catastrophe", in Susan Merrill Squier, (ed.) *Communities of the Air: Radio Century, Radio Culture* (Durham: Duke University Press, 2003), 63–75.

and taught widely on cultural and literary themes. In short, I come at this current project from two different directions—history and cultural studies. I am good at both the close reading of texts and the consideration of historical context. I have very broad knowledge of the time period and place, and am deeply immersed in the language. All these skills and more have been brought to bear here on radio. They also explain the sometimes idiosyncratic focus of the book: I am not your average historian of technology—or anything else. This is true of every author, but because I hope that this book will be read by many different constituencies, I need to lay my cards on the table in a particularly obvious way.

The Shape of Things to Come (in the Following Chapters)

The book proceeds more or less chronologically, though I have chosen to begin with the early 1920s, and then backtrack to consider some aspects of the fin de siècle and the First World War. After that, things move chronologically, though we occasionally stop and do a "deep dive" into a particular topic.

Chapter 2—"The Beginnings: Radio in the 1920s"—looks at the beginnings of broadcast radio in Germany and the extremely rapid development of a radio hobby culture located largely in clubs. It discusses the importance of the clubs for the spread of radio, and raises issues of class and public perception of the new media. It covers the beginnings of a real mass-based radio hobby culture in Germany, from 1923 to approximately 1927, shows how technology was domesticated by hobbyists, and describes the mix of fear and desire with which they were met by both government and business.

Chapter 3—"German Radio Before Broadcasting: Scientists, War, and Imperialism"—takes a step back, to the invention of radio as a communications technology. It pays particular attention to the First World War and its contribution to making the explosion of radio in the 1920s possible. It argues that the War not only spread technical knowledge but also set parameters for subsequent implementation of radio as a cultural medium, leaving a long and broad legacy for the beginnings of the radio hobby.

In Chap. 4, "Technology and the Radio Hobby Mature, 1927–1929", we can see the maturation of the hobby culture of radio in the late 1920s, in terms of both technological sophistication and organizational complexity.

It shows how "ham radio" developed into a distinct practice and describes the parallel worlds of middle-class and working-class hobby organizations. It also looks at the importance of radio in bringing the outside world into the living room, and raises the subject of the role of women in the radio hobby. Nevertheless, the chapter begins with a discussion of the lasting shadow of the First World War and the conflict over the language to be used to discuss the new medium and hobby of radio.

Chapter 5 is the longest chapter. In "The Nazification of the Radio Clubs, 1929–1935", I show the development of the radio hobby from roughly 1931 to 1934. This is the crucial period of the Great Depression, the rise of National Socialism, its coming to power, and the "*Gleichschaltung*" or Nazification of the entire public sphere, leading to the near extinction of all private associations. Well before 1933, the hobby clubs were infiltrated by pro-Nazi elements, and the entire debate around radio was dominated by the growing political and economic crisis. The coarsening of the political debate prepared the ground for Nazi violence. Once in power, the Nazis attempted to remake society and destroy or Nazify the entire public sphere in the process. This destroyed all but a small part of the radio hobby. Yet changes in manufacturing and design of radios at just this time would have led to a decline in the radio hobby in any case, though certainly not as fast. Once again, both technology and society together set the context for radio use.

Chapter 6, "The Radio Hobby in the Service of National Socialism, 1935–1945", looks at the remnants of the radio hobby during the Nazi Era and the Second World War and shows how the Nazis feared hobbyists but still needed their expertise, allowing a handful of "ham radio" enthusiasts limited scope to pursue their interests, so long as they supported the regime and contributed to rearmament. Because of both technical change and political repression, the other radio clubs disappeared, but ham radio remained active and even grew during the Third Reich, albeit in a highly regimented form. As it turned out, even the Nazis needed hobbyists.

The final Chap. 7, "The Radio Hobby Comes in from the Cold, 1945–1955", covers the surprising rebirth of a smaller yet still substantial radio hobby between 1946 and 1955, during Allied occupation and the early years of a divided Germany. Despite Cold War efforts to use hobbyists in both German states, an autonomous radio hobby persevered, though it was by then mainly limited to ham radio. Even in the dark days of privation and occupation just after the Second World War, some hobbyists remained surprisingly dedicated to their hobby and to a personal engagement with technology.

Not least, the conclusion again raises explicit comparison between the hobby culture of radio and the modern fascination with computers, the internet, and the maker movement. Hobbies and hobbyists matter, but so does the social and political context within which they exist.

Engagement with Theory

This book sits at the intersection of several disciplines: history of technology, history of private associations, history of media, political and social history, sociology, and even literature insofar as I often use close readings of key texts and take language seriously. It has been my goal in this book to wear theory lightly. Because the book draws from so many fields, the theory which informs it is very eclectic. The author views theory in a very utilitarian way, as a set of tools rather than creeds. As an unreconstructed historian, I see the sources as always paramount and theory only as a tool to order and understand them. Nevertheless, any scholarly work today has to detail its theoretical influences, and so here are mine.

In my notions about technology and the social, I have been heavily influenced by the classic book by Wiebe E. Bijker, Thomas P. Hughes, and Trevor Pinch, *The Social Construction of Technological Systems*.[24] There are good reasons why the "school bus book" is still one of the best-selling publications of Massachusetts Institute of Technology (MIT) press, and even after 30 years, it is insightful and provocative. The ideas about the social construction of technological systems (SCOT),[25] which are elaborated by the editors and contributors to this volume have guided me in my own thinking about how clubs and individuals relate to the technology of radio. My work here clearly places the social at the center, and has much less to say about the technological than most work in SCOT, but the influence is still important. Similarly, the short and graceful *Technology Matters: Questions to Live With*, by David E. Nye is a model for thinking about the

[24] Wiebe E. Bijker, Thomas P. Hughes, and Trevor Pinch (eds.), *The Social Construction of Technological Systems. New Directions in the Sociology and History of Technology* Anniversary Edition (Cambridge, MA: MIT University Press, 2012, 1987).

[25] Note that this school of thought is not at all the same as the "post-modern" "Social Construction of Science", which caused such heated discussions in the late 1990s and early 2000s.

relationship between the human and technology.[26] In a very different vein, and in ways I still don't fully grasp, my thinking about media and modernity is also influenced by the work of the German media scholars Friedrich Kittler and Jochen Hoerisch.[27]

In looking at radio as both a technology and a set of social practices, Kristin Haring,[28] Corey Ross,[29] and Kate Lacey[30] have been strong influences. Haring certainly influences the ways in which I think about hobbies, while Lacey has been a great influence in my thinking about both formal and informal communities around broadcast radio. Both Haring and Lacey are also models of studies which look at how technology and gender intersect. Inge Marßolek has been another major influence on the ways I think about radio and class, gender and power. Pam Potter and Celia Applegate have both been a strong influence on my thinking about both forms of media and the role of private associations.[31] Susan Squier supported my very early work on amateur radio and gender in the US, and her feedback was invaluable as I first struggled with grasping radio.[32]

My ideas about both radio and private associations as manifestations and builders of the public sphere are, of course, fundamentally influenced by Jürgen Habermas.[33] My understanding of the ways people act in associations is also colored by Michel de Certeau[34] and Pierre Bourdieu.[35] Celia Applegate

[26] David E. Nye, *Technology Matters: Questions to Live With* (Cambridge, MA: MIT University Press, 2007).

[27] Friedrich A. Kittler, *Gramophone, Film, Typewriter* (Translated, with an Introduction by Geoffrey Winthrop-Young and Michael Wutz) (Stanford: Stanford University Press, 1999); Jochen Hörisch, *Eine Geschichte der Medien. Von der Oblate zum Internet.* (Frankfurt a/M.: Suhrkamp, 2001, 2004).

[28] Kristen Haring, *Ham Radio's Technical Culture* (Cambridge: MIT University Press, 2007).

[29] Corey Ross, *Media and the Making of Modern Germany: Mass Communications, Society, and Politics from the Empire to the Third Reich* (Oxford: Oxford University Press, 2010.

[30] Kate Lacey, *Feminine Frequencies: Gender, German Radio, and the Public Sphere, 1923–1945* (Social History, Popular Culture, and Politics in Germany) (Ann Arbor: University of Michigan Press, 1996). See also Kate Lacey, *Listening Publics: The Politics and Experience of Listening in the Media Age* (Cambridge, Oxford, Boston: Polity Press, 2013).

[31] Pamela Potter and Celia Applegate (eds.) *Music and German National Identity* (Chicago: University of Chicago Press, 2002).

[32] Susan Merrill Squier (ed.) *Communities of the Air: Introducing the Radio World* (Durham and London: Duke University Press, 2003).

[33] Habermas, *Public Sphere.*

[34] Michel de Certeau, *The Practice of Everyday Life* (Berkeley and Los Angeles: University of California Press, 1984).

[35] Particularly the notion of habitus. Pierre Bordieu, *Outline of a Theory of Practice* (Cambridge Studies in Social and Cultural Anthropology 16) (Cambridge: Cambridge University Press, 1977 (1972)).

had a profound influence on my thinking about private associations, both in conversations and in her superb book *Bach in Berlin*.[36] Also very important has been the work of Lorenz Pfeiffer and Hajo Bernett on the *Gleichschaltung* of German sports clubs.[37] Nevertheless, despite all these important intellectual influences, my most important understanding of civil associations is drawn from my own participation in them, in the US, France, and Germany. Most formative was my close observation of the way in which a contemporary Berlin sports club addressed its own history at the occasion of the 100th anniversary of its founding.[38] My conversations with Dr. Jochen Laufer during the writing of the club history, when I was a guest member of the club in the years 2005–2006, were foundational for my understanding of German club culture and history. When pondering the role of parties and celebration in associations and the possible relationship to technology, I looked first to Bakhtin.[39] I can't help but think there is more to learn on that score, but as my undergraduate students constantly remind me, parties are hugely important institutions. Colleagues such as Prof. Bill Fischer, Prof. Giulia Pacini, Prof. Florence Martin, and Prof. Charley McGovern have also patiently listened to my tales of dancing radio nerds, and helped me understand what was going on beyond the obvious fun of dancing and drinking.

When doing "thick description" of what went on in the radio hobby at various times, and in the close reading I give certain texts ("Achim's

[36] Celia Applegate, *Bach in Berlin: Nation and Culture in Mendelssohn's Revival of the Saint Matthew Passion* (Cornell: Cornell University Press, 2014). See also Celia Applegate, *A Nation of Provincials: The German Idea of Heimat* (University of California Press, 1990) and Celia Applegate, *The Necessity of Music: Variations on a German Theme* (German and European Studies) (Toronto, Buffalo, London: University of Toronto Press, 2017).

[37] Particularly Lorenz Pfeiffer, " 'unser Verein ist judenfrei'—Die Rolle der deutschen Turn—und Sportbewegung in dem politischen und gesellschaftlichen Wandlungsprozess nach dem 30. Januar 1933", in *Historical Social Research / Historische Sozialforschung* 32, no. 1 (119) 2007, *Sport und Diktatur: Zur politischen und sozialen Rolle des Sports in den deutschen Diktaturen des 20. Jahrhunderts/Sports and Dictatorship: On the Political and Social Role of Sports in the German Dictatorships of the 20th Century*, 92–109. Hajo Bernett, "*Der deutsche Sport im Jahre 1933*", in *Stadion* 7 (1981) 225–283; Hajo Bernett, *Weg des Sports in die nationalsozialistische Diktitur. Die Entstehung des deutschen (nationalszialistischen) Reichsbundes für Leibesübungen* (Schondorf: Hoffman,1983) 44–47; Hajo Bernett, "die Zerschlagung des deutschen Arbeitersports durch die nationalsozialistische Revolution", in:*Sportwissenschaft* 13 (1983) 349–373.

[38] Peter Sturm et al (eds.), *100 Jahre Ruderklub am Wannsee. Festschrift, 13. September 1906–2006* (Berlin: Ruderklub am Wannsee, 2006).

[39] Mikhail Bakhtin, *Rabelais and His World* (Bloomington, Indiana University Press, 2009).

Receiver", the 1936 address by Adm. Gebhardt at the 1936 Deutscher Amateur-Sende-und Empfangsdienst (DASD) annual membership meeting), I am influenced by Stephen Greenblatt and the "New Historicism".[40] Long ago, I heard him speak to a National Endowment for the Humanities Summer Seminar led by Prof. Anton Kaes at the University of California, Berkeley. In a more general sense, I must acknowledge the strong influence over the years of not only Tony Kaes but also my fellow seminar participants Prof. Vibs Petersen, Prof. Mary-Beth O'Brien, Prof. Glenn Cuomo, and Prof. Stephen Brockman, all of whom very patiently introduced me to a thing called "cultural studies". Their scholarship and our constant discussion of our work have been invaluable.

On a very different side of my scholarship, a number of scholars have influenced my thinking about the Third Reich in general, and the *Gleichschaltung* in particular. Along with the pioneering work of Karl Dietrich Bracher, Wolfgang Sauer, and Gerhard Schulz,[41] my discussions with and readings of Profs. Stephen Hördler,[42] Sven Reichardt,[43] Richard Bessel,[44] Mathilde Jamin,[45] Robert Frank,[46] Eleanor Hancock,[47] Andreas Werner,[48] Daniel Siemens,[49] Wolfgang Pusy and Hagen Schulze[50] have all

[40] Above all, Stephen Greenblatt, *Renaissance Self-Fashioning: From More to Shakespeare* (Chicago: Chicago University Press, 2005).

[41] Karl Dietrich Bracher, Wolfgang Sauer and Gerhard Schulz, *Die Nationalsozialistische Machtergreifung. Studien zur Errichtung des Totalitären Herrschaftssystem in Deutschland 1933/1934* (Wiesbaden: Springer, 1960).

[42] Stefan Hördler (ed.), *SA-Terror als Herrschaftssicherung: "Köpenicker Blutwoche" und öffentliche Gewalt im Nationalsozialismus* (Berlin: Metropol Verlag, 2013).

[43] Sven Reichardt, *Faschistische Kampfbünde. Gewalt und Gemeinschaft im italienischen Squadrismus und in der deutschen SA* (Industrielle Welt/Schriftenreihe des Arbeitskreises für moderne Sozialgeschichte) (Vienna, Cologne, Weimar: Böhlau Verlag, 2009).

[44] Richard Bessel, *Political Violence and the Rise of Nazism. The Storm Troopers in Eastern Germany 1925–1934* (New Haven, Yale University Press, 1984).

[45] Mathilde Jamin, *Zwischen den Klassen. Zur Sozialstruktur der SA-Führerschaft* (Wuppertal, Peter Hammer Verlag, 1984).

[46] Robert H. Frank, "Hitler and the National Socialist Coalition 1924–1932", PhD. Diss., Johns Hopkins University, 1969.

[47] Eleanor Hancock, *Ernst Röhm: Hitler's SA Chief of Staff* (New York: Palgrave Macmillan, 2008).

[48] Andreas Werner, "SA: 'Wehrverband', 'Parteitruppe', oder 'Revolutionsarmee'? Studien zur Geschichte der SA und der NSDAP 1920–1933", PhD. Diss., Friedrich-Alexander-Universität zu Erlangen-Nürnberg, 1964.

[49] Daniel Siemens, *Stormtroopers: A New History of Hitler's Brownshirts* (New Haven and London: Yale University Press, 2017); Daniel Siemens, *Horst Wessel: Tod und Verklaerung eines Nationalsozialisten* (Munich: Siedler Verlag, 2009).

[50] Hagen Schulze, *Freikorps und Republik 1918–1920* (Militärgeschichtliche Studien 8), (Boppard a/R., Harald Boldt Verlag, 1969).

profoundly shaped my thinking about National Socialism, and the *Gleichschaltung* in particular. (This is a short list; if you are not on it, sorry!) Of course, I owe a huge intellectual debt to my *Doktorvater* Prof. Robert L. Koehl, who first taught me about different and creative ways to think about National Socialism and its relationship to power.[51] In a different but equally profound way, I have been greatly influenced by the towering figure of Prof. Gerhard Weinberg, who, among his many gifts, is always able to bring a discussion down to what is essential, and who always reminds us that when there is state violence, there are also victims, and they need to be heard.

People and Institutions to Thank

Beyond these fundamental theoretical influences, I owe a debt to very many individuals who have helped and challenged me along the way. None of them are even remotely responsible for any shortcomings of this book or its author.

Wolf Harranth, the head of the Dokufunk Archiv in Vienna, the noted media scholar Prof. Inge Marßolek, and Prof. Kristin Haring have been profoundly helpful in discussions about the radio hobby, and have all been very supportive of my project.

Aside from these scholars, my work has benefited from numerous discussions with many others, including the French and Francophone Literature and Film and Media scholars Prof. Giulia Pacini, Prof. Michael LeRuth, Prof. Maryse Fauvel, and Prof. Florence Martin, the Sociologist and Peace Studies scholar Dr. Robin Crews, the Historian of National Socialism Prof. Katrin Paehler, and the Germanist Dr. Sylvia Klötzer, all of whom I count among my friends as well as being intellectual influences. Prof. Martin and Dr. Crews provided me with extremely useful criticism and comment on a rough draft of this book, and I owe everyone here for their stimulating and insightful criticism and support over many years. The book has further profited from conversations with the History of Science Reading Group at the College of William and Mary, particularly the physicists Prof. Gene Tracy and Vice-Provost Prof. Dennis Manos, the historian of science Prof. Nick Popper, and the chemist, writer, and artist Prof. Carey Bagdassarian. Prof. Bill Fischer in Anthropology, and Prof. Charlie

[51] Robert L. Koehl, "Feudal Aspects of National Socialism", in *American Political Science Review* 54, no. 4 (December 1960), 921–933.

McGovern in American Studies both gave sage advice about aspects of the book, including the importance of sound and the role of celebration. I must also thank the Franco-American anthropologist Prof. Christy Shields-Argelès for information and ideas on the same topic.

The Society of the History of Technology (SHOT) has been a very welcoming and stimulating home for a lateral newcomer. I must thank all of its members who have introduced me to the History of Technology and initiated me into the secrets of Social Construction of Technology (SCOT), Actor-Network Theory (ANT), and Large Technological Systems (LTS), and caused me to think for the first time about "thingness" and "materiality". I owe everyone in SHOT a great intellectual and personal debt, not least for concrete practice in the intersection of technology and festival. The special interest groups within SHOT on electrical technology and communications, the "Jovians" and "Mercurians" have been particularly helpful. I owe a particular intellectual debt to Dr. Elizabeth Bruton, Curator of Engineering and Technology at the Science Museum, London, and also Profs. Michael Krysko, Jennifer S. Light, Heidi Tworek, and Derek Vaillant with whom I have appeared on SHOT panels and who have all critiqued and encouraged my work.

In a similar way, I owe a very great deal to the German Studies Association (GSA) and its members. The GSA has always been my professional "village", the place where I go to draw inspiration, see friends, and the only place in the world where everyone speaks my language. Much of the work in this book received its first airing as papers at the GSA. I owe a debt to the following participants in GSA panels, who all participated in the discussion of my work: Profs. Manuela Achilles, Ute Chamberlin, Christoph Classen, Elizabeth A Drummond, Andrew Evans, Bryan Ganaway, Martin Gutman, Thomas Pegelow Kaplan, Inge Marßolek, Heather Perry, Mark Russell, Aeleah Soine, Adam Stanley, Julia Timpe, Heidi Tworek, Susanne Ungar, and Jonathan Wiesen. I am very fortunate to have the GSA as my primary professional association. Without its open community and strong emphasis on interdisciplinarity, work such as this would be unthinkable. Associations matter.

As all academic books, this one was peer reviewed by the publisher, Palgrave Macmillan, by two specialists in the field. As is the custom, I will never know who these experts were, but I am deeply grateful for their hard work and helpful criticism. This has become a much better work with their help. I also owe a great debt to the editorial team at Palgrave Macmillan. Megan Laddusaw was kind enough to offer me a contract and helped in the background more than I can say. Christine Pardue handled the editing and showed great patience in helping bash my awkward prose into shape. Both were an absolute delight to work with. M. Vipin Kumar of SPi Global supervised the final production, and was a model of efficiency. Every author should have such great support.

Sources for this book have come from many and often obscure places. I owe a very great debt to the staff of the Berliner Staatsbibliothek and Swem Library at the College of William and Mary, and in particular to the skilled and patient staff of the Inter-Library Loan Office at Swem. I am particularly grateful to the Berlin Staatsbibliothek and to the captain and crew of the *CMA CGM Florida* for providing me with absolutely optimal writing conditions over the years. Further, I have been helped by the staff of many archives, including the US National Archives II in Maryland, the German Federal Archives (Bundesarchiv) in Lichterfelde, the Bundesarchiv-Militärarchiv in Freiburg, the Berlin Museum für Kommunikation, the Hamburg Museum für Kommunikation (now closed), the Deutsches Rundfunkarchiv in both Babelsberg and Frankfurt, the Deutscher Amateur Radio Club (DARC) Distrikt E Archive, Hamburg, and particularly its director Gerd Hoyer, DJ1GE. I have already mentioned Prof. Wolf Harranth, OE1WHC, Curator of the Wiener Dokumentationsarchiv zur Erforschung der Geschichte des Funkwesens und der elektronischen Medien—Internationales Kuratorium QSL Collection (Dokufunk), for his endless wisdom and good cheer. This book would have been impossible in its current form without the many unique documents collected in the Dokufunk archive. I own Dr. Harranth and his staff very special thanks.

The German and US amateur radio associations, respectively, the DARC (Axel Voigt, DO1ELL) and American Radio Relay League (ARRL; Steve Mansfield N1MZA (SK), then ARRL Legislative and Public Affairs Manager), both provided very helpful information. My thanks to Dr. Thomas Boghardt and his colleagues, Dr. Nick Schlosser and Dr. Kathy Nawyn, for their help in navigating the intricacies of the US Military Government in Germany and in tracking down the probably American benefactors of illegal German amateurs in 1946. These three dedicated civil servants answered the e-mail of a complete stranger, and selflessly gave me the benefit of their advice and knowledge. Long ago, Bob Eldridge VK6A7, Jeff Jeffrey VK6AJ, and Jim Muiter N6TP commented on my ideas when I was just beginning. Prof. Dr. H.-J. Kowalski of the Beuth-Schule gave me information about the university's archive.

Finally, nothing gets done in the Academy without funding, and I have several individuals and institutions to thank for their generosity. These include Mr. and Mrs. Robert D. Pavey, who helped fund my Class of 1964 Term Associate Professorship, the Office of the Dean of Arts & Sciences at the College of William & Mary, and particularly its head, Dean Kate Conley, and finally the Skunkworks Fund of the Center for the Liberal Arts at the College of William & Mary.

In the course of this research I have come to admire deeply many of the hobbyists, tinkerers, and hams whom I have never met but whose passion shone through the dry documents nearly a century later. My thanks for leaving enough documents to tell their story to later generations. I hope I got it right.

Last of all, I owe an immeasurable debt to my partner and spouse, Prof. Maryse Fauvel. She kept my "eyes on the prize" over many years, and made sure somebody had my back. I owe everything to her.

CHAPTER 2

The Beginnings: Radio in the 1920s

> To be a radio-amateur means revealing to oneself how and why radios work, not as a profession, but out of a love for the thing itself. It means, building on this, trying to improve the performance of equipment not only in respect to range but also regarding its qualitative performance. (…) Currently, temporarily, and maybe forever, radio equipment is just not as easy to use as a watch, a motor, or a telephone. Something secretive, something mysterious remains attached to it. Radios give their highest performance only when they are used with empathy and tenderness. Radio wants to remain a sonorous mystery with many open questions and riddles. Being a radio-amateur means teasing the last secrets out of the equipment.[1]

[1] Dr. Ernst Tross "Franz Wolff, ein vorbildlicher Radio-Amateur", in: *Der Radio-Amateur* 2, No. 22 (September 19, 1924) 572–574. "*Ein Radio-Amateur sein, heißt sich darüber Rechenschaft zu geben, wie und warum die Apparate arbeiten, und zwar nicht aus Profession, sondern aus Liebe zur Sache, und davon ausgehend, zu versuchen, die Leistungen der Apparate zu steigern nicht nur in Hinsicht auf Reichweite, sondern auch in Hinsicht der qualitativen Leistung*". (…) *Es ist nun einmal vorläufig so, und vielleicht bleibt es so, daß ein Radio-Apparat nicht so einfach wie eine Uhr, ein Motor oder ein Telephon zu bedienen ist. Es haftet ihm etwas Geheimnisvolles, etwas Mysteriöses an. Er gibt seine Höchtsleistung nur bei verständnisvoller, liebevoller Behandung her, er will ein tönendes Mysterium bleiben mit vielen Fragen und Rätseln, und Radio-Amateur sein, heißt dem Apparat die letzten Geheimnisse entlocken*".

B. B. Campbell, *The Radio Hobby, Private Associations, and the Challenge of Modernity in Germany*, Palgrave Studies in the History of Science and Technology, https://doi.org/10.1007/978-3-030-26534-2_2

Imagine the scene: following instructions out of a magazine, a young German boy[2] builds a rudimentary crystal radio receiver[3] out of a few simple parts, mostly taken from the family toolbox. Suddenly, out of his army surplus earphones comes a voice from a faraway major city or an even farther away foreign country. The radio worked! In an instant, everything is changed. The voices would now be there around the clock. Notions of space, privacy, national security, and social agency were irrevocably warped. The radio age had arrived. Welcome to the twentieth century.[4]

Here is why this is noteworthy: all of it is taking place on somebody's kitchen table, and no authority figures were involved (no teachers, priests, officers, professors, policemen, foremen, or saints). The relatively simple technology was accessible to everyone who could read or listen. This is the real revolution of radio as a hobby. What had previously been the domain of armies and navies, physicists, major corporations, and a handful of genius inventors was now, literally, child's play. The time between the discovery of a way to pass messages by imposing order on electromagnetic waves, and the passing of this knowledge into the hands of ordinary people, was extraordinarily short, a matter of a handful of years. After this, the

[2] As we see below, most radio hobbyists were men, though a not insignificant number of women were also involved. See Chap. 3.

[3] A "crystal radio" is one of the simplest receivers possible. It consists of a "detector" (diode) made of a metallic crystal, such as galena, an antenna, and a ground. The detector resonates (vibrates) with electromagnetic energy emitted by the antenna of a radio transmitter. Noteworthy is that these are "passive" receivers, meaning that they do not require any form of amplification (though it could also be added) and thus need no electricity. It is possible to make all of the parts at home, though most builders of crystal receivers choose to purchase at least the headphones. Commercially manufactured receivers of this type, as well as home-built ones, were widespread in the early 1920s, and were still produced and used in Germany into the 1930s and 1940s. See: United States Army Signal Corps and National Bureau of Standards, *The Principles Underlying Radio Communications*, Second Edition, Radio Communications Pamphlet Nr. 40 (Washington: Government Printing Office, 1922), 434–439. See also: Reichsverband Deutscher Funkhändler, *Neuester Illustrierter Radio-Katalog 1928–1929* with *Nachtrag zum Illustrierten Radio-Katalog mit Preisberichtigungen 1929/30* (Berlin: Kreditgenossenschaft des Reichsverbandes Deutscher Funkhändler, 1928). The catalog contains several pages of crystal radios and an additional several-page section on crystal radio parts for home builders.

[4] The voices are still there, at least as of the writing of this current book (2018). More and more shortwave broadcasters are shutting down, in favor of internet operations. Broadcast radio using AM and FM seems more robust, yet is also under threat from internet music streaming services and digital music listening devices (today, often combined with telephones).

world invaded the family kitchen directly, and the reach of someone sitting at the kitchen table extended around the world.

The story of radio as a hobby in Germany began more or less simultaneously with the beginnings of broadcast radio in Germany in 1923, which is to say, rather late. This is in stark contrast with the US and England, where radio as a hobby came long before broadcasting, and was well established before the First World War.[5] On the other hand, developments in Germany are quite similar to the broad development of radio in other parts of continental Europe, such as France, Sweden, the Netherlands, Italy, and Austria, where radio as a hobby was also weak or nonexistent before the end of the First World War. This does not mean that the radio hobby in Germany was caused by or limited to broadcast radio, simply that it did not exist on a large enough scale to be noticed until broadcast radio was a part of the public conscious and imagination. The development of the radio hobby in Germany was certainly also influenced by larger inter- and transnational developments, yet it only became a mass movement as a by-product of the state-initiated and state-controlled introduction of broadcast radio. This social time scale is quite different from one which might be driven by the technology alone, which, as we shall see in the next chapter, dates from the end of the nineteenth century. But we start in this chapter with the 1920s, a time when the radio hobby was born and grew rapidly, a time when "anything goes".

[5] As early as 1910, there were nearly 500 amateur transmitters. See the list in: Wireless Association of America, *Wireless Book of the Wireless Association of America*, Second Annual (New York: Modern Electrics Publication, 1910), 10–16.

The only academic study of amateur (ham) radio in the US is: Kristen Haring, *Ham Radio's Technical Culture* (Cambridge: MIT University Press, 2007) but see also:

Clinton B. DeSoto *200 Meters and Down: The Story of Amateur Radio* (West Hartford, CT: American Radio Relay League, 1981), and, for Great Britain, Elaine Richards, G4LFM *Centenary: 100 Years Working for Amateur Radio* (Bedford: Radio Society of Great Britain, 2013). The work of Susan J. Douglas is indispensable on the history of broadcasting in the US, and she is quite sensitive to the role played by amateur radio. See: *Inventing American Broadcasting, 1899–1922* (Johns Hopkins Studies in the History of Technology) (Baltimore: Johns Hopkins University Press, 1987) and *Listening In: Radio and the American Imagination* (Minneapolis: the University of Minnesota Press, 2004). See also: Michelle Hilmes, *Radio Voices: American Broadcasting 1922–1952* (Minneapolis: University of Minnesota Press, 1997) and Tom Lewis, *Empire of the Air: The Men Who Made Radio* (New York: Harper Collins, 1991). On Great Britain, see: Andrew Crisell, *An Introductory History of British Broadcasting* (2 ed.) (New York: Routledge, 2002).

For a stimulating comparison, see Michelle Hilmes, *Network Nations: A Transnational History of British and American Broadcasting* (New York: Routledge 2012).

THE "DOG THAT DIDN'T BARK": WHY GERMANY HAD NO RADIO HOBBY BEFORE 1923

Just as for Sherlock Holmes, in the story of the radio hobby in Germany, there is a "dog which didn't bark".[6] In Germany, there was simply no widespread popular radio hobby until the government began broadcasting.[7] As mentioned above, this is in stark contrast to developments in other major industrialized countries. The contrast is strong enough to stand out immediately. In the US, there was a broad and active radio hobby (including hobby transmitters) well before the First World War, and long before the invention of broadcasting in the early 1920s.[8] Developments in Britain, though at a smaller scale, were similar. By the eve of the First World War, there was a widespread radio hobby in Britain, and the first association of British radio amateurs was founded in July 1913.[9]

In Germany, this was not at all the case, despite the fact that Germany was a major industrial power, had a highly educated population, and was at the forefront of the scientific discovery of electrical and radio science. There were a small handful of German scientists who experimented with radio before 1923,[10] and some former military radiomen did begin to

[6] This is of course a reference to Sir Arthur Conan Doyle's short story "The Adventure of Silver Blaze", first published in *The Strand Magazine*, Dec. 1892–Dec. 1893, and in 1894 in: *The Memoirs of Sherlock Holmes* (New York, London: Harper & Brothers, 1894).

[7] The "Radio-Verein e.V. Coburg, Studiengesellschaft für Elektrotechnik und Funkentelegraphie" was founded in the Fall of 1919, and is likely the oldest radio hobby club in Germany. "Briefkasten", in: *Der Radio-Amateur. Zeitschrift für Freunde der Drahtlosen Telephonie und Telegraphie. Organ des Deutschen Radio-Clubs* 2, No. 6 (April 30, 1924), 174. There may have been some scientific experimentation in Stuttgart and Berlin, but nothing like a hobby. W. F. Körner identifies a handful of early "hams" before 1923, including one, Richard Dargatz, in Charlottenburg, who applied for a permit to have a private radio transmitter in 1919, only to receive a categorical refusal from the Postal authorities. All of the men Körner identifies transmitted without permits and quickly entered into contact with one another. See: *Geschichte des Amateurfunks, 1909–1963* (Hamburg: Verlag Rojahn, 1963), 13–19. The point is that in all of these cases, early radio enthusiasts remained small in number, isolated, and without any larger social resonance or identity.

[8] DeSoto *200 Meters*; Douglas *Inventing American Broadcasting*, 187–215.

[9] Richards, *Centenary*, p. 1.

[10] Prof. Wolfgang Harranth holds that the nucleus of the future hobby may be located in the young graduate students and engineers who built radio receivers for the post office and private companies in 1921 and 1922, before the beginning of broadcasting but after the postal service in Germany had developed a subscription-based radio news service. Though this is plausible, there are few records, and in any case, these men were paid radio technicians

come together and form associations from 1919 onward, but there was no radio hobby. German radio enthusiasts at the beginning of the radio age themselves struggled to explain this fact. One set of explanations comes in a 1924 article by Dr. Eugen Nesper, a man active in the scientific study of radio and one of the strongest advocates for radio as a hobby. According to Nesper, German radio enthusiasts knew they were far behind the US and Great Britain.[11] Nesper, in the name of German radio hobbyists in general, cited several reasons. First, the German Postal Ministry had not authorized amateur transmission and did not even authorize ordinary citizens to own a radio receiver until late 1923. Second was the allegation that Germans don't occupy themselves with technology as people do in the US, where kids build their own little machines. In other words, they claimed German society lacked the presence of other technical hobby cultures, which might have served as models and catalysts for the spontaneous growth of a radio hobby. Moreover, Nesper criticized Germany for not making radio part of the school curriculum before the war, whereas radio was ostensibly already in schools in the US, Britain, and France.[12]

This was all true, or at least plausible, but also all wide of the mark. The postal authorities were quite slow and reluctant to grant private transmitter licenses, and this issue remained a sore point for German radio hobbyists into the late 1940s, as we shall see. It is certainly also true that the German postal authorities were extremely jealous and protective of their monopoly over all forms of electronic communications, much more so than in other countries. This may be due in part to the authoritarian nature of the German Empire and particularly the closeness between the postal authorities and the military. This closeness was a constant from the Empire, to the Weimar Republic, to the Third Reich, and owes as much to the particular German conception of the civil service as an agent of the ruler as it does to the special strategic nature of modern communications media, be they letter, telegraph, telephone, or radio. But it was also due to a deep fear in Germany that private receivers would inevitably lead to a violation of the privacy and secrecy of all communications. The postal authorities in Germany had a laudable commitment to the right of citizens to personal

and students and thus not amateurs. Oral discussion with the author, November 21 and 26, 2018. On the subscription-based service, see Lerg, *Rundfunk*, chapters 1 and 2.

[11] Dr. Eugen Nesper, "Zur Einführung", *Der Radio-Amateur* 2, No. 1 (August 23, 1924), 1–2

[12] Ibid.

privacy in any communications, which passed through their hands—though the specific reasons for this probably had less grounding in human rights and more in commercial reasons: the German Post Office made a great deal of money supplying economic information to businesses on a subscription basis, and did not want to risk either this or its monopoly on letter and telegraph transmission.[13] Though Nesper's critique is clearly marked by the trauma of the First World War and the fear that Germany was falling behind its competitors, it leaves unmentioned the other national trauma of (socialist) revolution and civil war. Private receivers, let alone private transmitters, were consistently viewed with suspicion by the Interior Ministry and police. There was great fear that the Left would use radio technology against the Weimar state in the event of internal unrest or foreign attack.[14] On the other hand, blaming the Post Office was also partly beside the point: lack of legal authorization did not stop radio enthusiasts in the US from transmitting before the First World War, even though it was also true that legal authorization came relatively quickly in the US, nor did it later stop German radio amateurs from illegal transmission in the 1920s, 1930s, and even 1940s. Moreover, radio was definitely not a common part of the school curriculum in the US before the early 1920s, and it is doubtful that France or Great Britain was any different, whereas Germany had a strong school system, and world-leading universities and technical schools, even if there may have been a relative lack of courses on radio science.

Instead, one of the key differences between Germany and the US was simply the lack of public "buzz" around radio. In particular, magazines, aimed at a popular audience in the US and Britain, were full of articles about the new medium of radio, and niche magazines aimed specifically at radio hobbyists were quickly founded.[15] In Germany, before the First

[13] The German Postal Service had a number of financial interests to protect. Aside from the threat radio might pose to traditional letter service, the Postal Service also controlled the Imperial Telegraph Service (*Reichstelegrafenamt*); telegrams were expensive, and the service was quite lucrative. It used radio links between telegraph stations, on occasion. In addition, the Postal Service ran a subscription-only business news service via the (wired) telephone system, which was also quite profitable, and which used radio links for speed. There was thus great fear in the Postal Service that radio receivers in civilian hands would be able to listen in on these services and simultaneously destroy the guarantee of privacy and the profitable monopoly over certain kinds of information upon which much of the financial stability of the Postal Ministry depended.

[14] See Lerg, *Rundfunkpolitik*, 93–108.

[15] The following radio magazines were published in Britain and the US *before 1920*: in Great Britain: *RadCom* (1913), *Wireless World* (1913), *The Marconigraph* (1911), and

World War, this public consciousness of radio simply did not exist at the same level. On the other hand, after the First World War, and once broadcasting was authorized, radio very quickly gained public attention. Not only did radio-oriented publications explode onto the scene in Germany in 1923, 1924, and 1925,[16] but radio also became a topic in existing newspapers and magazines. To put it another way, the major reason for the lack of a radio hobby in technically advanced Germany before the First World War was the lack of a social space for it. The relatively more authoritarian and closed nature of Imperial Germany played a role,[17] but the key was the lack of a public consciousness of the existence and accessibility of radio to ordinary citizens. Nesper was certainly correct that the lack of other technically based hobbies retarded the German adoption of radio as a hobby substantially, but he ignored the fact that it was less the lack of hobbyists than the lack of widespread public knowledge that technology could be a hobby.

There is yet another very basic reason why the widespread development of a radio hobby in Germany came only after the introduction of broadcast radio: first, there had to be something to listen to, before large numbers of people could become enthusiastic about building receivers, and there had then to be a community of people not only interested in using the new technology but in understanding it and using it in creative ways. Without this community to supply basic technical information and to put some signals on the air to hear beyond the commercial and maritime traffic, there was just not enough going on to create the "buzz", which would

Radio Review (1919); in the US: *Modern Electrics* (1908), *QST* (1915), *Radio News* (1919), *Pacific Radio News* (1919), *Wireless Age* (1913), *Electrical Experimenter* (1913), *Popular Electricity in Plain English* (1908), *Everyday Engineering* (1915), and *Electrician and Mechanic* (1910). Many more radio and electronics magazines were founded in the 1920s, and articles about radio appeared in a much wider range of magazines aimed at tinkerers, do-it-yourselfers (DIY), or boys. Scanned back issues of many of these radio magazines are available at: http://www.americanradiohistory.com/.

[16] The earliest German-language radio hobby magazine was *Der Radioamateur*, which began publication in late 1923. Popular German radio magazines from the 1920s included *Der Radio-Amateur* (1923), *Funk* (1924), *Die Funk-Stunde* (1924), *Radio* (1924), *Radio-Umschau* (1924), *Der Funk Bastler* (1924), *Die Norag* (1924), *Der Rundfunk-Hörer* (Munich, 1924), *Z.-J.-Funk* (1925), *Radio-Woche* (1925), *Süddeutscher Rundfunk* (1925), Noru: *Norddeutsche Rundschau für Funk und Film* (1926), *Funk-Woche* (1926), *Funkschau* (1929), and *Funk und Schall* (1929).

[17] In this context, it is hard not to think of the current debate in the US over "net neutrality". Innovation definitely flourishes best in open societies, but this is only part of the story.

attract others to the hobby in large numbers.[18] The introduction of broadcast radio in Germany created the “buzz”. Once it got started, the radio hobby in Germany was able to build on two key foundations: the well-developed German system of technical education and the introduction of thousands of soldiers to telegraphy and the basics of electrical theory during the Great War.[19] This gave many a crucial foundation of technical knowledge, which allowed them to understand the new science of radio and hope to use it in creative ways. But the simple knowledge was never enough; to have a hobby required a community of hobbyists, and some excitement. There wasn’t much of a hobby in simply listening to the radio, but there was in building new radios or improving existing ones or even in exchanging messages with other hobbyists half a world away. Technological hobbies are about doing creative things with technology, not just passive listening. They require some kind of a community to thrive.

But from the beginning, for all its spontaneity once it got going, the radio hobby in Germany was profoundly influenced by state regulations and state needs. The German case may well be unique in some aspects. Because of the way the German government—and this meant primarily the Postal Ministry—designed the broadcast radio system and regulated radio in Germany, nascent hobby clubs were brought in as partners, but also immediately brought to heel and formed in a way dictated by the Postal Ministry for the needs of its (broadcast) radio plans. The Postal Ministry actually supported the creation of radio clubs for hobbyists as a means of creating enthusiasm (a market) for broadcast radio and as a way to spread the necessarily technological understanding of radio so that the problem of interference could be mastered, a precondition for the success of broadcasting, given the kind of equipment available in 1923. Creation of a market for broadcast radio also meant creation of the radio receivers necessary to listen to radio, and here again, the clubs served to instruct individuals on how to build their own receivers, a real necessity, given the high cost and low quality of early commercially produced receivers in Germany.[20] Finally, the clubs were forced to take on much of the work of

[18] This still partially begs the question, since a radio hobby did develop in both the US and Great Britain well before broadcasting. In both places, hobbyists listened to maritime and other commercial traffic, the military, and to each other, since in both countries radio hobbyists quickly began to transmit. Certainly, the long coastlines of both countries made maritime listening relatively easy.

[19] See the following chapter.

[20] See Table 2.1 for information on costs.

administering examinations, which were preconditions for obtaining a license to own a radio receiver, thus sparing the Postal Ministry a great deal of cost and effort, yet still leaving it in control.[21]

The radio revolution created a market for commercially produced radios, but also more generally for radio and electrical parts to feed the do-it-yourself (DIY) movement. This helped the electrical industry in Germany to come out of its post-First World War depression, and it even led to a tremendous and rapid expansion of the industry, already a world leader, and made the necessary parts for radio equipment widely available. Out of the home-built radio movement then came the minority of radio hobbyists who also wanted to transmit rather than simply receive. We have much more to say about the clubs, and the home-building hobbyists, but for now, let's not get ahead of ourselves.

Hans Bredow, the Invention of Broadcast Radio, and the Radio Hobby in Germany

Because it happened a bit later than in other parts of the world, the introduction of a state-run broadcast model of radio in Germany was certainly influenced by external developments, not least in the US and Britain, as well as France and Belgium. Indeed, the architects of the German system claimed to have studied foreign models carefully before designing their own, uniquely German vision or radio, although there was also strong continuity with the earlier German model of a state telegraph monopoly. Be this as it may, the German version of broadcast radio was largely the vision of one man, Hans Bredow, who was an Under-Secretary (*Ministerialdirektor*) in the Postal Ministry tasked with overseeing wireless communications.[22]

[21] Note the particular constellation of circumstances at work: Germany in the 1920s was a liberal democracy, but its institutions were still heavily influenced by earlier conservative authoritarian traditions dating from the German Empire. The state was also quite limited in its economic resources, given the need to pay back war loans, deal with the aftereffects of the war, and the hyperinflation of the early 1920s.

[22] For information on Bredow, see: *Wer ist Wer?* XI (Berlin, Ariani Verlag, 1951), 69. See also: Dr. Gehne: "Staatssekretär Dr. Bredow über den Rundfunk", pp. 880–884, and W. Grunicke: "Die Organisation des deutschen Rundfunks" pp. 886–888 both in: *Der Radio-Amateur 2*, "1. Messeheft", No. 33 (December 5, 1924) and "Staatssekretär Dr. Bredow über den deutschen Rundfunk" II. Teil "Stellungnahme zur Amateurbewegung", *Der Radio-Amateur* 2, "2. Messeheft", No. 34 (December 12, 1924), 906–908.

Bredow was an engineer and businessman, who came to play a leading role in the Telefunken corporation before the First World War. Today, we would call the Telefunken of the early twentieth century a major part of the German "military-industrial complex".[23] It made most of its money selling early radio equipment to the German military and merchant marine. Closely tied to the German government, it was an international company with business ties to, and patent agreements with, the leading radio-electrical companies of the world, such as the Marconi Corporation. Moreover, it was a key partner of the German government efforts before the Second World War to create a world radio network linking German colonies to the mother country independent of the British-controlled underwater telegraph lines.[24] It is significant that Bredow had this background: it gave him a large-scale vision, a strong business background, but also tied him to a corporate model of a strong, monopolistic state working with a handful of leading national industrial firms with international connections.

Bredow's concept of radio was at once visionary and narrow. He foresaw the importance and popularity of a broadcast model of radio for Germany very early on,[25] and stuck to his vision with tenacity. He developed a plan for a network of strictly state-controlled and owned yet regionally based broadcast radio stations, whose content would be chosen by semi-independent regional program authorities. It would take the regionalized and strongly federal structure of Germany into account, and yet would maintain a strict state monopoly over radio. The large amounts of capital needed to get started would come from carefully selected groups of private investors close to government, but would be invested in the regional programming corporations. Programming would focus on high

[23] The term came into prominence after it figured prominently in the Farewell Address by US President Dwight D. Eisenhower, January 17, 1961.

[24] See, for example, Michael Friedewald, *Die "Tönenden Funken". Geschichte eines frühen drahtlosen Kommunikationssystems, 1905–1914*, Aachener Beiträge zur Wissenschafts- und Technikgeschichte des 20. Jahrhunderts 2 (Berlin: Diepholz/GNT Verlag, 1999), 13; James W. Carey, "Technology and Ideology: The Case of the Telegraph". *Prospects* 8 (1983), 303–25 and Heidi J.S. Evans (Tworek), " 'The Path to Freedom'? Transocean and Wireless Telegraphy, 1914–1922", *Historical Social Research* 35, No. 1 (2010), 209–236.

[25] In a possibly apocryphal story, Bredow played phonograph music over the radio while at the front in the First World War. Gerhart Goebel, "Der Deutsche Rundfunk bis zum Inkrafttreten des Kopenhagener Wellenplans", Sonderdruck aus dem *Archiv für das Post- und Fernmeldewesen* 2, No. 6 (August 1950), 357–358.

culture and education, and overt political content would be banned.[26] It would be financed through a monthly license fee for each receiver, and ownership of radio receivers would be strictly controlled and supervised.

As a businessman, Bredow understood that none of this would work without a market for radio programs and radio receivers, and this meant creating strong public interest and support. He developed a very clever, and a very German means of creating public excitement through the encouragement of independent clubs of radio enthusiasts, tapping into the strong German culture of clubs and private associations. These private clubs would work to spread the word about radio, and would also teach the public enough technical knowledge to be able to operate what were, in the beginning, fairly complex or at least, very "user-unfriendly" receivers, which could create strong interference if not properly used. The clubs were thus intended to carry out much of the necessary education and publicity needed to make broadcast radio viable and to take on a share of the regulation of radio, without any extra expense for the state (which was financially weak after the First World War). This was a stroke of marketing genius.

Despite the importance of radio clubs in Bredow's plan, he mistrusted both the general public and the clubs, and always strove to limit their political influence and power, generally with great success. Bredow's vision for radio was that of a child of the German empire schooled in the military-industrial complex. The state knew what was best for the citizen, and demanded a total monopoly over communications media. But Bredow was also a modern businessman, who knew the public could and should be mobilized to create new markets. A radio hobby would almost certainly have developed in Germany without Bredow's intervention: a non-professional (hobby) engagement with technology in a larger sense (though not radio) already existed in Germany,[27] and the radio hobby certainly developed in neighboring European states without German-style state

[26] The emphasis on the political neutrality of state radio also conveniently preserved the political status quo and left state dominance of the media unquestioned.

[27] Many other forms of technology enthusiasm were characteristic of the 1920s, in both Germany and elsewhere. Aviation, automobiles and motorcycles, and luxury liners are all examples. See, for example, Bernhard Rieger, *Technology and the Culture of Modernity in Britain and Germany, 1890–1945*, New Studies in European History (Cambridge: Cambridge University Press, 2009) and Peter Fritzsche, *A Nation of Flyers: German Aviation and the Popular Imagination* (Cambridge: Harvard University Press, 1994).

direction and manipulation.[28] And yet, with his model for the development of broadcast radio, Bredow created the radio hobby in Germany as a mass movement in a very short period of time and set patterns which would persist long after he was no longer in charge.

Radio and the Law

The way radio law was written in the first few years of broadcasting in Germany made the hobby clubs essential partners. Legal requirements more or less brought the clubs into existence as a mass movement. But when the law changed in 1925, the clubs, though forced to struggle a bit, actually weathered the change nicely, for the simple reason that they had built up their own momentum and reason for being.

Bredow's plan was both modern and clever. Yet it was implemented by a state structure and bureaucracy inherited from the German Empire. Even after the German government decided to introduce broadcast radio in Germany, the authorities in the Postal Ministry and other state representatives resisted the ability of ordinary Germans to own radio receivers. When broadcasting was first introduced, the ownership of receivers was tightly restricted, and the conditions were only gradually loosened in a series of laws and internal implementation guidelines. At first, only receivers which had been technically tested and approved by the Postal Ministry were allowed to be used. Commercially produced models had to be sealed, and no user-modification was allowed.[29] Moreover, one or more permits were required in order to own and use a receiver. First, a "Radio Permit" (*Radio Genehmigungsurkunde*) had to be obtained. These were at first quite expensive,[30] and only German citizens whom the police deemed reliable were entitled to have one.[31] The basic receiver permit,

[28] There are no academically satisfying studies of amateur radio in Europe outside of England. Note that the main German radio journals such as *Funk* or *CQ* were filled with regular reports on developments in neighboring countries and the US.

[29] Rechtsanwalt Franz Landsberg, "Der deutsche Rundfunk", *Der Radio-Amateur* 1, No. 4 (November 1923), 83–88.

[30] The original cost or a radio permit was 60 Gold Marks, a huge sum. Remember, too, that broadcast radio was introduced at the height of the hyperinflation, which followed the First World War. In May 1924, after a currency reform and the return of stability, the cost was reduced to RM 2 per month (RM 24 per year). This price remained stable into the 1970s.

[31] There were exceptions for citizens of countries which allowed German citizens to own radios.

once issued, also entailed the yearly (later monthly) payment of the basic broadcast radio fee or tax, which had to be paid as long as the receiver was owned. Ownership of a receiver (let alone its use) without a permit was illegal, and could lead not only to a fine and confiscation of the receiver but even to prison in certain cases. Postal authorities were entitled to search any premises where an illegal radio was suspected at any time, and without a warrant.[32] Second, the ownership and use of receivers which used one or more tubes to amplify the received signal (meaning any receiver beyond the most technically simple crystal detector) required the owner to hold a second permit, the so-called Audion Research Permission (*Audionversuchserlaubnis*, AVE). This permit was only issued after the applicant had passed a strict technical test on radio theory and practical use.[33]

In the very earliest period after the introduction of broadcasting (October 1923–March 1924), the rules governing radio ownership were issued by the Postal Ministry itself as administrative regulations, with only the Imperial Telegraph Law of 1892 as a legal basis.[34] After broadcast radio had existed for less than a year, the law was formalized and strengthened, but not through the regular parliamentary procedure. Instead, the *Verordnung zum Schutz des Funkverkehrs* of March 8, 1924 (Decree on the Protection of Radio Traffic of March 8, 1924)[35] was issued by the German president as an emergency decree based on article 48 of the constitution, without a parliamentary vote. It provided a clear legal basis for the way the Postal Ministry had already decided to regulate radio. It provided a more modern definition of what a radio was and reduced the Broadcast Radio Fee to RM 2 per month. It allowed home-built crystal receivers to be used without direct technical approval of the Postal Ministry, but required that

[32] "Der deutsche Rundfunk" von Rechtsanwalt Franz Landsberg, in: *Der Radio-Amateur* 1, Nr. 4 (November 1923), 83–88.

[33] Amtsblatt des Reichspostministeriums Nr. 46, included in: *Der Radio-Amateur* 2, No. 8 (May 28, 1924) and Obertelegrapheninspektor Ernst Schulze, "Die neuen Rundfunkbestimmungen", *Der Radio-Amateur* 2, No. 11 (July 4, 1924), 289–292.

[34] Stud. Jur. H. Steiniger, "Die Grundgedanken des Reichstelegraphengesetzes und der Telefunkennovelle", *Der Radio-Amateur* 1, No. 1 (August 23, 1924), 18–19.

[35] Der Reichspräsident [Ebert], "Verordnung zum Schutze des Funkverkehrs. Vom 8. März 1924", *Deutscher Reichsanzeiger* und *Preußischer Staatsanzeiger*, No. 66 (March 8, 1924). For background on the law and the complex negotiations behind the scenes between the Post and other government ministries, see Barch, record group R/4701/10858 Reichspostministerium Z 1/17936 Akten Betr. Funkgesetznovelle 1922, and record group R/4701/8673 Reichspostministerium Sitzungsberichte R.F.K. Geh. Registratur Z, Band 1 1919–1925.

anyone wishing to build a crystal radio (instead of buying a manufactured one) become a member of an approved radio hobby club and receive a permit issued by the club. It further stipulated that the approved radio hobby clubs could also administer the Postal Ministry test and issue to its members the *Audionversuchserlaubnis* required for ownership of a receiver with amplification (tubes). Commercially produced radios still had to have their design approved by the Postal Ministry, and could not be sold to someone who did not already have a radio permit. Punishment for possession of an unlicensed radio or even for attempting to build a radio without first holding a permit remained draconian and was, in fact, now increased in an attempt to reduce the already considerable number of illegal receivers then in use. Nevertheless, the law also allowed for the first time the issue of a limited number of permits to own and operate a radio transmitter for experimental purposes, a practical necessity if industry were to be able to manufacture, test, and develop radios, and scientists to gain greater understanding of how radio worked. A handful of the permits were also issued to radio clubs (as opposed to companies), and even a few well-known scientists), under the assumption that they would also engage in technical experiments.[36] Whatever the intention, this also opened the door to ham radio, a fact the postal authorities certainly came to regret. As we shall see below, the issue of legalizing amateur transmission in Germany would remain highly contested until after the Second World War.

What this meant was that in the first two years of broadcast radio in Germany, ownership of a radio receiver was possible, but it was quite costly and hedged in with all sorts of rules and restrictions. To simply own a radio, one had to hold a *Radio Genehmigungsurkunde*; to build a crystal receiver, one needed a *Detektorenversuchserlaubnis*, and to own or build a radio with tubes, one needed an *Audionversuchserlaubnis*. These latter were issued only after successfully passing a rather strict technical exam. Ownership of a (legal) receiver further required membership in a radio hobby club. This was a great benefit for the clubs, which saw their membership explode as the fad for radio grew. They were suddenly state

[36] Amtsblatt des Reichspostministeriums Nr. 46, included in: *Der Radio-Amateur* 2, No. 8 (May 28, 1924) and Obertelegrapheninspektor Ernst Schulze, "Die neuen Rundfunkbestimmungen" in: *Der Radio-Amateur* 2, No. 11 (July 4, 1924), 289–292 and Justizrat Dr. Felix Szkolny, "Rechtsprechung. Radio-rechtliche Zeit- und Streitfragen" (part 1) in: *Der Radio-Amateur* 2, No. 34 "2. Messeheft" (December 12, 1924), 935ff and part 2, in: *Der Radio-Amateur* 3, No. 4 (January 23, 1925), 97–99. See also: "Mitteilungen", in: *Der Radio-Amateur* 2, No. 29 (November 7, 1924), 778–782.

partners and were placed in charge of the technical education of the German people. True, this arrangement was intended simply to save the Postal Ministry money and resources, but it gave the hobby clubs a real mission. Yet, on the other hand, the law indirectly favored the well-to-do and the residents of urban areas. Rural areas often did not have and could not support a radio hobby club of their own, which greatly restricted the spread of radio to rural areas. Moreover, not only were the permits and the monthly radio fee expensive in themselves, but membership in a hobby club also brought the cost of dues and a subscription to the club magazine or newsletter. Legal radio ownership in Germany in the 1920s and early 1930s was always an expensive proposition, but in 1923, 1924, and early 1925, it was particularly expensive and difficult. The result was predictable: thousands of Germans opted for the least expensive solution, which was illegal radio ownership. A vast illegal DIY radio culture developed almost overnight.

The Postal Ministry worked hard to limit illegal radio use. As we have seen, even the attempt to build a radio without first holding a permit was a crime, and the postal authorities were often very heavy handed in their enforcement of the law.[37]

It made regular use of its powers to search private premises without warning or warrant. Many were caught, and the Postal Ministry worked hard to publicize the numbers in order to dissuade others.[38] Though we can never know for sure how many illegal radios existed, it is certain that the police and postal authorities were unable to stop illegal radio ownership. The reasons were simple: given the state of radio technology in the 1920s and into the 1930s, it was technically very difficult to determine the location of an illegal radio transmitter when it was used carefully, and it was next to impossible to discover an illegal receiver when used correctly. Those who were caught were mostly denounced by neighbors or by radio parts sellers, or were betrayed by owning a visible external antenna. Based

[37] For example, see the case of A.M. Uelzen, whose unfinished crystal radio was confiscated by the postal authorities. A.M. Uelzen, "Briefkasten. Beschlagnahme einer Empfangsanlage", *Der Radio-Amateur* 2, No. 20 (September 5, 1924), 526–527.

[38] See, for example, "Warnung für Schwarzhörer!", in the rubric "Verschiedenes", in:

Der Radio-Amateur 3, No. 15 (April 10, 1925), 375. The article announced that recently 78 people have been sentenced to fines of up to RM200 for "building and use of radio sets without a permit" ("wegen Herstellung und Betrieb nicht genehmigter Funkanlagen"). In every case the radios were also confiscated.

on the sales of the parts necessary to construct a simple crystal radio receiver, one contemporary estimate held that there were 25 illegal receivers for a single legal one.[39]

This early system was too cumbersome and impractical to last. Over the Spring of 1925, rules for radio permits were slowly relaxed in a series of rulings from the Postal Ministry.[40] In the summer, the Postal Ministry announced a major change in the law in the *Bekanntmachung über den Unterhaltungsrundfunk vom 24. August 1925*, which went into effect on September 1, 1925. A permit and monthly fee were still needed to own a radio receiver, but both the Crystal Radio Permit and the Tube Radio Experimental Permit were dropped, as was the requirement that radio owners be members of a radio club. Any sort of radio could now be used, as long as it didn't cause interference.[41] With a few minor changes, this basic structure remained in effect into the 1950s and beyond.[42]

The climb-down embodied in the 1925 law did not mean that the Postal Ministry suddenly trusted the public; it was simply a recognition that radios in private hands did not bring the kind of anarchy and chaos that many in the bureaucracy feared. It was also at least a tacit recognition that the clubs had done their job of educating the public. The problem of interference did not go away, but it remained manageable.[43] The change in the law also weakened the radio clubs, just at a time when they were becoming increasingly aggressive in their negotiations with the Postal Ministry.[44] It is not known if this played a role in the decision by the Postal

[39] Ansgar Diller, "Aus den Kinderjahren des Rundfunks. Wie Post und Rundfunk vor fünfzig Jahren auf Schwarzhörerfang gingen", *epd/Kirche und Rundfunk*, 56, (August 24,1974), 4. as cited in Fuge, "Schwarzhörer".

[40] "Unterhaltungsrundfunk", *Der Radio-Amateur* 3, No. 15 (April 10, 1925), 26.

[41] Dr. Paul Gehne, "Von der Audionversuchserlaubnis zur Sendeerlaubnis", *Der Radio-Amateur* 3, No. 29 (July 17, 1925), 719–721, and "Neuregelung des Unterhaltungsrundfunks" in the rubric "Kleine Funkmitteilungen", *Der Radio-Amateur* 3, No. 35 (August 28, 1926), 31–32.

[42] "Neue Bestimmungen über den Rundfunk. Gültig vom 1. Mai.1930" reprinted from the *Amtsblatt des Reichspostministerium* Nr. 32 vom 15.4.1930 in: *Funk Bastler* 17 (April 25, 1930).

[43] The progress of radio receiver design also helped, since simple regenerative receivers, which were the state of the art in 1923 or 1924 and which were so prone to causing interference when used improperly, were rapidly superseded by other designs by 1925 or 1926, which were not so prone to causing interference.

[44] Not least, the major radio clubs began to act as spokesmen for the ham radio enthusiasts in their midst, and increasingly agitated for the more widespread introduction of private transmitters—this despite their own often mixed relationship with the transmitting amateurs.

Ministry to dilute the law, but it certainly had this effect. Without the need to be a member of a club in order to obtain a radio license, club membership dropped dramatically in late 1925.[45]

Yet once the genie of radio had been "let out of the bottle", it was difficult to control. Bredow's understanding that the role of the public would be limited and subject to strict, state-policed limits soon came into conflict with the dynamic of the radio hobby and the wishes of countless hobbyists, who were well informed about developments in the rest of the world, and who did not want Germany to take a backseat to other countries in the development of the new technology. Despite the mistrust of the public, the fact that Bredow made a role for hobbyists in his plans at all was crucial in the development of radio as a hobby in Germany in the 1920s. It encouraged private enthusiasm for radio, helped create a network of radio clubs, and genuinely helped spread knowledge of the technology of radio. And despite the near-contempt with which Bredow and his colleagues in the Postal Ministry often held the amateurs, Bredow's plan gave the clubs a strong raison d'être. Once they were made (very junior) partners of the state, they couldn't be ignored completely or even suppressed later on. And once people caught radio fever, their enthusiasm was genuine and hard for the state to curb. The German model of broadcast radio created the context within which popular enthusiasm for radio could exist, even far beyond the intended top-down model. Though the radio mania of 1923, 1924, and early 1925 could not be sustained, the radio hobby and its clubs emerged from the founding period as established entities, which could not be ignored when it came to framing media policy.

There were very good reasons why the home-building of radio receivers became so popular so quickly. The self-assembly of complex and cutting-edge electronic instruments quickly became an end in and of itself for hobbyists, not just in Germany. In a symbolic but very tangible fashion, there was no better way to appropriate and control technological modernity for private individuals. Home-building made the potentially frightening and abstract new world of radio personal, domestic, and controllable. It fit well with a male emphasis on tools and building, and with

For example, see Dr. Paul Gehne, "Von der Audionversuchserlaubnis zur Sendeerlaubnis", *Der Radio-Amateur* 3, No. 29 (July 17, 1925), 719–721.

[45] For example, the Wilmersdorf chapter of the Deutscher Radio-Klub (DRC) in Berlin went from a membership of 547 to a low of 279 by September 1925. "Mitteilungen", *Der Radio-Amateur* 3, Nr. 39 (September 25, 1925), 27–31.

a very German national identification with engineering and precision technology.[46] It also fit into an existing middle-class discourse of technical progress and education as a solution to perceived national and personal decline. But equally as important, the DIY radio movement was a financial imperative in the early years of radio in Germany. Early manufactured radios were extremely expensive in the 1920s, and the German economy (and thus individual purchasing power) was very weak in the wake of the lost war. One must remember the economic context of instability and relative impoverishment of Germany in the early 1920s. Broadcast radio was founded in the midst of the hyperinflation of 1923, in which large parts of the German middle class lost not only its financial foundation but often its moral bearings as well.[47] The economy recovered a bit in the late 1920s, only to collapse during the Great Depression, so that the economic imperative to home-build radios also lasted much longer in Germany than in other countries like the US. This fueled the radio hobby to a considerable extent (Table 2.1).

Note that radios were first sold without tubes, which had to be purchased separately and added another 30–40% to the price of the radio. Radio tubes had a limited life span even with the best of handling, and thus needed to be replaced periodically. Radios also needed earphones or a loudspeaker (these were rarely built-in until the early 1930s), both of which were additional expenses. And until well into the 1920s, most radios were powered by batteries (made necessary by the design of radio receivers of the time, as well as the fact that not everyone had a reliable source of electricity in the home at a regulated voltage). Batteries were expensive, often messy, and even dangerous, and needed periodic recharging. Radios which ran on (alternating) household current became common only in the early 1930s, and, even then, electrification in Germany

[46] Note that building competitions judged not only the technical merits of design but the precision and cleanliness of the construction. See Haring, *Ham Radio*, on the male nature of the radio hobby.

[47] On the inflation in Germany, see: Gerald D. Feldman, *The Great Disorder: Politics, Economics, and Society in the German Inflation, 1914–1924* (Oxford and New York: Oxford Univ. Press, 1996); Adam Fergusson, *When Money Dies: the Nightmare of Deficit Spending, Devaluation, and Hyperinflation in Weimar Germany* (New York: Public Affairs, 2010); Bernd Widdig, *Culture and inflation in Weimar* (Berkeley: University of California Press, 2001); Frederick Taylor, *The Downfall of Money: Germany's Hyperinflation and the Destruction of the Middle Class* (London: Bloomsbury, 2013); Theo Balderston, *Economics and Politics in the Weimar Republic* (New Studies in Economic and Social History, 45) (Cambridge: Cambridge University Press, 2002).

Table 2.1 Sample prices for manufactured radio receivers in Germany[a]

Year	*1923–1924*[b]	*1925–1927*	*1928–1932*	*1933–1934*
Simple crystal radio receiver	ca. 20	4.80[c]–12[d]	9[e]–18[f]	
1–3 tube receiver	ca. 90	20–195[g]	34[h]–210[i] (+ 2 tubes)	76[j]
Luxury receiver	ca. 150–200	300[k]–669[l]	375[m]–2400[n] (+7 tubes)	
Shortwave receiver	N/A	200[o]	135.50[p]–789[q] (+6 tubes)	

[a]All prices in Reichsmarks. This table can only give a general idea of relative prices. There was great variation in price and performance between models. Most radios were sold without tubes (adding roughly RM 8 per tube), and tubes eventually burnt out and needed replacement. Early radios ran on batteries, which had to be purchased separately (ca. RM 3–5 ea.), and needed regular recharging. They also required headphones (RM 5–12.50). By about 1927, radios which ran off of AC or DC household current became more and more common, though such radios usually had a roughly RM 30–40 price premium attached. By this point, radio receivers beyond simple crystal radios included a built-in loudspeaker (Though some radios required purchase of a separate loudspeaker, RM 30–250.00). Luxury radios were often available in wood cabinets with a variety of finish and wood available. They sometimes also included a built-in phonograph and speaker. Remember, too, that the radio subscription was RM 2 per month per radio. In short, prices could vary widely, but were generally expensive, compared to wages.

[b]Dr. Eugen Nesper, "Welcher Empfangapparat kommt für den Radio-Interessenten in Betracht?", *Der Radio-Amateur* 3, No. 31 (November 21, 1924): 814–818. The price given for a crystal radio includes twin headphones.

[c]Radio Bauer, 1925–1927 p. 4 Detektor Nr. 1 RM 4.80. (Since no brand was given, this was probably produced by or for Radio Bauer).

[d]Ibid. Detektor Nr. 6, RM 12.00. Again, this was probably a receiver made by or for Radio Bauer.

[e]Radio Diehr, 1928–1929, p. 28. Nora DA50 crystal receiver, RM 9.00.

[f]Ibid. DeTeWe DE3 crystal receiver, RM 18.00.

[g]Radio Bauer 1925–1927, p. 10. Radiosonanz "Osoflex" 3-tube receiver RM 195. A set of tuning coils for various bands was an additional RM 33.

[h]Radio Diehr 1928–1929, p. 29. Lorenz "Volksfreund" with two tubes, RM 34. (Transformer required at additional cost).

[i]Radio Diehr, 1928–1929, p. 32. Mende E RM 45, 210.00.

[j]*Volksempfänger* VE301 (3 tubes). In 1938, the DKE 38 (2 tubes) was released at RM 35.00.

[k]Radio Bauer 1927, p. 10. Telefunken 3/26 RM 295.00.

[l]Radio Bauer 1927, Interimsliste, p. 3. "Superhut IV", complete with batteries, frame antenna and tubes, RM 669.00.

[m]Radio Diehr 1928–1929, pp. 39–40. Lange Neutrodyn 5-tube with built-in tuning coils, RM 375.00.

[n]Radio Diehr 1928/1929, p. 43. Siemens & Halske Neutrogerät in Mahogny Cabinet, with frame antenna and loudspeaker, but without tubes or batteries, RM 2400.00.

[o]Radio Bauer 1927, p. 9. "Radiofrequenz" RM 200.00.

[p]Radio Bauer 1927, p. 33. 3-tube Saba KE, RM 135.50.

[q]Radio Bauer, 1927, p. 48. 6-tube Reico RF 366 RM 789.00, without tubes.

was by no means complete. All this means that radio was expensive to buy and expensive to run when German broadcasting first began. Moreover, in the early years after the introduction of broadcast radio, hundreds of German firms entered into a radio frenzy and brought radios to market which were often of poor quality and design, yet were sold with unrealistic promises of performance. Aside from the cost savings, the home builder thus often had a clear advantage in quality, at least until 1927 or 1928. Home builders could assimilate the latest technical developments from abroad more rapidly than industry could, meaning that until the early 1930s, homebuilt receivers were often technically more advanced.[48]

To be a consumer of broadcast radio in the early 1920s anywhere in the world thus often meant also being a producer of the tools with which to do so. This was true in the US and Great Britain, and it was even more true in Germany. Today, radios are ubiquitous and banal machines that any child can use and which are so cheap that most people own several without even really being conscious of it. But at the time, the commercial market for consumer radio receivers was just developing, there were few economies of scale possible, and radios were treated as luxury devices. Access to the new medium, even for an act as simple as listening to broadcast radio, required the acquisition of considerable technological knowledge, and usually the self-production of a receiver. This situation lasted in Germany into the early 1930s. This helped make radio a hobby in Germany on a large scale. For economic reasons, the radio revolution was also a DIY revolution. Because of the strong DIY component, for many years, in Germany, there was a great deal of overlap between what later developed into separate communities of hobbyists. In the 1920s, listening to radio always meant having enough technical understanding of radio to be able to tune the receiver—not a simple undertaking with early regenerative radios (which fed some of the amplified signal provided by one or more vacuum tubes back into the circuit, and which thus could easily be pushed into the transmission of noise, which would block others' reception across a potentially wide area), and usually also meant building one's own receiver as well. There was fairly little difference between the community of simple listeners, the community of audio enthusiasts, and the community of

[48] "Schlechte Radio-Werbung", under the rubric "Briefkasten", *Der Radio-Amateur* 2, No. 15 (August 1, 1924), 395–396; Dr. M Arndt, "Die Krise in der Radio-Industrie", *Der Radio-Amateur* 2, No. 16 (August 8, 1924), 412–413. (These articles echo reports in trade-based radio magazines of the same time.)

do-it-yourselfers interested in the building of radios—the German term is *Radio-Bastler*—and the "ham" amateur radio enthusiasts who wished to transmit as well as receive. Until roughly 1927, there was no real distinction between the four groups, and they remained in close contact long thereafter. This was also true in the US and everywhere else, save that in the US and Britain, the distinction between these four groups became apparent much earlier, and amateur radio enthusiasts (transmitters or "hams", as opposed to "just listeners") in particular were a clearly separate (and self-defined) group much earlier. Separate amateur radio organizations for these (transmitting) hobbyists did not form in Germany until 1925, whereas in the US, the American Radio Relay League (ARRL) was founded as early as 1914, and it is estimated that as many as 480 private transmitters existed in the US by 1910.[49]

The Radio Revolution of the 1920s

With the invention of broadcasting to give people something to listen to, and the mandate for clubs to teach the basics of radio and help people gain the license needed to own a radio, a real radio revolution broke out all over Germany. People suddenly became obsessed with the new medium, even without knowing what it was all about. Ironically, in the first few years of broadcasting, in Germany, there was very little German (as opposed to foreign) radio available outside of Berlin and, soon, a handful of other big cities. Both broadcasting practice and radio technology were in their infancy, but radio was suddenly a powerful symbol of modernity and what a later age would call globalization. Clubs sprang up everywhere, first, in the big cities (where the first German broadcast stations were located, and thus where people could hear radio signals with comparatively simple receivers), then in ever-smaller towns over the course of the 1920s. The German electrical industry seized on the radio frenzy, seeing the production of radio receivers, kits, and part as a solution for their systemic depression after the Second World War. Radio "stars" were just as well known,

[49] Körner, *Aamateurfunk*, 4, citing the Second Annual *Wireless Book of the Wireless Association of America* (New York: Modern Electrics Publication, 1910), 10–16. By the end of 1913, roughly 2000 amateur transmission licenses had been issued in the US, and a large number of others transmitted without formal licenses. De Soto, *2000 Meters*, 34. In general, see also 16–37.

glamorous, and influential as the "stars" of cinema at the time, and both were the darlings of popular print media.

Berlin, the site of the very first broadcast transmitter and long a center of the electrical industry, was the first place where radio clubs got started on a large scale.[50] But not by much: centers of radio enthusiasm formed almost immediately around other big cities. The key was not only size but the presence of technically educated people (usually men, given the German educational system), and, above all, being close to one of what were then only a handful of transmitters. Stuttgart, Munich, and Hamburg were also among the first centers of radio hobby activity along with Berlin. Much of this early expansion came before the end of 1924, scarcely a year after the beginning of broadcasts. From the big cities, clubs spread next to medium-sized cities, again especially those where there was a strong tech-educated population.[51] Then, as transmitters became more powerful, radio clubs spread out into smaller towns. Correspondingly, in the bigger cities, clubs founded chapters in more and more parts of the cities. This expansion also took roughly a year.

The most important of the radio hobby clubs, the "Deutscher Radio-Klub" (DRC), was founded in the Spring of 1923 in Berlin, even before broadcasting began.[52] (By October 1923, the spelling changed to "Club".)[53] Nearly simultaneously, other radio clubs were founded in other big cities, like Munich[54] or Frankfurt/Main. In August 1923, the first issue of "*Der Radio-Amateur*" (*The Radio Hobbyist*) appeared. It was the organ of the DRC and the first technical radio hobby magazine in

[50] The first radio club in Germany was formed in Coburg in northern Bavaria in May 1923. Dokufunk Archive, "Radioverein Coburg, Studiengesellschaft für Elektrotechnik und Funkentelegraphie, Minutes of the General Assembly of 25. May 1923". In May 1928, the club became part of the Süddeutscher Radioklub. Dokufunk Archive, "An das Amtsgericht Coburg, Abteilung Vereinsregister of 7 May 1928".

[51] Stuttgart was then closer to being a middle-sized city, but the presence of a technical university and its location on the western border of Germany helped make it one of the first-wave cities. Coburg was much smaller than Stuttgart but also had a technical university.

[52] "Mitteilungen des Deutschen Radioklubs e.V.", *Der Radio-Amateur* 1, No. 3 (October 1923), 61–64. According to its first President, Dr. Eugen Nesper, the planning for the club went back to December 1922. Dr. Eugen Nesper, *Der Radio-Amateur. Radio-Telephonie. ein Lehr- und Hilfsbuch für die Radio-Amateure aller Länder* (Sechste, bedeutend vermehrte und verbesserte Auflage) (Berlin: Verlag Julius Springer, 1925), 25.

[53] *Der Radio-Amateur* 1, No. 3(October 1923), 64.

[54] "Gründung des Bayrischen Radioklubs", *Der Radio-Amateur* 1, No. 3 (October 1923), 64.

Germany.[55] It was the most important of the radio hobby magazines in German and also one of the longest-lived, ending publication only in 1944, when wartime created paper shortages and most magazines and newspapers were closed down. Other hobby magazines followed. In addition, by early 1924, most newspapers were including a special column on radio.[56]

In early 1924, the Postal Ministry moved to organize the burgeoning radio clubs as adjuncts of the new radio law. On January 24, 1924, it convoked a meeting of the major radio clubs in Berlin in order to tell them about the new rules which were soon issued in the form of the "Verordnung zum Schutz des Funkverkehrs" of March 8, 1924 (Decree on the Protection of Radio Traffic of March 8, 1924).[57] Present at the meeting were the German Radio Club (Deutscher Radioklub [sic], Berlin), the Southwest German Radio Club (Südwestdeutscher Radioklub, Frankfurt a/M.), the Hamburg Radio Club (Hamburger Radioklub, Hamburg), the North German Radio Club (Norddeutscher Radioklub, Bremen), the Württemberg Radio Club (Württembergischer Radioklub, Stuttgart), the Southern German Radio Club (Süddeutscher Radioklub, Munich), the Schleswig-Holstein Radio Club (Schleswig-Holsteinischer Radioklub, Kiel), the Dresden Radio Club (Radioklub Dresden), the Leipzig Radio Union (Radiovereinigung Leipzig), and the Coburg Radio Association (Radioverein Coburg).[58] These were the major regional clubs, which had come together one day earlier to form the "German Radio Cartel" (*Deutsches Funkkartell*). Its headquarters was first in Hamburg.[59] By this

[55] The first issue of *Der Radio-Amateur* appeared August 23, 1924. *Der Radio-Amateur* 1, No. 1 (August 23, 1924).

[56] P. Wassman, "Allerlei Neuigkeiten", *Der Radio-Amateur* 2, No. 2 (February 1924), 61.

[57] Der Reichspräsident [Ebert], "Verordnung zum Schutze des Funkverkehrs. Vom 8. März 1924", *Deutscher Reichsanzeiger* und *Preußischer Staatsanzeiger*, 1924, Nr. 66 (March 8, 1924). For background on the law and the complex negotiations behind the scenes between the Postal and other government ministries see: Barch. record group R/4701/10858 Reichspostministerium Z 1/17936 Akten Betr. Funkgesetznovelle 1922, and record group R/4701/8673 Reichspostministerium Sitzungsberichte R.F.K. Geh. Registratur Z, Band 1 1919–1925.

[58] See: Dokufunk Archive, Reichspostministerium, "Niederschrift der Besprechung am 24 Januar [1924] im Großen Sitzungssaales des Reichspostministeriums" (The original came from the papers of the Coburg Radio Association.).

[59] Not all members of the Cartel were present at the January 24 meeting, only the bigger regional clubs. Member organizations included: Funkverband Niederdeutschland e.V., Hamburg, Deutscher Radio-Club e.V., Berlin, Hamburger Radio-Klub e.V., Norddeutscher

time, many more clubs existed, but these were the big ones, from whom the Postal Ministry could expect both leadership and discipline.[60] The postal authorities had pushed for the formation of the Cartel because it did not want to have to deal with an ever-shifting number of small local and regional clubs and wanted to establish a clear transmission belt for communicating its wishes. The clubs, on the other hand, realized that the formation of the Cartel would strengthen their hand in negotiations with the authorities. Consolidation was thus an imperative for both sides, given the explosive growth of radio and the sheer size of the radio movement. The Postal Ministry made the clubs, and their members become central actors in the introduction of broadcasting and the spread of radio in Germany. But rather than being the passive actors the Postal Ministry expected, the clubs began to lobby and even politely push back against regulations they deemed excessive.[61]

Meanwhile, the formation of more and more local radio clubs continued in early 1924, as radio fever hit Germany. On January 11, the "Radio-Klub Görlitz" (Görlitz Radio Club) was founded. Some 70 people immediately joined.[62] By the end of January 1924 (and possibly earlier), a radio club had been founded in the industrial town of Chemnitz, with a

Radio-Club e.V. Hamburg, Radio-Club Lübeck e.V., Schleswig-Holsteinischer Radio-Klub e.V., Kiel, Mecklenburgischer Radio-Verein e.V., Rostock, Verein der Funkfreunde e.V., Schwerin, Mecklenburg-Strelitzer Radio-Klub e.V., Neustrelitz, Funkverein Wismar e.V., Wismar, Radio-Klub Bremen e.V., Bremen, Radio-Klub Unterweser e.V., Bremerhaven, Oldenburger Radio-Klub e.V., Oldenburg, Radio-Vereinigung Wilhelmshaven-Rüstringen, Wilhelmshaven, Emder Funkverein e.V., Emden, Osnabrücker Radio-Klub e.V., Osnabrück, Funkverband Weser-Ems, Bremen, Mitteldeutscher Radio-Verband e.V., Leipzig, Oberdeutscher Funkverband e.V. Stuttgart, Ostdeutscher Radio-Klub e.V., Königsburg, Pommerscher Radio-Klub e.V. Stettin, Süddeutscher Radio-Klub e.V., München, Gesellschaft von Freunden der Radiotelephonie und Telegraphie Südwestdeutscher Radio-Klub e.V., Frankfurt/M, Verein der Funkfreunde Schlesiens e.V., Breslau and Westdeutscher Funkverband e.V., Münster. Nesper, *Der Radio-Amateur*, 26–33. "e.V." indicates that the club was legally incorporated as a club according to German law. Note the strength of port cities in this list, as well as the larger regional metropoles.

[60] "Tagung der Deutschen Radio-Klubs in Berlin", *Der Radio-Amateur* 2, No. 2 (February. 1924), 62–63.

[61] See the transcript of this first big meeting between the Post and the big regional clubs: Dokufunk Archive, "Niederschrift der Besprechung am 24 Januar [1924] im Großen Sitzungssaales des Reichspostministeriums".

[62] "Klubsachen", *Der Radio-Amateur* 2, No. 5 (April 16, 1924), 154. Again, both university and business supported the founding of the club.

branch in nearby Leipzig.[63] On February 5, 1924, the "Radioclub Kassel" was founded in that same city. Thirty members immediately joined.[64] On March 9, 1924, the "Radio-Club Herne" was founded in that city in Westphalia. What was notable about it, aside from the early date of its founding, was that it was the first radio hobby club in the occupied territories—where radio ownership was still illegal. Forty members joined at the founding.[65] On March 18, the "Oberschlesische Funktechnische Gesellschaft" ("Upper Silesian Radio-Technical Society") was founded in Gleiwitz, around a core group of industrial engineers in Upper Silesia.[66] At roughly the same time, students and faculty of the university of Münster founded the "Verband der Rheinisch-Westfälischen Funkfreunde" ("Association of Rheinish-Westfalian Radio Hobbyists").[67] Also in March 1924, the "Gesellschaft der Funkerfreunde in Hannover" was founded there by a group of business leaders.[68] In late March or early April, the Rheinisch-Westfälischer Radio-Club (Hagen i. W) was founded. Once again, over 70 people immediately joined.[69]

These examples demonstrate the dynamics behind the radio revolution in Germany. Within a year of the introduction of broadcasting, radio had clearly been identified as the "next big thing". Thousands flocked to join the hobby clubs, which spread rapidly across Germany. It is clear from this short sketch of the early days of radio hobby clubs in Germany that there was a very strong presence of professionals from the electrical industry, and scientists and engineers from the universities and technical schools. "Old elites" from business and the civil service were also highly influential. This should not be a surprise: all clubs liked to have as many notables in the club leadership as possible, for prestige reasons. Nor should it be a surprise that those with a professional or academic reason to be interested in radio should flock to the radio hobby clubs. Even today, many engineers, communications professionals, and scientists are amateur radio

[63] P. Wassman, "Allerlei Neuigkeiten", *Der Radio-Amateur* 2, No. 2 (February 1924), 61.

[64] "Briefkasten", *Der Radio-Amateur*, 2, No. 6 (April 30, 1924), 174.

[65] "Radio-Club Herne", in: *Der Radio-Amateur* 2, No. 9 (June 11, 1924), 251.

[66] "Briefkasten", *Der Radio-Amateur*, 2, No. 6 (April 30, 1924), p. 174.

[67] "Gründung eines Amateurverbandes für den rheinisch-westfälischen Bezirk", *Der Radio-Amateur* 2, No. 6 (April 30, 1924), 174.

[68] "Gründungsversammlung der Gesellschaft der Funkerfreunde in Hannover", Der Radio-Amateur 2, No. 6 (April 30, 1924), 174. It was founded by the Handelskammer Hannover and Verkehrsverband Niedersachsen (two regional business groups) on March 31, 1924.

[69] "Klubsachen", *Der Radio-Amateur* 2, No. 5 (April 16, 1924), 154.

operators—those in technical fields are natural members of a technical hobby. Within the German context, the network of technical universities, with their emphasis on engineering and practical technology, were particularly important in spreading knowledge of radio and providing many of the early members of the radio hobby.[70] If there is a surprise here at all, it is that radio fever extended well beyond the "usual suspects" of notables and technical professionals, to include all levels of society. The Postal Ministry refused to let clubs be dominated by business interests, be allowed to issue radio permits, and administer the Postal Ministry tests without supervision, in part simply to keep the hobby clubs from growing too powerful through business ties.[71] But the large numbers of men (and a few women) who joined the radio hobby clubs in their early days were clearly not just from the social elites. On the other hand, the highly technical nature of the exam needed to obtain the *Audionversuchserlaubnis* offered a great advantage to anyone who had a strong background in math and physics, such as one might obtain in an elite *Gymnasium* (college-preparatory high school)—or to workers in electrical fields.

Non-elites had to work harder to have access to the new medium, but radio fever and the radio hobby were never limited to just the educated middle classes in Germany. People from all classes were fascinated by radio. Networks of specifically working-class clubs developed almost as quickly as those dominated by the wealthy and educated. The growth of a separate working-class network of clubs linked to the Social Democratic Party (and later a separate, parallel network linked to the Communist Party) helped non-elites to gain access to the world of radio. For example, the "Worker's Radio Club" ("Arbeiter-Radio-Klub", ARB) was founded in Berlin in the Spring of 1924, just as the middle-class radio clubs were being founded. It rapidly spread all over Germany.[72] This should be no surprise; Germany

[70] This is a topic much larger than this book. The German university system was and still is very hierarchical. The universities—those institutions of higher education, which carried the university title in their names—carried the highest prestige, but tended to focus on theoretical rather than applied knowledge, which was left to the larger numbers of the less prestigious "Institutes of Higher Technical Education" (*Technische Hochschule*, etc.). From the late nineteenth century, the higher technical schools battled the universities for recognition, mirroring an even larger social conflict between the upper and "middle" middle classes.

[71] "Tagung der Deutschen Radio-Klubs in Berlin", *Der Radio-Amateur* 2, No. 2 (February 1924), 62–63.

[72] Peter Dahl, *Arbeitersender und Volksempfänger: proletarische Radio-Bewegung und burgerlicher Rundfunk bis 1945* Frankfurt/aM.: Syndikat, 1978. On the Arbeiter-Radio-Klub founding, see pp. 39ff.

had a strong electrical industry and large numbers of highly skilled workers. These men and women were potential radio enthusiasts, for they combined the technical skills needed to build radios with the financial need to build rather than buy. Despite the highly technical nature of the first radio ownership exam, it was not only middle-class men and women who had the requisite technical background. The German radio hobby quickly developed parallel middle-class and working-class clubs.

This separation between middle-class and working-class clubs was an entrenched characteristic in German society, dating from the second half of the nineteenth century. In most towns, all kinds of clubs—choirs, gardening clubs, sports teams—existed in parallel middle-class and working-class versions. Radio was no different, though by the 1920s, in the aftermath of the First World War, the divisions between the clubs began to soften, and particularly in the newer technologically based clubs, individuals could cross class lines and might even be members of both a working-class and a middle-class radio club at the same time. The two types of clubs also often worked together in their common interest.[73] Nevertheless, this fundamental division in German society was not overcome until after the Second World War, and still casts a ghostly presence on club life today. We have more to say on this later.

The spread of the hobby clubs continued in 1924 and 1925, as regional cartels of clubs formed. By June 1924, the national Deutscher Radio-Club brought together its own Berlin organization with eight other regional clubs, each with its own network of local chapters.[74] Similar consolidation happened elsewhere as well; other regional club networks were formed as local clubs joined together in regional unions.[75] Strong local clubs,

[73] See below.

[74] "Klubmitteilungen", in: *Der Radio-Amateur* 2, No. 10 (June 25, 1924), 274–276. The member clubs included the Deutscher Radio-Club e.V. Berlin, the Funkverband Niederdeutschland, Hamburg, the Rheinisch-westfälischer Funk-Verband, the Ostdeutscher Radio-Club, Königsberg, the Mitteldeutscher Funkverband, Leipzig, the Schlesischer Radio-Club, Breslau, the Badisch-Württembergischer Radio-Club, the Südwestdeutscher Radio-Club, Frankfurt a.M., and the Süddeutscher Radio-Club, Munich.

[75] For example, the "Rheinisch-Westfälischer Funkverband e.V." was created on May 25, 1924, as the union of all radio clubs in the Rhineland-Westphalia region. Similarly, on May 18, 1924, all of the radio clubs in the (then) two federal states of Baden and Württemberg joined together, though they had trouble finding an appropriate name, since the regional adjectives "Süddeutsche" (South German) and "Südwestdeutsche" (Southwest German) had already been taken by regional associations in Munich and Frankfurt, respectively. And the radio clubs in Bremen, Oldenburg, Wilhemshaven-Rüstringen, Bremerhaven-Geestemünde-Lehe,

particularly in the big cities, were also expanding by forming more and more local neighborhood chapters.[76] By early July 1924, the DRC had 20 chapters in Berlin alone.[77] And independent local clubs continued to be formed.[78] Membership in the radio clubs swelled across Germany. The clubs grew daily and were in the process of consolidating their organization and integrating themselves into the social fabric of Germany.

Consolidation continued the following year. On July 28, 1925, the Radio Cartel held its annual meeting in Munich. Here, it voted to join with the *Funktechnischer Verein* ("Radio-Technical Association", FTV), the largest radio club not yet within the cartel, to form the *Deutscher Funktechnischer Verband* ("German Radio-Technical Union", DFTV).[79] This new

Emden, and Delmenhorst all came together to found the "Funkverband Weser-Ems". "Klubmitteilungen", *Der Radio-Amateur* 2, No. 10 (June 25, 1924), 274–276.

[76] For example, within the Deutsche Radio-Club, Berlin, a new local chapter was formed on April 25 in the well-to-do Berlin district of Steglitz and another in the district of Lichterfelde. By June 1924, other chapters had been founded by employees of the "Priteg" (Berliner Privat-Telephon-Gesellschaft) and another by students of the Beuth-Schule, an engineering school. This latter was a purely student-faculty group, but already had 200 members by June 1924. It soon had its own lab and offered courses in radio technology taught by junior faculty members. "Klubmitteilungen", *Der Radio-Amateur* 2, No. 10 (June 25, 1924), 274–276. Most of the chapters in and around Berlin drew heavily from students of the several Berlin universities, and employees of the large electrical industry, for which Berlin was a major center.

[77] This does not include Potsdam, which had its own chapter.

[78] For example, the Pommerscher Radio-Club e.V. Stettin, which, by June 1924, had 235 members. It had its own lab and lecture hall, where it held eight-day classes in radio technology. It had joined the Deutsche Funkkartell in Hamburg and was in the process of building a network of local chapters, including one specifically for pupils of local schools. By June, dues had been raised from 1 to 1.5 Gold Marks, which included a subscription to *Der Radio-Amateur*. Together with the Association of German Radio Dealers, it was already lobbying the Postal Ministry to obtain a local broadcast transmitter for Stettin. As with most of the clubs at this time, the club leadership was firmly in the hands of notables from science, industry, and the civil service. Somewhat unusual was the open participation of senior German Army officers in the club. "Klubmitteilungen", *Der Radio-Amateur* 2, No. 10 (June 25, 1924), 274–276.

[79] "Tagung des Deutschen Funkkartells in München am 28 Juli 1925", *Der Funk-Amateur* 3, (July 31, 1925), p. 1 of insert; "Die Gründung des Deutschen Funktechnischen Verbandes", *Der Radio-Amateur*, 3, No. 32 (August 7, 1925), 792–795; "Arbeit, Aufgaben und Ziele der Deutschen Vereine von Funkfreunden (= Tätigkeitsbericht of Generalsekretär des Deutschen Funk-Kartell, Friedrich Schmidt, auf 28.Juli.1925 meeting in Mü)", *Der Radio-Amateur* 3, No. 32 (September 11, 1925), 795–797; "Mitteilungen des Deutschen Funktechnischen Verbandes e.V., Berlin", *Der Radio-Amateur* 3, No. 37 (September 11, 1925), 915–916.

organization served as the umbrella organization for all the German radio clubs until 1934. From now on, it would be the contact point between the Postal Ministry and the world of the hobby clubs, and would be the main negotiator with the Postal Ministry.[80] It was led by men who had built the radio hobby movement from its earliest days, and its leadership contained a large number of those who were particularly interested in transmitting. Predictably, this led to conflict with the Postal Ministry.

The rapid growth of both radio ownership and radio clubs is worth a deeper look here (Fig. 2.1). In Germany, radio listeners went from basically zero in early 1923 to at least 1,205,310 as of April 1, 1926.[81,82]

In 1930, 40,000 crowded into the Great German Radio Exposition in Berlin, and there were 2.38 million (legal, licensed) radio receivers in Germany.[83] The development in all the industrial countries was broadly similar (and often much faster!). This was a genuine technological revolution, and because of it, radio truly was at the time the most prominent symbol of modernity. In July 1923, broadcasting did not exist in Germany, and there could only have been a handful of receivers and transmitters in all of Germany outside of the government and the military, mostly in uni-

[80] There is some indication that the fusion of the *Funktechnischer Verein* (FTV) and the Funkkartell was pushed by von Bredow, who obviously had an interest in having a single, influential representative of the hobby world with whom to work. This is plausible. See: "Mitteilungen des Deutschen Funktechnischen Verbandes e.V., Berlin", *Der Radio-Amateur* 3, No. 37 (September 11, 1925), 915–916.

[81] Winfried B. Lerg, *Rundfunkpolitik in der Weimarer Republik* (Munich: DTV, 1980), 116. These official figures are only for registered, legal paying subscribers. They certainly underestimate the true number of radios in existence, as the Reichspost admitted. Lerg, *Rundfunkpolitik*, 104. In any case, given that radio listening was quite often a family or communal activity in the 1920s and 1930s, the number of radio listeners might have been as much as four to five times the official figure.

[82] Growth of radio receiver ownership in Germany. Figures from Hans Peter Richter, "Die Entwicklung der Rundfunk-Teilnehmerzahlen für das Deutsche Reich und die Bundesrepublik nach den Angaben der Post für den Stichtag 1 April", *Rundfunk und Fernsehen* 3 (1955): 398. Note that this table actually counts only those paying the Radio Fee. This is a good proxy for radio ownership, but it tends to underestimate the total numbers, possibly by a wide margin.

[83] "Eröffnung der Großen Deutschen Funkausstellung und Phonoschau 1930", *Funk* No. 35 (29. August, 1930), Hauptteil, 175. The Opening took place on August 22. The US ambassador was there, as were representatives from the Italian, Turkish and Czech embassies, as well as lots of prominent Germans. Albert Einstein spoke. On the number of radios, see: Kurt Wagenführ (ed.), *Jahrbuch Welt-Rundfunk 1937/1938* (Heidelberg-Berlin: Kurt Vowinckel Verlag, 1938), 52–53.

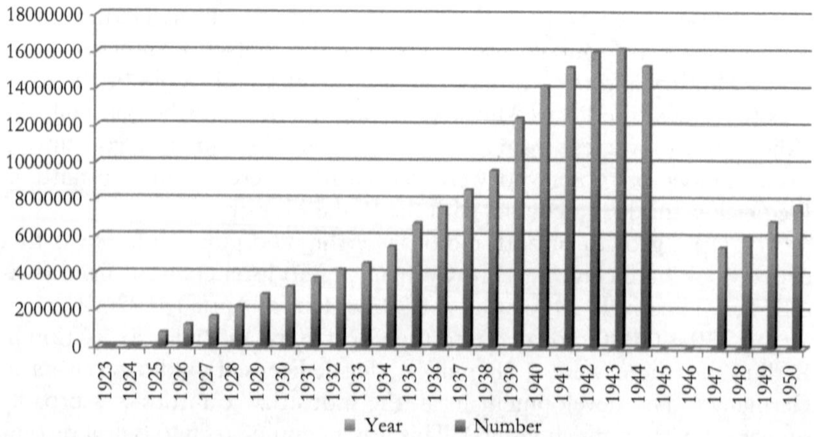

Fig. 2.1 Number of German radio subscriptions, 1923–1950

versity laboratories. By the first half of 1925, radio mania had struck Germany. There was a growing network of German state broadcast transmitters, and radio clubs were springing up across the entire country. In just over a year, a fearfully complex technology which had been the provenance of an educated elite in the government, military, and a handful of university programs had taken the population by storm. In just over a year, radio, if not yet banal, had become everyday, and accessible to people of all walks of life, regardless of education. This is an extraordinary development.

Bredow's careful planning was responsible in part, but it would have been useless if it had not been for three other elements: "radio mania" (widespread popular enthusiasm), the availability of a few basic parts and components (produced by a well-developed electrical industry, seeking new markets after the war), and the structure of the radio hobby clubs to teach technical knowledge and provide an infrastructure for do-it-yourselfers ("maker spaces").

Spreading the Gospel: Radio Clubs, Party Animals, and the Roaring Twenties

The clubs were the key element. The media could create "buzz" around radio, the Postal Ministry could plan all it wanted, and the electrical industry could try its best to flog its products, but without the clubs to mobilize individuals, pass on technical knowledge about how radios worked and how they could be built, and provide physical space and tools for the do-it-yourselfers, radio would not have been nearly as successful or as widespread.

The most straightforward role of the clubs was to not only spread the radio gospel but to spread technical knowledge about radio. This was the task Bredow had assigned them, and for the first two years or so after the introduction of broadcast radio in Germany, the legal structure demanded that people have contact with a club in order to obtain a permit to own a radio receiver. To obtain the permit, there was a rather strict exam to pass, which was so technical that in practice, one had to first attend classes for several weeks. Since the state did not have the resources to provide training classes or even an infrastructure for holding thousands of exams all over the country, the new radio hobby clubs were by law given the job of administering the exam.[84] They also provided most of the preparatory classes, though these could also be offered by other organizations such as the network of adult education schools (*Volkshochschulen*) or even some schools.[85] Remember, at this stage (1923–1925), permits were needed to simply own and operate a radio receiver. Anyone who wanted to have a radio had to have at least fleeting contact with a radio club. Many became members because the clubs provided a lot of benefits beyond administering the ownership permit exam.

[84] "Verordnung zum Schutze des Funkverkehrs. Vom 8. März 1924". *Deutscher Reichsanzeiger* und *Preußischer Staatsanzeiger*, 1924, No. 66 of March 8, 1924. See also: Amtsblatt des Reichspostministeriums Nr. 46, included in: *Der Radio-Amateur* 2, No. 8 (May 28, 1924).

[85] In 1924 and early 1925, holding classes was one of the main activities of the radio hobby clubs. For example, the "Funkverein Dresden e.V." kicked off a series of classes for the AVE with a talk by its Chairman, Dr. rer. Techn. Max Sende, on the history of Radio Tubes. "Mitteilungen befreudeter Radio-Clubs", *Der Radio-Amateur* 2, No. 14 (25 July 1924), 373–374. In the same issue, the DRC published a model syllabus for AVE classes. "Die Funkausbildung zur Erlangung der Audionversuchserlaubnis. Lehrplan für die Unterrichtkurse des Deutschen Radio-Clubs", ibid., 371–373.

The technical knowledge required for the early radio permits was overly complex and included a good stiff dose of mathematics. It was a system designed by the engineers and bureaucrats in the Postal Ministry and only lasted until August 1925. The basics for building and operating a radio receiver were much simpler than the knowledge required to pass the exam. They still needed to be learned, and existing institutions (outside of the new radio clubs) were not able to teach them. Even after the end of the onerous AVE test and permit requirement, the clubs were still one of the main sources of radio information. They continued to offer classes in various facets of radio. They were not only the obvious place to which a beginner could turn, they were also the place where those with the basics could deepen their knowledge. Many, certainly the majority, simply wanted to listen to a radio and were willing to study a bit in order to be able to build one (if only for financial reasons), but large numbers of people became interested in the technology itself. The clubs were the obvious place to deepen and perfect technical abilities and understanding—at least for those outside of graduate programs in physics or professional schools.

The radio clubs were very active in spreading the word. They sponsored public and private lectures on radio technology, designed both to attract new members and deepen the understanding of current ones.[86] They published journals with extensive technical articles.[87] They maintained libraries of key technical reference works[88] and back issues of journals (usually including at least the principle German and often foreign club journals).[89]

[86] For example, the Berlin-Lichtenberg chapter of the DRC held a lecture in November 1925 on radio during the war, presented by a Herr Bieleit. "Mitteilungen", "Die Vortragsfolgen" insert, *Der Radio-Amateur* 3, No. 47 (November 20, 1925), 24–29. In another example, in September 1925, the Berlin-Charlottenburg chapter of the DRC presented a talk by an (unnamed) radio hobbyist, who spoke on tips and tricks of the hobby. "Mitteilungen", "Die Vortragsfolge", insert, *Der Radio-Amateur* 3, No. 37 (September 11, 1925), 27–31.

[87] For example, *Der Radio-Amateur*, the club journal of the DRC, was one of the major radio-hobby publications in Germany. As we shall see below, the club journals became increasingly more technical in content as time went on.

[88] Every club seems to have had at least a small library collection, and the larger chapters often had fairly extensive collections. Remember that this is well before the internet, so libraries were important. While public libraries also quickly obtained popular reference books, the club libraries often had more, certainly outside of the bigger cities. Moreover, clubs usually had examples of foreign radio journals and magazines in their libraries, making it possible for club members to keep up to date with international trends.

[89] For example, in 1926, the local "Wannseebahn" chapter of the DRC (with some 500 members) had a library with copies of a selection of foreign radio journals. "Das Clubleben

They negotiated discounts and organized the purchase and sale of basic radio components for members.[90] Moreover, they provided a physical space where members could come together and actually build radios. At the most rudimentary, this might just be the back room of a bar where the club held its regular membership meetings.[91] But the bigger club chapters often built workshops and stocked them with tools. Some of these workshops were fairly sophisticated, and in some cases, the electrical industry itself provided hobbyists with workspace.[92]

As any real estate agent will tell you, physical space is not a trivial thing. Remember that housing was a major preoccupation in Germany (as elsewhere) in the 1920s and 1930s.[93] Many people still lived in very cramped and even unhealthy conditions. When living conditions were this tight, it is clear that there often wasn't much space for hobbies at home. Similarly, wages and the general standard of living were much lower in the 1920s and 1930s than today, meaning that most households could not afford even rudimentary tool kits. The clubs served a very important function by providing public space and by pooling resources to purchase basic tools

in der Ortsgruppe Wannseebahn (Sitz Steglitz) des Deutschen Radio-Clubs e.V.", *Der Radio-Amateur* 4, No. 12 (March 19, 1926), 235–238.

[90] For example, as early as July 1924, the DRC in Berlin had negotiated discounts from nine different Berlin firms. "Mitteilungen des Deutschen Radio-Clubs e.V., Berlin", "Einkaufserleichterung für unsere Mitglieder!", *Der Radio-Amateur* 2, No. 14 (July 25, 1924), 368–371. Because many of the specialized parts used to build shortwave equipment were not always commonly available, the Deutscher Amateur- Sende- und Empfangs-Dienst (DASD) (the association of German "hams"), even had a mail-order department, from which members could order parts and components.

[91] For example, in 1926, the local "Wannseebahn" chapter of the DRC (with some 500 members) met in the basement of the "Patzenhofer" bar, where it maintained not only meeting rooms but also a library and a lab. "Das Clubleben in der Ortsgruppe Wannseebahn (Sitz Steglitz) des Deutschen Radio-Clubs e.V.", *Der Radio-Amateur* 4, No. 12 (March 19, 1926), 235–238.

[92] An example of a club with a well-equipped workspace was the Radio-Vereinigung Dessau e.V. "Aus den Vereinen u. Verbänden", *Funk* No. 20 (May 16, 1930) Programmteil, 2–4. An industrialist offers fully equipped maker space in his factory: "Mitteilungen des Deutschen Radio-Clubs e.V., Berlin", *Der Radio-Amateur* 3, No. 1 (January 2, 1925), 17–21.

[93] See, for example, Tanja Poppelreuter, *Das neue Bauen für den neuen Menschen: zur Wandlung und Wirkung des Menschenbildes in der Architektur der 1920er Jahre in Deutschland* (Hildesheim/Zürich/New York: Olms, 2007) and Karl Christian Führer, *Mieter, Hausbesitzer, Staat und Wohnungsmarkt. Wohnungsmangel und Wohnungszwangswirtschaft in Deutschland 1914–1960* (Vierteljahrschrift für Sozial- und Wirtschaftsgeschichte, Beiheft 119) (Stuttgart: Franz Steiner, 1995).

and reference materials. The "maker spaces" of the 1920s and 1930s were extremely important as the incubators of innovation and the teachers of knowledge about radio.

Clubs held regular construction sessions under the supervision of experienced and skilled members, sometimes engineers who actually worked in the radio industry, sometimes graduate students in engineering or the physical sciences.[94] In this way, the clubs were an important conduit of knowledge between universities, industry, and the general public. This exchange of knowledge was by no means one-way: technical innovation often came from the radio hobbyists themselves. Clubs held public displays of their latest products, and even held competitions for the best and most technically innovative home constructions.[95] At a time when radio design and construction was literally being made up from scratch, workers in industry often gained inspiration or copied ideas which first came out of the clubs.[96] This was a major reason that many electrical manufacturers sponsored radio hobby clubs for their employees.[97] Similarly, most clubs

[94] For example, the Charlottenburg local chapter of the DRC had so many members wishing to participate in DIY evenings in early Fall 1924 that it had to form three separate groups, even though participation cost RM 3 per month. "Mitteilungen des Deutschen Radio-Clubs e.V., Berlin", *Der Radio-Amateur* 2, No. 20 (September 5, 1924), 527–53.

[95] Individual club chapters and individual clubs often held contests or the best built radio receivers, and the yearly German Radio Exposition held in Berlin always featured a display of the best amateur constructions held by the clubs. For example, the Hamburg Radio-Klub e.V. held a competition for the best crystal radio at the German Radio Exposition in Hamburg. in the early summer of 1924. Among the four winners, the third place was won by a woman, Frl. Anita Berg from Hamburg (others were #1 Reinhard Meyer, Lübeck, Rudolf Hansen, Hamburg (Staatl. Lehranstalten), and #4 A. Falke, Hamburg. "Klubmitteilungen", *Der Radio-Amateur* 2, No. 13 (July 18, 1924)), 346–348. On the display of home-built radios at the Great German Radio Exposition, see: "Mitteilungen des DFTV", *Der Radio-Amateur* 3, No. 39 (September 25, 1925), 32. The 1925 display included radios built by members of the Southern German Radio Club, the Radio Association of Lower Germany, the Upper German Radio Association, the DRC, and the Radio-Technical Association (all regional associations of clubs). These displays were regular features of the Great German Radio Exposition.

[96] Engineers from the electronics industry were always careful observers of the displays of home-built radios held by the major clubs, particularly during the annual German Radio Exposition in Berlin.

[97] For example, by June 1924, a chapter of the DRC had been founded by employees of the "Priteg" (Berliner Privat-Telephon-Gesellschaft) and another by students of the Beuth-Schule, an engineering school. This latter was a purely student-faculty group, but already had 200 members by June 1924. It soon had its own lab and offered courses in radio technology taught by junior faculty members. "Klubmitteilungen", *Der Radio-Amateur* 2, No. 10 (June 25, 1924), 274–276. The Postal Ministry itself also had its own internal radio hobby club.

owned a fairly up-to-date advanced receiver, and some even owned one of the few legal transmitters. This meant that clubs were a place where people could go to listen to the radio, view (and even copy) radios which were more advanced than they or their neighbors might possess, and generally get "infected" by the radio bug. They could go to the clubhouse and experience radio firsthand, again often including transmitters. This exercised a powerful draw. Radio was magic. Clubs were nodes where people became excited about the new medium.

Radio as a mass medium was being invented in the 1920s, both as a social practice (and the clubs played a big role there, as we see), but also as a technology. Radio design on a technical level advanced at a breakneck pace in the first half of the twentieth century and never faster than in the 1920s and 1930s. Hobby clubs were as interested in this, and as eager to contribute to it, as industry was. (Remember, they were often the same people.) Today we understand that informal contacts between interested individuals are one key to innovation. Tech companies today, in particular, are all eager to facilitate random creative interactions and to provide space which is conducive to such ad hoc interactions.[98] The creation of such spaces has proliferated at the author's home university over the last decade, be they cafés or multipurpose meeting rooms. New courses, new research, and a great deal of informal teaching have come out of such spaces. Radio clubs and their various special activities played a similar role in the Weimar Republic. They were, by definition, ad hoc assemblies of people around a common interest. That the common interest was defined as a hobby or free-time activity is secondary, particularly since there was little other space (outside, perhaps, of trade fairs and public lectures) for the kind of unstructured meeting and exchange of ideas that the clubs provided.

The role of radio hobby clubs in mobilizing people behind and for radio was crucial. It is clear how they could be places where those who already were fascinated by radio and already had some technical knowledge—like engineering students or electrical industry scientists—could meet and exchange ideas. But they were even more important in spreading the fascination of radio to non-technical individuals. In this regard,

[98] On the importance of informal meetings, see, for example, Anne-Laure Fayard and John Weeks, "Photocopiers and Water-coolers: The Affordances of Informal Interaction", *Organization Studies* 28 No. 5, (May 1, 2007), 605–634 and Stephanie E. Hampton and John N. Parker, "Collaboration and Productivity in Scientific Synthesis", *BioScience* 61, No. 11 (November 1, 2011), 900–910.

structure and tradition were very important. Germans knew clubs. The building of clubs and membership in clubs (again, the German word is *Verein*) might even be termed a quintessentially German activity. Private clubs of all kinds had played a major role in the formation of a German national conscience in the nineteenth century, and were even then very widespread.[99] They were and still are a mainstay of German social life and a key part of the public sphere. They were so important to German society that their structure was closely regulated by law.[100] As a result, Germans understood clubs. They were familiar with their form, their structure, their purpose, and their function. They knew how to build them, how to run them, and how to act as members. There was a highly developed and widespread club landscape in place long before the radio boom of the 1920s. So naturally, radio enthusiasts immediately formed radio clubs, even without the prompting of Bredow and the Postal Ministry, as important as this was in giving the clubs an official role. This is more significant than it seems. Clubs were familiar structures, even when—or particularly when—they focused on a new medium and an unfamiliar technology. The sheer familiarity of the club structure made the new technology of radio less threatening, and made it much easier to establish radio as a set of social practices. Radio hobbyists didn't have to reinvent the wheel while they were inventing radio. All they had to do was put what they were doing in the context of a club. The clubs provided a social structure within which the new technology of radio could find a social space. And this made the technology much more accessible to ordinary Germans.

Since clubs already had a firm place in German society, the newness of radio could more easily become anchored in existing social structures. While the young technology enthusiasts (dare I say, "nerds"?) could enthuse to their hearts content, the club structure was familiar and comforting for others, both inside and outside the hobby clubs. The presence of notables in the club leadership gave them social respectability, an important and often overlooked factor in the social acceptance of any new tech-

[99] In this regard, see, above all, the work of Celia Applegate. For example, *Bach in Berlin: nation and culture in Mendelssohn's revival of the St. Matthew Passion* (Ithaca, N.Y.: Cornell University Press 2005).

[100] Paragraphs 20–78 of Book 1 of the *Bürgerliches Gesetzbuch* (BGB) cover laws regarding private associations (*Vereine*). This compendium of German civil law was first issued in 1896, and has been continually updated and amended since then. Bundesministerium der Justiz, *Bürgerliches Gesetzbuch*, available at: https://www.gesetze-im-internet.de/bgb/BJNR001950896.html#BJNR001950896BJNG000502377. Accessed August 14, 2017.

nology.[101] The clubs worked in many other ways to make radio an everyday part of the social landscape. Not least, whatever else clubs provided, German clubs all had a wider social function beyond their immediate focus. They served as perhaps the major focal point for the social life of their members. All provided a range of social activities for their member's free time. And remember, radio first became a hobby in Germany in the "roaring twenties".[102] Social activities beyond the narrow focus of the clubs are always important in binding members to them, but in the 1920s, they took on particular importance.

This is certainly borne out by the activities of the radio hobby clubs. When reading their journals from the 1920s and early 1930s, one thing which sticks out is not only the number of major parties they announce, but their scale and duration. The bigger clubs could naturally hold the biggest parties, and in the case of Berlin, they often did so with the aid and participation of the local broadcast authority and with the presence of the cream of the nation's radio stars. Parties were always a part of the life of any club in Germany. But a party is not just a party. Particularly in the case of a technological hobby like radio, club parties played an essential role in popularizing and normalizing the technology.

Some radio club parties were fairly simple: In January 1925, the "Wannseebahn" chapter of the DRC—one of the wealthiest and most scientifically oriented of the Berlin neighborhood chapters—held a regular general meeting. Wives and girlfriends of the members were invited, and after the technical part of the evening and club business, the meeting became a party, with dancing to music from the radio. Members had so much fun that the club promised another such meeting soon.[103] This is pretty tame, as parties go. But other radio parties were much more spec-

[101] For example, when the Rheinisch-Westfälischer Radio-Club (Hagen i. W) was founded in the Spring of 1924, the President (1. Vorsitzender) was a Dr. Henrich, the Vice-President (2. Vorsitzender) was a high official of the Post Office and a nobleman (Telegraphendirektor von der Heyde), and the secretary also had a Ph.D. (Dr. Hammers). "Klubsachen", *Der Radio-Amateur* 2, No. 5 (April 16, 1924), 154.

[102] "Roaring twenties" is actually a US and British term. In France, the corresponding concept is *les annees folles*. Because the German experience of the postwar period was a bit different, so is the term: *die wilden zwanziger*. Nevertheless, the central idea—wild parties, lots of alcohol, new medias (jazz, radio, film)—is definitely a part of the German concept. There is simply more of an edge and less innocence in the retrospective German conception of the period.

[103] "Mitteilungen", *Der Radio-Amateur* 3, No. 5 (January 30, 1925), 131–136. In fact, it did soon organize another, even larger party, on March 15, 1925, this time with both live

tacular affairs, with all the features one might expect from a 1920s party: stars, dance bands, alcohol, and glitter. For example, in April of 1925, the "North" chapter of the DRC held a "Radio Ball". The popular Voß dance band played (live), and the well-known radio artists Willi Weiss and his wife, Mizzy (Miezi) Peery, both sang, while she played the piano. All the men who attended were given a bottle of "regenerative oil" (*Rückkopplungsöl*—probably schnapps) as gifts, and all the women were given a bag of party favors. The party lasted until 5:00 AM.[104] And there were even still bigger affairs. In early May, the "Wannseebahn" chapter celebrated the anniversary of its founding with a huge party, featuring the following artists: Gerda Görwitz (a story teller and raconteuse), the Schuhplattler und Yodel Group of the Bavaria-Club, Jochen Albrecht and Senta Wenderoth (dancers), Willy Rosen (poet and composer), and a number of well-known musical artists from the Berlin Broadcasting Station (Berliner Funkstunde): Charlotte Freyer (also of the Berlin Staatsoper), Eugen Transky (a chamber singer), Willi Weiß (a tenor), accompanied on the grand piano by Miezi Peery, Franz Baumann (an opera singer), Bruno Seidler-Winkler (a bandmaster), and "last but not least", Alfred Braun, perhaps the biggest radio star in Berlin. There were Vienna waltzes, modern jazz, talks and spoken presentations, music from the members of the club, and comedy skits with Austrian characters. The party ended at 6:00 AM with the last song by Willy Weiß ("Ja, der Sonnenschein"), as thousands of paper "radio waves" fell from the ceiling.[105] This was a grand ball worthy of Hollywood, even if the Austrian/Bavarian theme and the amount of sausage consumed were particularly German. Another popular type of club party, particularly in Berlin, but also in other areas with access to large bodies of water, such as Hamburg, Bremen, or Frankfurt, were the popular overnight cruise parties: a club would hire a cruise boat on a local river or lake for an all-night party with live bands, radio music (!),

music and radio music. See: "Mitteilungen", *Der Radio-Amateur* 3, No. 11 (March 13, 1925), 25–29.

[104] "Mitteilungen", *Der Radio-Amateur* 3, No. 18 (May 1, 1925), 27–31.

[105] "Mitteilungen", "Die Vortragsfolge", *Der Radio-Amateur* 4, No. 20 (May 14, 1926), 26–29. The song was the German version of "Yearning (Just for You)" by Benny Davis and Joe Burke. It was widely performed and recorded in Germany. Note that the song was performed here in German just a year after it appeared in the US. Many of these early radio stars were Jewish, and suffered greatly under the Nazis as symbols of modernity and urbanity. There is a need for an academic analysis of the first German radio stars.

dancing, food, other light entertainment (amateur skits and songs were popular), and, of course, drink.

Even in the relatively brief accounts published in the club journals, it is clear that these were serious parties. Both cruises and big balls would typically last into the very early hours of the morning, and there was heavy drinking. While the big, highly organized radio balls in big cities naturally catch our attention, it must be said that small local club branches also held regular parties. And it should be emphasized that, as radio enthusiasts, they all had access to a key part of any party: music. While the gala balls all featured live music and appearances by national stars, nearly all of these parties, even the smallest, featured music from the radio or from phonographs.[106] And music also meant dancing, and dancing inevitably meant the opposite sex. Being able to offer music, whether live, captured from the airwaves, or recorded, was a definite advantage for the radio clubs. The rabbit raiser's or gardening clubs couldn't compete.

Parties are not trivial activities.[107] They play an important role in any society or organization as a means of enhancing cohesion and as a release from the day-to-day struggle for existence. They also are important for fund-raising and for raising the social profile of the clubs. They put a public face on the club, helped create an identity for it among the public at large, and allowed the clubs to display the attractiveness of members and membership. Remember, the notables who were the honored guests always got the table next to the stage, and were mentioned in the newspapers the next day. But parties in this context also had other functions. First, they helped to attract new members and interest to existing clubs, which otherwise might become too narrow in composition. More importantly, parties helped dispel any lingering anxiety about the new technology, and, indeed, helped make it attractive and "sexy". The radio clubs quickly adopted the term *Funkball* ("Radio Ball") for their big parties, literally, "spark" balls, or parties—*Funk* in German also signifies radio. "Sudden, discontinuous arcs of energy between conductors through the air or another dielectric" might have sounded either vaguely alarming or boringly technical, but "sparks" were bright, effervescent, and shocking ("like a glimpse of stocking"). The

[106] Music from phonographs was not yet ubiquitous in society, yet was very common at Radio club functions, not least because radio enthusiasts were often "hi-fi" enthusiasts as well.

[107] See Mikhail Bakhtin, *Rabelais and His World* (Bloomington, Indiana University Press, 2009).

use of many other (often) tongue-in-cheek technical references around the parties, like the "radio wave" confetti mentioned above, is not just an expression of in-group language. It served to popularize and demystify radio technology, and thus to make radio popular among a wider audience. And heaven knows, anything might go at a party named after sparks and filled with electricity. What else could set a flapper's heart so alight?[108]

In this way, technology was made sexy instead of scary. Sex sells, as they say, and Radio Balls were deliberately sexy. All efforts were made to attract member's spouses, girlfriends, siblings, and people from the public at large—particularly if they were young and female. While people of all ages belonged to the radio clubs, they naturally tended to attract a disproportionate number of young, single men. By inviting the public to parties, women became part of the mix in larger numbers, and they helped to satisfy that most human and basic of urges: dancing.

Radio in the 1920s and early 1930s was one of the most important conduits for modern music everywhere, and in the 1920s, this meant jazz.[109] A small town or village would probably not be able to attract one of the big national or international jazz bands, but the local radio club could have access to the latest music over the radio—if not from the generally conservative German broadcasts, then from French or British ones. In a certain sense, that meant jazz music was one of the most important vectors (as well as symbols of) modernity itself. After all, in the US, we call the 1920s "the jazz age". Jazz, modern popular music of all types, was an important symbol of modernity at the time, for it was a major marker separating the past from the present. It was exciting, a little naughty, and resolutely modern. In Germany, it was marked not only as modern but also as foreign, since it came to Germany from the US, Britain, and France, all enemies in the recent war. Jazz in Germany was thus never innocent because it was seen as not only originating outside of Germany but precisely among Germany's recent enemies. Jazz was always controversial in Germany, and traditionalists (mostly, but not exclusively on the right) associated it with all sorts of fears of cultural pollution and decline, even miscegenation. After all, the common right-wing term for it in German at

[108] The allusion is, of course, to the Cole Porter hit song of 1934 "Anything Goes", from the musical of the same name.

[109] Today, we might rather say "swing", or distinguish between various different types of jazz. The term here is used in its widest sense, to mean all sorts of modern, non-traditional music.

the time was "nigger music".[110] The polemic around modern music (and in a larger sense, modernity) would later come to poison the debate around radio by the early 1930s. But the music was and remained oh so attractive, and in the 1920s, it helped associate radio and the radio clubs with modernity.[111] The presence of jazz at radio balls and on the airwaves, the explicit link between radio and modernity made radio and the radio hobby very attractive. It promised an alternative to the drab and puritan world of wartime and postwar privation. it promised association with other Western countries to a nation made an international pariah by the war. And it promised not only modernity but a future, and one which would likely be prosperous.[112] Radio seemed directly linked with the future, and the radio hobby clubs basked in this association.

When we think of the 1920s today (at least in the US and Britain), we think of a time when upper-class pursuits seemed within the reach of everybody, or at least of the middle class. From golf to ocean cruises, non-elites could at least dream of one day participating in activities previously reserved for their social betters. The lives of the rich and famous (be they gangsters, royalty, or film and radio stars) were both a preoccupation and a symbol of the "roaring twenties". Al Capone was a cultural symbol on both sides of the Atlantic, Marlene Dietrich in both places a major star, and if American audiences devoured *The Great Gatsby*, German-speaking audiences were no less taken by Vicki Baum's *Grand Hotel*.[113] Radio partook of this same glamour. Radio had its own stars and star system, whose

[110] In German, *Negermusik*, is a single word. There is considerable research into jazz in the Weimar Republic. In general, see Jonathan O. Wipplinger, *The Jazz Republic: Music, Race, and American Culture in Weimar Germany* (Ann Arbor: University of Michigan Press, 2017); Michael H. Kater, "The Jazz Experience in Weimar Germany", *German History* 4, No. 2 (April, 1988) p. 145; Jürgen Wilhelm Walter Heinrichs, " 'Blackness in Weimar': 1920s German Art Practice and American Jazz & Dance" Ph.D. Dissertation, Yale University, 1998; Joshua Stemfeld, "Jazz Echoes: The Cultural and Sociopolitical Reception of Jazz in Weimar and Nazi Berlin, 1925–1939" Ph.D. Dissertation, University of California, Los Angeles, 2007, and J. Bradford Robinson, "The jazz essays of Theodor Adorno: some thoughts on jazz reception in Weimar Germany", *Popular Music* 13, No. 1 (January 1994), 1–25.

[111] Though the Nazis officially banned jazz, in practice, it continued to be played and enjoyed by millions, though officially, what people danced to in the Third Reich was called "German Dance Music". Michael H. Kater, *Different Drummers: Jazz and the Culture of Nazi Germany* (New York: Oxford University Press, 1992).

[112] For the economic promise of radio, see below.

[113] F. Scott Fitzgerald, *The Great Gatsby* (New York: Charles Scribner's and Sons, 1925). Vicki Baum, *Menschen im Hotel* (1929). *Menschen im Hotel* was first published in 1929 as a serial novel in the *Berliner Illustrierte Zeitung* 38, No. 13–26 (April 30 to June 29, 1929),

lives were bared in the newspapers and in new magazines dedicated to them. Radios in Germany were also very expensive, and were marketed in Germany as luxury items.[114] While the big luxury radios were usually out of reach for the average radio enthusiast just as for the average citizen, all could aspire to the status of owning a radio. Radio was not only sexy in the 1920s, it was glamorous. Attendees at a radio ball could act out their upper-class aspirations for an evening. It might be a rented tuxedo, the music might come out of a home-built radio, but the booze and the girls (or boys) were real. Parties are neither innocent nor trivial. They were an important part of the German radio hobby, and helped spread the gospel of radio. Radio helped people imagine another way of life parallel to the everyday. By mid-1924 or early 1925, radio was firmly established, and clubs took on a routine form with all the attributes of the German culture of private associations or *Vereine.*

Radio and Middle-Class Anxiety: Technology and Social Aspirations in the 1920s Germany

The interwar period in both Europe and the US was a time in which modern technologies took on a mass character, and technological modernity swung with a jazz beat. This was certainly the case in Germany, but it also was a worldwide phenomenon. One reason for the mass spread of modern technologies in the industrialized world, and certainly in Germany, was the close link made between technology and the aspirations, dreams, and fears of the middle classes. Like the scene in the 1970s American film "The Graduate", in which the main character is given the mysterious job advice: "plastics", important new technologies in the 1920s and 1930s were wishfully seen as holding the prospect of material and social advancement. This gave new technologies an importance far greater than the actual work they did. The rise of the working class, economic disruption caused by the aftereffects of the First World War (and the new technologies themselves), and economic rationalization were making the social and economic status of large parts of the middle class increasingly tenuous. Meanwhile, many

and later by Ullstein Verlag as a novel. It was translated into English, and later made into a film in the US in 1932.

[114] As just one example, see the two-page advertisement for a Seibt radio receiver in *Radio* 5, No. 5 (August 10, 1927). The entire issue carried many more ads than usual, in the run-up to the Great German Radio Exposition.

in the working class, whose lives had always been tenuous, began to see a path, if not out of the working class, then at least toward greater prosperity and security. Mastery of new technologies was seen as just such a path to security and social advancement. The new technology of radio was linked to aspirations of the middle class in Germany in the 1920s. To demonstrate this, we can look at a fictional story entitled "Achim's Receiver" (*Achims Empfänger*) which was serialized in five issues of the journal *Der Radio-Amateur* ("The Radio-Hobbyist") in the first half of 1924, less than a year after broadcast radio was launched in Germany.[115] New technologies often become mirrors for the dreams, aspirations, and fears of social groups that are economically under pressure and keen on maintaining or improving their social position—that is the significance of the "plastics" scene in the film "The Graduate". Popular enthusiasm for the new technology of radio thus grew in part from the belief that it would provide the key to a larger social and economic dilemma faced mainly by the middle class. In all of this, the class-based fears and hopes were projected not onto the content of broadcast radio, but rather onto the technology itself. It is not a question of hearing the right words over the radio, but rather of "owning" the ability to manipulate the technology of radio.

Here is the story: Achim is a very normal boy. He likes hiking and enjoys school, particularly history and geography. But then come foreign languages, and he starts to do poorly. Outside of school, he is friends with another boy, who has a miniature train set, and loves trains and their complicated signal systems. But Achim, instead, is drawn to the "secrets" of radio,[116] and is excited at the prospect that a radio station[117] will be built in the neighborhood. He begins to save money to buy the parts to build a radio, and takes over the family tool box, to the distress and absolute incomprehension of the women in the family. His desire to build a radio becomes an obsession, which completely changes his character and gives him a new sureness and self-discipline. He begins to get up early, and stays up late after dinner. His life suddenly gets a new purpose. Achim had discovered his profession: "wireless telegraphy and telephony".

[115] Hans Ostwald, "Achims Empfänger", *Der Radio-Amateur* 2, No. 2 (February 1924):47–48; No. 3 (March 1924):86–88; No. 4 (April 1924):112–114; No. 5 (April 16, 1924): 141–143; No. 6 (April 30, 1924): 165–166.

[116] "*Er fühlte sich zu geheimnisvolleren Vorgängen hingezogen*".

[117] Oddly, the original German uses a term which specifies that this was a reception station, probably a mistake by the author.

The degree of the transformation Achim goes through is astounding: "And [he] got back so much strength and self-confidence that he lost all his earlier shyness and sluggardliness. Only now did he really feel at home in the world, it seemed to him that he only now had a real connection to humanity".[118] This makes it clear that we are not dealing with an ordinary story of a boy who finds a new hobby. It is not just a simple coming of age drama, it is a radical transformation *of character* through modern technology.

Achim had not been doing well at school, so Achim's father has to speak with Dr. Färber, Achim's teacher. Achim's teachers were concerned, for if he didn't improve in foreign languages, he would be held back, and have to repeat the school year. Achim's father rails at the blind application of rules, despairs that his son might not be able to sit for the university entrance exams (*Abitur*), and searches for a proper path for Achim. He understands that Achim hates soulless book learning, but is instead drawn to practical tasks. And yet, his father cautions him that he still has to learn all his academic subjects, if he is going to "get ahead"[119] and obtain a "prestigious profession".[120] Just when things look dark for Achim and his ambitious father, well… Achim builds his radio. And uses it. He suddenly starts to make rapid improvements in English, since there was so much of it on the air, and even begins to improve his French. Together with some younger teachers—Dr. Winterburg, the science teacher, and Dr. König, the history teacher—Achim's father persuades the principal to allow Achim to move on to the next class at the end of the year. Radio saved the day for little Achim.

What is remarkable about this story is the clarity with which it reflects the fears and desires of many middle- and lower-middle-class, even working-class Germans in the 1920s, and the fact that it appeared in an otherwise highly technical radio magazine. To understand the need to believe in the radical transformative power of modern technology, we need to remember the context of the early 1920s in Germany. The loss of the First World War led to a great deal of soul-searching and uncertainty in Germany, and particularly in the noble and educated middle-class circles, which had been the backbone of the old empire. The period of revolution and flux immediately

[118] "*Und empfing so viel Kraft und Sicherheit zurück, daß er alle seine frühere Scheu und Unlebendigkeit verlor. Er fühlte sich erst jetzt wieder so richtig heimisch auf der Welt, kam sich vor, als habe er erst jetzt richtigen Anschluß an die Menschheit.*"

[119] "*Wer da richtig vorankommen will, der muß sich in alles einarbeiten. Und wen das nur möglich ist, wenn man gewisse Bedingungen erfüllt—dann muß man sie eben erfüllen*".

[120] "*ein höherer Beruf*", p. 143.

after the war confirmed many in their fears that parliamentary democracy was unequal the challenge, and that the country was adrift and rudderless. The German revolution and its aftermath also raised the specter of communism and the threat it seemed to represent to private property—a cornerstone of middle-class existence. Even worse for the confidence of the middle class, the Empire's failure to properly finance the war through taxation instead of borrowing led to massive hyperinflation by 1923—exactly when this article about Achim was being written—which gutted the savings of the middle class and led to a massive crisis in values.[121] Finally, the 1920s seemed to usher in a new age of consumerism and frivolity, which were, yet again, major threats to the hard-working middle class, threats which were dangerous precisely because they were so seductive. Is it any wonder, in this context, that even the dry pages of a technical journal would reflect middle-class angst about the future?

The story favors classic middle-class values: individualism, self-motivation, creativity (within socially accepted limits), and a kind of service ethic: "And that the tasks, which such a person must fulfill, must be solved for the common good, for humanity. He never solves them just for himself. No, he is actually only the agent of others".[122] There is even a whiff of strict middle-class conformity in this quote and maybe, just maybe, a foreshadowing of Nazi notions of "people's community" (*Volksgemeinschaft*). All this is highly characteristic of middle-class, "modern" values, particularly in the German context of the early 1920s.

The story of Achim's radio is certainly a polemic against the existing educational system, which is portrayed here as old fashioned, hidebound, dominated by useless knowledge, too abstract, and unable to motivate modern youth. There is a clear fear expressed in the story that the traditional educational system, as controlled by the educated elites, was not only inappropriate for a world where technical subjects would be necessary but also threatened to stand in the way of social advancement or even lead to a loss of status. Poor Achim with all those useless foreign languages. If he had to repeat a class, if he did not do well (because the school was unable to motivate him, of course), then he might not be allowed to take

[121] The most important work on the German inflation is: Gerald D. Feldman, *The Great Disorder: Politics, Economics and Society in the German Inflation, 1914–1924* (Oxford and New York: Oxford University Press, 1997).

[122] "*Und daß die Aufgaben, die ein solcher Mensch zu erfüllen hat, für die Allgemeinheit, für die Menschlichkeit gelöst werden müßen. Er löst sie nicht nur für sich selbst. Nein, er ist eigentlich nur rein Beauftragter der anderen.*"

the university entrance exam, and might be forced into leaving school and beginning an apprenticeship in a manual trade, leading to a déclassé existence. On the other hand, some of the younger teachers were not quite so strict, and showed the potential to try to reform the school system to motivate clever boys like Achim, if the head of the school, old Dr. Färber, would only let them. This educational critique actually reflects a much larger conflict within the German middle class between the graduates of the prestigious universities and those who were trained in the less prestigious (but more concrete and practical) technical universities and institutes, as much as it reflects anxieties about social decline.[123]

Aside from more generalizable middle-class values, which were not specific to Germany in the 1920s, the story thus reflects the generational conflict between the pre-war, wartime, and postwar generations which was characteristic of that time and place. Achim, a child of the post-First World War world, is clearly in rebellion against his elders (if not specifically against his father: he is, after all, a good, obedient boy—this is wish-projection, not reality), but he chafes under the dead learning of the older generation represented by many of his teachers, representatives of the old, pre-war order. Standing by Achim are not only his father but the younger teachers—inevitably the members of the "front-line generation" (in 1924, anyone old enough to be a father of a pre-teenager had to have been of military age in the last war). The struggle is one for the survival of the (technological and highly competitive) modern world, for technological modernity itself, and Achim is at the forefront.

More than this, the story argues that creative people should not be held to the same mundane standard as the rest. Those who are particularly gifted in one area need to do what they are good at, and should not be held back by rules designed for the average person. It argues that life is short, and the creative part of it has to be used to the fullest and not wasted on empty requirements. Each person has a responsibility to society to use his/her talents to the fullest. Is it much of a stretch to hear an echo of Nietzsche here or the cult of the genius so dear to the Romantics? And yet, there are clear bounds to the exercise of creativity: it must be practical, it must serve the common good, and it must lead to a profession. This is

[123] See the work by Kees Gispen, *New Profession, Old Order: Engineers and German Society, 1815–1914* (Cambridge: Cambridge University Press, 2002) and *Poems in Steel: National Socialism and the Politics of Inventing from Weimar to Bonn,* Monographs in German History 6 (Cambridge and New York: Berghahn Books, 2002).

not praise of creativity and individualism in the shape of the bohemian artist or wild genius, it is instead the voice of the applied technologist railing at the theoretician.

Of course, the story is also highly gendered. It is Achim's radio, and not his sister Anna's. The "female beings" in his family hated the tool and spare parts box with all its bits of wire, spare screws, and so on.[124] Anna just couldn't understand what he wanted with all that old stuff: "Of course the girl just couldn't have a clear thought in her head about what this 'junk' might be good for. Of course she had the female notions of useless scrap metal and stuff that was just in the way".[125] The message here is clear: tech was for men, not women, whose brains just couldn't grasp any of it.

But above all, what turns around dreamy, creative Achim is not the best efforts of his teachers, the pleading of his father, or the fear for his future, but rather his radio, and the nearly single-minded enthusiasm for technology which it engendered. This not only has the power to fix his personal problem (one which, in earlier times, would be dealt with through either the bible or the strap) but also the larger systemic problems associated with the school and the hidebound teachers who control it. It isn't that Achim becomes smarter, the changes wrought by the magic radio were above all to his character: he begins to get up earlier and stay up later, he shows drive and focus, and he even starts to do better in French! Radio makes Achim a better person. His story is a classic middle-class morality story jazzed up with modern technology, Horatio Alger with an antenna.

The clear preference for practical and technological subjects, and the disdain for typical humanities subjects like foreign languages (even though they are clearly shown to be, in fact, highly useful in radio), echoes our current cultural myths about technical education in very familiar ways.[126] Good education is seen as practical and, indeed, practicality (defined as technical, creative and leading to a career of high social status)[127] is portrayed as both masculine and healthy, and above all, active, rather than passive: as Achim's

[124] *Der Radio-Amateur* 2, No. 2 (February 1924).

[125] "*Das Mädchen konnte natürlich keinen ordentlichen Gedanken fassen, wozu dieser "Kram" wohl gut war. Die hatte eben die weibliche Vorstellungen von überflüssigem Altmetall und Bruch, der im Wege war*". *Der Radio-Amateur* 3, No. 3, (March 1924).

[126] This passage echoes very contemporary concerns in an age when STEM (Science, Technology, Engineering and Medicine) fields are all the rage. Foreign languages, what good are they? Kids should learn to code instead, or so the argument goes. The author has been teaching in a department of Modern Languages and Literatures for the last 20 years, and metaphorically, Achim's father has been around a lot.

[127] "*ein höherer Beruf*", p. 143.

father assures him, " A real boy at your age isn't just satisfied anymore with mere learning".[128] Yet these sentiments are also in conflict with the ethos of the educated German middle class, the "Bildungsbürgertum", which, while it has valued individualism and hard work, had also favored education in the arts and the humanities, and had both disdain and alarm at the growing importance of technology and science. This rift is best symbolized by the greater status accorded in the Empire to humanistic schools (Gymnasium) over technical ones, and traditional universities (which taught subjects such as law, theology and classics) over mere "Institutions of Technical Higher Education" ("Technische Hochschulen").

If "Achim's Radio" reflects middle-class fears and anxieties, this is only half of what is going on. The article is also a volley in the ongoing fight for recognition of those with a technical and scientific education—above all, of those with an education in the *applied* sciences. Remember, Achim had trouble in school precisely because he had a practical turn of mind, and disliked empty, theoretical learning. And foreign languages did him in—until he learned through his radio that they had a practical application. Moreover, the love of (applied) technology and the particular fears of social and economic loss of status reflected in the article resonated particularly well with the lower middle class but also with portions of the working class as well, who also increasingly needed technical knowledge more than brute strength. After all, the lower middle class and working classes were the most at risk economically and, therefore, the most willing to seek new professional opportunities in a new technology like radio. So, what is mirrored in "Achim's Radio" are not only the fears and aspirations of the German middle classes in a particularly trying time but also an ongoing civil war within the middle class itself, one which extended into the upper reaches of the working class as well.

Interesting in this context is the effect of Achim's radio on his father. On the one hand, the father is described as an expert in politics and economics, and therefore likely quite a practical man. He certainly understands the risk to Achim if he does not do well in school. But with a touch of the 1920s glamor, the father actually makes his living as a film writer—he is thus very modern, in the sense of working in film, which like radio was a medium which symbolized both glamor and modernity. The father is also described,

[128] "*Ein ordentlicher Junge in deinem Alter benügt sich auch nicht mehr mit dem bloßen Lernen*" p. 143.

by his dress, as an "artist",[129] leaving us with an improbable mix of features: economics, film, and bohemian. Yet by redeeming Achim, the father, too, is also redeemed by radio. Both men are shown the importance of understanding modern technology, not just as a medium (writing film), but for its own sake.

As a piece of fiction in an otherwise sober and scientific publication, "Achim's Radio" is certainly an outlier. And, in fact, this is what makes it such a valuable document—it so clearly falls outside the norm for this publication that we have to take notice. Yet, if its fictional nature makes it a particularly clear case of projection onto technology, many of the factual articles in the same journal and others tell a similar story. In fact, the themes of "Achim's Radio" are echoed in the wider radio hobby press. Radio is often portrayed as contributing to education in general, and education in technical subjects in particular.[130]

Radio was also seen as facilitating the acquisition of a well-paying job. New, skilled, and relatively prestigious jobs were created by radio, such as radio operator, radio engineer, radio repair technician, radio dealer, or radio performer.[131] While they were actually quite hard to get, they held the imagination of members of a middle class threatened by economic conditions. And in some instances, radio hobby clubs actually worked to obtain jobs for their members, particularly when they were associated with technical institutes or universities.[132] In other cases, radio was seen as a tool for fighting unemployment during the Great Depression.[133] The point

[129] *Der Radio-Amateur* 2, No. 2 (February 1924), 47.

[130] As an example, Ober-Regierungsrat Dr. Paul Gehne, "Popularisierung der Technik. Der Rundfunk als technischer Kulturfaktor", *Funk Bastler. Fachblatt des deutschen Funktechnischen Verbandes E.V.* No. 13 (March 28, 1930), 209–210.

[131] See, for example: Dr.-Ing Max M Hausdorff, "Über Schulen für Radio", *Der Radio-Amateur* 1, No. 1 (August 23, 1924) and part 2: Dr. Ing Max M. Hausdorff, "Über Radioschulen (2. Folge)", *Der Radio-Amateur* 1, No. 3 (October 1923), 56. This article introduces the National Radio Institute in Washington, DC, and then gives a table of monthly salaries for those in the radio profession: $1500 (and full room and board) for ship's radio operators, radio repair technician $2000, broadcast radio operator $3000, and a radio salesman $5000 (including commissions).

[132] "Mitteilungen", *Der Radio-Amateur* 3, No. 28 (July 10, 1925), 25–28, which announces that the club branch in Charlottenburg now has a job brokerage (Stellenvermittlung) in its business office—especially since so many of its members were students at the Technische Universität, who could do quite valuable work for Industry.

[133] See, for example, "Der Winterhilfe des Rundfunks", *Funk. Die Wochenschrift des Funkwesens* 2, Hauptteil (January 8, 1932), 8. This article speaks of how broadcast radio was working with the "Winter Relief" organization (*Winterhilfe*) to set up groups of unemployed

is that the technology of radio, in its various forms, was seized upon by the middle class and parts of the working class as a solution to one if its chief worries: appropriate employment.

Finally, but certainly not least, technology was not only linked to potential employment, it was also glamorized in the 1920s, as we have seen above. This occurred even in radio publications, which were otherwise dry and technical. Radio was linked to a fantasy upper-class lifestyle of taste, style, and luxury: the image was Gatsby with headphones. Advertisements for radios often showed their owners as urban *mondänes*, sipping champagne and wearing tuxedos while listening to the radio.[134] This is, of course, more telling about advertising codes and norms in the 1920s than anything else, but, still important, for linking technology with the lifestyle of the rich and famous had to have been at least plausible at the time. Given the cost of manufactured radio receivers in Germany in the 1920s, advertisers understood that they had to address those with money. Much more telling, however, is the same kind of link between radio and upper-class, cultured, and urban lifestyles when it is made in advertisements for radio components such as vacuum tubes.[135] The dream of an upper-class lifestyle sold well precisely to the kind of reasonably well-educated and self-motivated people who were part of the radio hobby. In the advertisements in radio publications, radio is linked directly to the world of educated high culture or, in German, *Bildung*. Those interested in radio were also seen as interested in high culture. Again, this was partly a reflection of the kind of programming provided on German broadcasting, but it is also indicative of the aspirations of the radio hobby public. For example, a 1932 ad for the electronics company Grawor features "Three Acknowledged Masters"—Beethoven, Goethe, and the Grawor "Perkeo Dynamo" loudspeaker.[136] And, indeed, members of radio clubs in the 1920s certainly sought out high culture. Some clubs even made efforts to provide

artists and musicians to work in broadcasting. (Note that the so-called Winterhilfe, which is often seen as a Nazi invention, existed long before the Nazis took power.) See also H. Schnell, "Der Funkerberuf", *Funk Bastler*, No. 50 (December 12, 1930), 797–798. Note that this article warns readers that the profession is saturated, and is not accepting new trainees.

[134] Advertisement for Radiofrequenz G.m.b.H., *Der Radio-Amateur* 2, No. 9 (June 11, 1924), inside rear cover.

[135] See the "Valvo" advertisement for vacuum tubes, *Funk*, No. 10 (März 4, 1932), Programmteil, 3. The ad makes specific reference to jazz music.

[136] Advertisement for Grawor: "Drei anerkannte Meister", *Funk*, No. 18 (April 29, 1932), Programmteil, inside front cover.

discounted tickets to the theater.[137] Radio clubs and radio advertising thus both reflected and reinforced the (often aspirational) middle-class values of their members, whatever their actual class.

The debate over the cultural role of broadcast radio in Germany in the 1920s was quite heated, but to nearly all concerned it was self-evident that radio would be a tool for spreading high culture to the masses.[138] It is clear from both the debate and from the way radio was portrayed in the part of the press aimed directly at radio enthusiasts that it was seen—or at least being sold—as a way for the cultured middle class to express and embrace its values, and for those who aspired to this status to join the club.

Radio technology was a screen upon which members of the middle class and particularly the lower middle class could project their greatest fears and their fondest hopes in Germany in the 1920s and early 1930s. Like the physical screen in the movie theaters, it was also a projection surface for at least parts of the working class as well. Bursting on the scene as a mass technology just at a time when economic weakness was posing a particular threat to middle-class security and status, it was seen by many to hold the potential of offering good and prestigious careers, social advancement, and continued status. These dreams were too much for any technology to fulfill in their entirety, but they were very real to those who dreamed them. The social promise of radio was a major aspect in the enthusiasm for radio which made the "radio revolution" of the early twentieth century.

[137] For example, the Deutsche Funkgesellschaft. See: "Aus den Vereinen", *Funk,* No. 18 (April 29, 1932) Programmteil, 1–4 which announces that the Deutsche Funkgesellschaft was now offering reductions in several Berlin theaters, including the Deutsches Nationaltheater, Theater des Westens, Theater im Admiralspalast, and so on.

[138] See, for example, Hans Bredow, *Vier Jahre Deutscher Rundfunk* (Berlin: Reichsdruckerei, 1927). For another contemporary view, see: "Die kulturelle Bedeutung des Rundfunks", *Der Radio-Amateur* 2, No. 33 (December 5, 1924) "1. Messeheft", 876–880. See also: Cory Ross, *Media and the Making of Modern Germany: Mass Communications, Society, and Politics from the Empire to the Third Reich* (Oxford: Oxford University Press, 2008), 85–86.

CHAPTER 3

German Radio Before Broadcasting: Scientists, War, and Imperialism

According to most histories, radio as a technology was invented by G. Marconi in 1895.[1] There are actually many other candidates for the first "inventor" of radio or the first to make a legitimate radio transmission.[2] All built on the basic science of electricity, acoustics, and general physics, which had been done (mainly in Germany) over the course of the nineteenth century. In a sense, the idea of radio was literally "in the air" at the time Marconi made his first transmissions. What distinguished Marconi from the other claimants to the title of inventor of radio was less his scientific acumen than his business ability and sense of showmanship. He was the first to commercialize radio, and built a business empire around it. Marconi and the companies he founded continued to advance the technology of radio both before and after the First World War, as did many others. These advances occurred in both the science of radio, the practical technology of making wireless transmissions, and in the social (meaning also political and economic) means of organizing its use. Governments, particularly navies, and businesses showed the greatest interest in the new

[1] The most recent, and definitive work on Marconi is: Marc Raboy, *Marconi: the Man Who Networked the World* (Oxford and New York: Oxford University Press, 2014).

[2] See Raboy, *Marconi*, Chapter 2. See also: G.R.M. Garratt, *The Early History of Radio, From Faraday to Marconi* (IET History of Technology Series 20) (London: the Institution of Engineering and Technology, 1994, 2006).

B. B. Campbell, *The Radio Hobby, Private Associations, and the Challenge of Modernity in Germany*, Palgrave Studies in the History of Science and Technology, https://doi.org/10.1007/978-3-030-26534-2_3

technology before the First World War, and poured huge amounts of capital into its use and development.

Radio would have looked and functioned (socially, if perhaps not technically) very differently if it had not been invented at this particular historical moment in the industrialized (and largely Western) world. One key element of this was the fact that the path for the development of radio as a medium of communications had already been prepared by earlier inventions and practices. Chief among them was (wired) telegraphy. The idea of long-distance communications was present long before the technology for it was perfected. As early as 1792, a semaphore (visual) telegraph (the Chappe Telegraph system) was invented and implemented in France during the Napoleonic wars.[3] Not long thereafter, a series of scientific discoveries and practical implementations led to the "invention" of a practical means of (wired) telegraphy by Samuel Morse in the US in 1832. This invention revolutionized warfare, business, and government. Soon, vast wired networks spanned the globe.[4]

Radio in its early years (in the form of wireless telegraphy) could function socially as simply an extension of wired telegraphy, which both aided its spread and acceptance, and limited the ways the new technology was used. By already understanding wired telegraphy, business and government could immediately grasp the utility of communications freed from wires. Radio or wireless fulfilled a need for rapid long-distance communications first awakened by telegraphy. Though at first it could not really compete with telegraphy over land, it fulfilled a very great unmet need for communications over water, where wired telegraphy was either impractical or tied to underwater cables, which could be cut in wartime and which, in any case, were largely dominated by British and US companies. It was no accident that navies and merchant shippers were the first and biggest customers for the new wireless technology. And it was also no accident that

[3] Note that after France, Prussia was one of the first major states to build lengthy semaphore telegraph lines.

[4] Independently of Morse, William Fothergill Cooke and Charles Wheatstone patented their own telegraph system in the same year in England, but it was much more cumbersome than Morse's. There were also a number of experimental wired telegraph systems before Morse and Cooke and Wheatstone. See: Tom Standage, *The Victorian Internet: The Remarkable Story of the Telegraph and the Nineteenth Century's On-line Pioneers* (New York: Bloomsbury, 2014). See also: Ken. Beauchamp, *History of Telegraphy, Its Technology and Application* (IET History of Technology Series 26) (London: the Institution of Engineering and Technology, 1994, 2006).

radio was quickly seized upon as a means of communication with the far-flung colonial empires of most industrialized countries, even if actual implementation struggled to meet expectations for many years. Like telegraphy before it, radio was intimately tied to European colonialism and imperialism.[5] Radio as a practical communications medium was invented at the high water mark of European imperialism, and was always closely related to it. As we will see later, the link between radio and far-off exotic places lasted a long time, indeed longer than many of the colonial empires themselves.

A need for long-distance communications—and the organizational, legal, and financial structures to carry it out—was not the only thing radio borrowed from its precursors. In its infancy, radio also borrowed its special language from telegraphy. Early radio was technically incapable of transmitting (modulating) the human voice and, therefore, used the same binary digital system to transmit information as wired telegraphy did: Morse code. Early radio was simply wireless telegraphy, and the term was a synonym for radio in its early years. Morse code rapidly became a world standard for wired telegraphy between 1840 and 1890. Morse code also immediately became the world standard for wireless telegraphy as well, and the first messages sent by Marconi were already in Morse code. But what became characteristic of early radio was not just the code; by itself, Morse code is simply a means of sending letters of the alphabet (first, the Roman alphabet, though later others were adapted to it). Those letters had to be combined to form words, which were of course dependent on the language of the sender. To become an international means of communication, to work as quickly and as efficiently as possible, and to standardize its use in the name of mutual intelligibility, wired telegraphy developed a complex set of abbreviations and short codes, which allowed telegraph operators to conduct their business and even to exchange rather complex sets of information independent of their native language. This language of telegraphy then became the basis for the international use of wireless telegraphy, which still lives on in the hands of hobbyists and spies, even though it has been dropped from commercial and most military communications since the

[5] Michael Friedewald, *Die "Tönenden Funken". Geschichte eines frühen drahtlosen Kommunikationssystems, 1905–1914*, Aachener Beitrage zur Wissenschafts-und Technikgeschichte des 20. Jahrhunderts 2 (Berlin: Diepholz/GNT Verlag, 1999); James W. Carey, "Technology and Ideology: The Case of the Telegraph", *Prospects* 8 (1983): 303–25 and Heidi J.S. Evans (Tworek), " 'The Path to Freedom'? Transocean and Wireless Telegraphy, 1914–1922", *Historical Social Research* 35, No. 1 (2010); 209–236.

1990s. The ready availability of this international language derived from (wired) telegraphy played a major role in the development of long-distance radio communications as a hobby. Using a simplified and adapted form of the telegraphic abbreviations, radio hobbyists (hams) around the world could communicate with each other irrespective of their native language, and non-hams could listen in. Morse code is still used by hobbyists today, although most have moved on to other modes of communication, such as voice or one of the many digital modes now available. At the birth of the radio hobby, Morse code gave particularly transmitting amateurs an international language, and gave all radio listeners access to the world of nautical and military transmissions by learning Morse code.

The First World War

The effects of the First World War on the development of radio technology and the social practice of radio were profound. The rapid development and expansion of radio in Germany (and throughout the continent of Europe) in the immediate postwar years would not have been possible without the experience of the First World War. Moreover, the way radio was organized and used socially was highly colored by that conflict down to mid-century at least.

The importance of the First World War for building the foundations of the radio revolution of the interwar period is cleverly demonstrated in the first section of Edgar Reitz's monumental film/mini-series "Heimat" ("Homeland").[6] We see a young man on a picnic with several of his neighbors and friends at the site of a romantic and completely rural castle ruin. The year is 1923, and we know him to be a veteran of the First World War. He has a home-built radio, and when he turns it on, the picnickers he is with hear a classical music concert from faraway Cologne, the first time any of them have heard a radio. Quite literally, radio brings the faraway, the urban, the foreign and, not accidentally, high-culture to the familiar world of the rural small town in the Hundsrück. Suddenly, the "*heimlich*" becomes "*unheimlich*": the familiar becomes foreign through the magic of the new medium of radio. It is no accident that the agent of the introduction of radio into the rural life of the Hundsrück in this clip from the film "Heimat"

[6] Edgar Reiz, "Heimat-eine Deutsche Chronik", Edgar Reitz Film (ERF), Sender Freies Berlin (SFB) and Westdeutscher Rundfunk (WDR) (1984).

The passage in question is roughly at the 57″ to 1:02″ mark in the first episode of the film, appropriately titled "Fernweh" ("Wanderlust").

is a returning soldier. Symbolically, the radio is ushering in the modern media age, which will now reach deeply into every village and town.[7] The man in the film with the radio later opens a radio store, and makes a living from the new medium—until he disappears, leaving the village for life in faraway America, the victim of the siren song of his own medium.

The Great War made technology ubiquitous in a way it never had been before: mass technologies for a mass war. This fundamentally changed people's (particularly men's) relationship to technology and technological modernity. It also set the stage for the postwar era, when many of the things which had been a part of the war became part of everyday civilian life, even for workers and farmers. After the war, automobiles, telephones, radios, mass-production, and organization all gradually—or not so gradually—seeped through and permeated all of society. This could happen not least because millions of men and women had become familiar with mass technology during the war.

If the war first made technology ubiquitous, it also made it an existential matter: technological advancements became, on a very fundamental level, matters of life or death: no one after the War could fail to understand that new technologies meant new weapons, new tools for both rule and ruin. It became ordinary and commonplace after the war for both governments and ordinary citizens to evaluate every new technology in terms of national defense and the potential to cause harm, whatever else was thought of it. This was true for radio just as for every other major new technology. The war led to a lasting militarization of civil society. The argument here is that the war influenced the perception of technology, not least by heightening the understanding that all technology had a military dimension. Radio was perceived and pursued in the post- or interwar period by both the state and private individuals as a strategic technology with a direct relationship to national defense and state power.

If the war broadened people's familiarity with technology and colored their perception of it, it also widened their perceptions of geography and space. Millions of soldiers became intimately familiar with the soil and geology of the battlefields, and hundreds of thousands mingled with it forever, their bodies buried in foreign fields. A soldier looks at space in a particular way, and this new look extended to civilians as well. Maps became ubiquitous in newspapers, movies, and posters. Ordinary people became conscious of distance and bearing in ways they rarely had before (though this process

[7] The radio shown in the film is an anachronism, since it uses radio design/technology characteristic of 1927 or 1928, rather than 1923 when the scene takes place.

was certainly begun by earlier colonialism, and even earlier still by the Christian practice of pilgrimage). The airplane (but also the cannon shell and mortar) even extended this spatial perception to the air, to "space", and people now conceived of space in three dimensions. The armies of the belligerents became the world's greatest travel agencies, and millions were exposed to travel and foreign cultures for the first time. Families in Little Rock or Pasewalk or Mandres received postcards and letters and packages from the Asia, Africa, or parts of Europe of which they had never heard before. After this experience of a new geography, "how ya gonna keep 'em, down on the farm (after they've seen Paree)?"[8] The war completed the process begun by Western imperialism and trade, by bringing the faraway into the intimacy of home. What souvenirs, letters, photo, and film did to make the foreign part of the everyday on a piece-by-piece basis, radio could do in real time. Without the widening of people's horizons and geographical perception brought about during the war, the fascination of people in the 1920s with the exotic (fed not least via radio, "direct from the south pole" and other exotic places) cannot be explained.[9] Both the hunger for the exotic, and the perception that safety required knowledge of what was going on far away, and as quickly as possible, helped fuel the radio revolution. And much of this is attributable to the Great War. The major point here is this: the war had a profound and lasting effect on people's perceptions, and these perceptions colored how the mass technologies of the postwar period were implemented and appropriated socially. Technology is not simply science, it is a social practice,[10] and the social practice of the 1920s and 1930s was fundamentally influenced by the Great War.

[8] This is a famous postwar song, published in 1918 and written by Joe Young and Sam M. Lewis with music by Walter Donaldson. http://libx.bsu.edu/cdm/ref/collection/ShtMus/id/725

[9] This could easily be the topic of a separate study. As "signposts", let me simply again mention the film "King Kong" (Merian C. Cooper, RKO Radio Pictures, 1933), or bring up other examples such as the novel "Murder on the Orient Express" (Agatha Christie, Glasgow: William Collins & Sons, 1934), the 1922 discovery of the tomb of Tutankhamen and following Egyptian craze or the Antarctic of Richard Byrd (1928–1930), with its direct radio broadcasts. See the following chapter.

[10] As examples of a much larger debate, see: Wiebe E. Bijker, Thomas P. Hughes and Trevor Pinch (eds.), *The Social Construction of Technological Systems: New Directions in the Sociology and History of Technology. Anniversary Edition* (Cambridge, Massachusetts and London: MIT University Press, 2012); for a more recent example, see: Jannis Kallinikos, Hans Hasselbladh, Attila Marton, "Governing social practice: Technology and institutional change", *Theory and Society* 42, No. 4 (July 2013): 395–421.

The First World War Not a Radio War, Yet Prepared the Way for the Radio Age

Radio technology was still in its infancy in 1900 or 1914, when war broke out. Not least because of the limited performance and fragility of radio technology at this time, all countries were slow to incorporate modern electrical communications technology—including radio—in their armed forces, with navies generally adopting wireless telegraphy (radio) faster than armies. Despite the importance of wired telegraphy during the American Civil War (1861–1865)[11] and the Austro-Prussian (1866) and Franco-Prussian Wars (1870–1871),[12] its adoption by the German military was comparatively slow. The success of the Japanese with using telegraph to direct artillery during the Russo-Japanese War (1904–1905), particularly in indirect fire, which advances in metallurgy and chemistry were increasingly making physically possible over greater and greater distances, provided European armies with a wake-up call. By 1914, the German Army[13] was among the world's most advanced in the number and quality of specialist communications troops, which were equipped for both wired and wireless telegraphy.[14]

Yet when the First World War broke out, the (comparatively) advanced state of both German technical education and the military's possession of advanced communications technologies did not necessarily translate into successful employment of the means at hand. The German Navy mainly

[11] See: David Hochfelder, *The Telegraph in America, 1832–1920* Johns Hopkins Studies in the History of Technology (Baltimore: Johns Hopkins University Press, 2012).

[12] Telegraph played a major role in the wars of German unification, including the Austro-Prussian and Franco-Prussian wars, but this involved, for the most part, civilian telegraph networks. The Prussian/German armed forces were comparatively slow to build their own military field networks and devise doctrine for their use until the early 1900s. Werner Niehaus, *Die Nachrichtentruppe, 1914 bis heute. Entstehung und Einsatz* (Stuttgart: Motorbuch Verlag, 1980): 12–19. See, in general, Hans Georg Kampe, *Nachrichtentruppe des Heeres und Deutsche Reichspost. Militärisches und staatliches Nachrichtenwesen in Deutschland, 1830 bis 1945* (Wandesruh bei Berlin: Project Verlag Dr. Erwin Meißler, 1999).

[13] Actually, one should say German "armies", since prior to 1919, the land forces of the German Army were made up of the separate armies from the Kingdoms of Prussia, Saxony, Bavaria, and Württemberg, though with far-reaching coordination in doctrine, training, and equipment. For a basic introduction, see: Gordon R. Craig, *The Politics of the Prussian Army* (Oxford: Oxford University Press, 1956) or Rudolf Absolon, *Die Wehrmacht im Dritten Reich*, Vol. I: *30. January 1933 bis 2. August 1934* (Boppard am Rhein: Harald Boldt Verlag, 1969): 1–15.

[14] Kampe, *Nachrichtentruppe*; Niehaus: *Nachrichtentruppe*.

stayed in port, and thus did not much depend on wireless communications, except to coordinate a comparative handful of overseas ships and merchant raiders (before they were destroyed), or the relative large but still limited number of submarines (whose radio traffic was used by the British to locate and destroy them). While long-distance communications with German colonies in Africa, or far-flung allies such as the Ottoman Empire, were successfully maintained despite British control or disruption of undersea telegraph cables, this was the only unequivocal German success when it came to the employment of wireless during the First World War. On the other hand, the German officer corps had a high social investment in what it saw as the fighting capabilities of the Army and the importance of individual character for the leadership of men under fire. This led to the fear that both telegraphy and radio would limit the initiative of subordinate commanders in battle. The fear was so acute that it had prevented the development of doctrine for the use of modern communications in the years leading up to war. In the early days of the fighting, the result was that the actual use of the modern communications means at the Army's disposal was chaotic. There is even some evidence that this failure to fully use the communications tools available played a major role in the ultimate failure to attain German objectives in the advance through Belgium and into France in the first, crucial two to three months of the war in 1914.[15]

As a result of both cultural and technical factors, the use of radio in the First World War by Germany (at least outside of the naval war) was quite limited,[16] and radio only became a part of mass culture after the war. Yet the slow, grinding evolution of doctrine for the use of wireless during the war, and the parallel and even slower effort to evolve the existing technology to make it lighter, smaller, less fragile, and more powerful, did make it clear even to lay observers by the end of the war that radio would be a part of all future warfare, and would play a major role in civil society. Despite

[15] Niehaus: *Nachrichtentruppe*. Of course, German problems with the employment of modern communications were relative; Germany's enemies were often no better, and sometimes, as in the case of Russia, much worse. In particular, Russian use of clear-text rather than encrypted radio transmissions to send orders to the field armies played a major role in initial German successes in the battles of Tannenberg and the Masurian Lakes.

[16] For a good introduction to the increased importance of modern communications media during the war, including radio, see the recent exhibition in the German Communications Museum, Berlin. Thomas Jander and Veit Didczuneit, *Netze des Krieges. Kommunikation 1914–1918* (Museumsstiftung für Post und Telekommunikation/Brandenburgisches Verlagshaus, 2014).

the limited scope for the use of radio during the war, by heightening consciousness of the importance and potential of technology and greatly increasing ordinary people's exposure to it, the war ushered in the postwar age of radio, at least in continental Europe. Indeed, even the immediate postwar era of revolutions was truly "rung in" by the infamous Bolshevik radio transmissions calling for world revolution to end the war in 1918.[17] The failure of radio to play a major part in the war in no way reflected the consciousness of radio's potential which came out of the war, nor does it give us an accurate look at how the war prepared the way for the later radio revolution in a technical and social sense.

Material Foundations

Despite the limited use of radio during the war, the conflict provided the material foundation for the postwar radio revolution. First, it led to a massive expansion of the already well-developed German electrical industry. This included both national champions such as Telefunken, Lorenz, AEG, or Siemens, but also countless smaller manufacturers and sub-contractors. The radio-boom of the 1920s was eagerly seized upon by hundreds of small business, which had expanded or developed during the war to supply electrical/communications/radio products. The virtual end of military contracts caused these firms to seek other outlets and customers, and the radio-boom came at just the right time. Expansion of industry and production also meant expansion of a skilled workforce with knowledge of electricity and radio. Many of these skilled workers would go on to postwar careers in the electrical/radio industry, or would themselves become radio hobbyists.

While much of the new production of electrical equipment of all kinds was consumed in combat, large stocks remained. Furthermore, the Versailles Treaty required the scrapping of vast additional amounts of military equipment in 1919 and the 1920s. Huge quantities of such material found their way into private hands or onto the commercial market after

[17] See Alexander Rabinowitch, *The Bolsheviks in Power: The First Year of Soviet Rule in Petrograd* (Bloomington: Indiana University Press, 2007): 164–165. Note that the Bolsheviks used radio to contact the Germans only because they had no other reliable means with which to do so; land telegraphy via Finland would theoretically have been possible, but the Bolsheviks had no guarantee that the often highly politicized telegraphers along the way would forward the message.

the war.[18] While hard to quantify, it is clear that neither companies nor private individuals could build radios on a large scale without some basic industrial supplies, such as wire or insulators, or without at least a few specific components, which were easier to buy than to make, such as vacuum tubes or Morse code keys. The presence of wartime products meant that costs were lowered for anyone building a radio in the early years of radio, even though it was still an expensive proposition.[19]

The war also led to great technical advances, and the acceleration of the transition from experimental technologies into practical ones. An excellent example of this is the vacuum tube,[20] which only went into serial production in Germany during the war. The German military was initially slow to use them, yet production soared in the latter part of the war, and since patents—including international ones—were suspended during wartime, the production of vacuum tubes was also spread out across many different companies in Germany.[21] Vacuum tubes were, of course, a fundamental building-block of the later radio revolution, until they were replaced by solid-state transistors in the 1950s, and permitted the sending of voice transmissions, as well as Morse code. The rapid increase in their production and the spread of such production beyond the original patent license

[18] For example, see the advertisement for radio material out of "army stock" (*Heeresbestände*) from the "Chrlottenburger Motorengesellschaft" *Radio*, No. 4 (July 25, 1923): 150 or the advertisement of the company Dürre & Bierstedt in Magdeburg. In: *Der Radio-Amateur* 2, No. 25 (October 10, 1924): 668 or: "Verwendung alter Heereskopfhörer" von Dr. G. Heußel, in: *Der Radio-Amateur* 4, No. 11 (March 12, 1926): 219–220.

See also: Curt Urban, "Der Erzieherische Wert des Radioamateurwesens", *Der Radio-Amateur* 1, No. 2 (September 1923): 33–35, which mentions the accessibility of surplus military equipment as an aid to would-be do-it-yourselfers.

[19] As someone who grew up with a deep fascination for the local army surplus store in the 1960s, this author would welcome a study of the social consequences of so much surplus material being available after the two world wars.

[20] The vacuum tube was first invented by Fleming in 1904, then (separately) made useful by von Lieben (1906) and DeForest (1907). It was used first as an amplifier in civilian wired telegraphy and telephony, particularly to amplify signals in very long lines. In this capacity, radio tubes were used in large numbers during the war. Only after their widespread use in wired telegraphy and telephony did the usefulness of vacuum tubes in radio receivers and transmitters become clear. See: H.J. Van Der Bijl, *The Thermionic Vacuum Tube and its Applications* (New York: McGraw Hill, 1920): xvii–xix. On the history of vacuum tubes, see: Massimo Guarnieri, "The Age of Vacuum Tubes: Early Devices and the Rise of Radio Communications [Historical]", *IEEE Industrial Electronics Magazine* 6, No. 1 (March 2012): 40–43.

[21] See: Hartmut Petzold, "Zur Entstehung der elektronischen Technologie in Deutschland und den USA. Der Beginn der Massenproduktion von Elektronenröhren 1912–1918", *Geschichte und Gesellschaft* 13, No. 3 *Sozialgeschichte der Technik* (1987): 340–367.

holding companies was a major factor in the radio-boom of the 1920s. Another example of technical advancements made during the war would be development of "low-loss" techniques in transmission and radio design.[22] The war stimulated much invention, which was crucial for the radio revolution of the 1920s.

Most important of all was simply the training of vast numbers of men in the basic technical subjects needed to create the radio revolution, such as simple electronics theory, basic practical electronics, and knowledge of Morse code. Specialized communications units existed in the Prussian army from 1899. In 1914, they numbered 550 officers and 5800 men.[23] Most of these specialized communications troops were still involved in wired telegraphy and telephony, though what few radios existed were also their responsibility.[24] An additional number of troops within Infantry, Artillery, and Engineer units outside of the Communications Troops were also trained in telegraphy and electronics (but not radio), since the communications troops only handled the construction of telegraph lines (or radio nets) between command units down to the divisional level; anything below that level was the responsibility of the unit itself. Communications troops connecting lower levels of command with the Division or smaller units were taken from the units themselves. On the opposite end of the chain of command, most telegraph lines within Germany itself were the responsibility of the Postal Service, and large numbers of postal telegraphy experts supplemented the comparatively small numbers of Army troops trained in telegraphy, even outside of Germany after the German advance into Allied territory.[25] Alongside the pre-war military communications assets of the German Army were those of the German Navy. While the Navy, of course, did not have much use for wired telegraphy, it did have a great interest in radio. By 1914, all major ships were equipped with radio transmitters and receivers, so that we can estimate that the Navy had some 2000 trained radio operators in 1914.[26]

Most of the Army or Navy troops trained in both wired and wireless telegraphy and telephony before the War would have been career professionals, so

[22] See: Dr. Eugen Nesper, "Nachste Ziele der Radiotechnik", *Der Radio-Amateur* 3, No. 35 (August 28, 1926): 856–857. "Low-loss" in this context means having low resistance and, therefore, low power loss in a circuit.

[23] Niehaus, *Nachrichtentruppe*, 24–25. It is not clear if this figure is only for the Prussian Army, or for the entire land forces of the German Empire (therefore, the Prussian, Saxon, Bavarian, and Württemberg armies combined.)

[24] Niehaus: *Nachrichtentruppe*, 21–24.

[25] See Kampe for a more detailed discussion of this organization.

[26] This is a very rough estimate.

that the relatively small numbers of such men before the War was compounded by the fact that their skills, while of a high standard, had little impact on the civilian world. This changed dramatically with the outbreak of war. In 1914, Germany had an active military establishment of 734,000 men.[27] By war's end, some 11 million men had served in the military. A very rough calculation suggests that by 1918, at least 400,000 men had been trained in basic radio (as it was understood at the time), Morse telegraphy, telephony, and the basics of electronics. The real number is surely much greater.[28]

[27] Absolon, Vol. I, 4.

[28] All estimates here are highly conservative. The estimate is based on the following calculation:

A. Army:

1. Army officers and enlisted men in the "Nachrichtentrupppen" (formerly "Telegraphentruppen") in 1918: 4, 381 officers and 185,000 men.
2. Communications troops organic to infantry divisions, roughly 1300 per division:
   (a) 200 with divisional staff
   (b) 130–150 communications troops per infantry regiment (@ two per division)
   (c) 17 in each battalion (@ three infantry battalions per regiment)
   (d) Field Artillery Regiment: assume similar number to Infantry regiment = 200 (one FA Regiment per division)

Total number of Infantry Divisions in German Army in 1918: 247
(247 × 1300 = 321,100 communications troops organic to infantry divisions)

In addition, there were 2250 independent Foot Artillery Batteries
(assume same number as battery of Field Artillery = 22 or 49,500 communications troops)

A total of 251 radio stations existed with airfields and aircraft units.
(assume 15 men ea. = 3765)

B. Navy:

520 ships (not counting East Asian gunboats, obsolete coastal defense ships or vessels too small to carry radios).
Assume at least ten men on average per ship trained in radio. = 5200 men
Further radio personnel with the:
Marinestation der Ostsee, Marinestation Kiel, Marinestation Friedrichsort, Marinestation der Nordsee, Marinestation Wilhelmshaven, Marinestation Geestemünde, Marinestation Helgoland, Marinestation Cuxhaven, Six Marinestationen in German colonies:
  assume at least 12 radio personnel per Marinestation = 144
Reichsmarineamt
Admiralstab der Marine (responsible for communications equipment)
  assume at least ten each = 20
Marinetelegraphenschule Lehe
  assume 50 men
(does not include radio personnel with navy division in Flanders or naval aircraft units)
Deutsche Militärgeschichte 1648–1939 (Munich: Bernard & Graefe 1983), Bd. 5, Abschnitt VIII *Deutsche Marinegeschichte der Neuzeit*, 263–273; 295; 308–310.

The presence of so many veterans with communications training clearly had a direct effect on the spread of radio and the radio hobby after the war. Even if they had not learned radio theory, they still had learned basic electronics (for many, certainly their first exposure to electricity), the workings of field telephones, and the language of Morse code.[29] The presence of so many men with basic technical skills—and even more importantly, a familiarity with long-distance communications, no matter in what form—formed an important material foundation for the radio revolution of the 1920s.[30]

Clearly, the First World War did not see the invention of technological modernity; by the dawn of the twentieth century, the industrialized world had already entered a technological age, and most of the major technologies that characterized life (such as radio) had already been invented. Some were even in fairly wide use, yet the age after the First World War is different, for only then did life become so thoroughly imbued with modern technologies, everywhere and in all classes. After the First World War, modern (electrical) technology was ubiquitous, and had a mass character. The war was a major catalyst for bringing this about, and colored the social forms and institutions within which technologies found use and expression.[31] It would be wonderful to write the story of radio while just focusing on the evolution of the technology, but sadly, the "ur-conflict of the 20th century" casts a long shadow over the social use and civil appropriation of radio in Germany.

= roughly 5214 men trained in radio with German Navy (real numbers surely higher; estimate does not include attrition).

Lower estimates may be found in the literature. See: Helmut Schanze: "Rundfunk, Medium und Masse. Voraussetzungen und Folge der Medialisierung nach dem 1. Weltkrieg", *Die Idee des Radios von den Anfängen in Europa und den USA bis 1933. Jahrbuch Medien und Geschichte 2004* (Konstanz: UVK Verlagsgesellschaft, 2004): 18ff, Schanze gives an estimate of 100,000 men trained as military radio operators, a figure surely too high for those trained specifically as radio operators, but plausible if it includes wired telegraphers.

[29] Even if the introduction of broadcast radio in 1923 meant that most people would listen to voice transmissions with their radios, Morse code was still highly important for identifying stations over the air and was a basic tool of ham radio. Articles abound in the radio hobby press well into the late 1920s urging radio listeners to learn Morse code.

[30] For example, when, at the beginning of broadcast radio in Germany, an "Audionversuchserlaubnis" was required to own and use a radio receiver, special advanced, accelerated classes were offered, which were limited to former military radio men. See: "Mitteilungen", *Der Radio-Amateur* 2, No. 26 (October 17, 1924): 691–694.

[31] The social forms for the employment of any given technology are, of course, quite a separate thing from the basic science behind it. Physics is one thing, but form is social. See, for example, Bijker, Hughes, and Pinch *Social Construction*.

CHAPTER 4

Technology and the Radio Hobby Mature, 1927–1929

> Hams expect that the authorities will give every possible support to the new technology, the spread of which lies primarily in the hands of the radio amateurs. [We demand this] particularly because of the tremendous national importance [of radio].[1]

The model for radio hobby clubs set in the early period continued to be followed in the latter 1920s. The clubs thrived, and those engaged in the radio hobby followed ever more specialized and diverse interests: do-it-yourself (DIY) construction, high-fidelity sound reproduction, phonograph recording, shortwave transmission, and even amateur television. Because the general pattern in the late 1920s was business as usual, this chapter focuses on several particular themes, which will enrich the story of the hobby.

[1] "Mitteilungen". "Der deutsche Sendedienst", in: *Der Radio-Amateur* 4 No. 8 (February 19, 1926) 31. "*Die Funkfreunde erwarten, daß die Behörden der Entwicklung der neuen Technik, deren Ausbreitung vornehmlich in den Händen der Funkfreunde liegt, in Ansehung ihrer ungemeinen nationalen Bedeutung jede mögliche Förderung zukommen lassen*".

B. B. Campbell, *The Radio Hobby, Private Associations, and the Challenge of Modernity in Germany*, Palgrave Studies in the History of Science and Technology, https://doi.org/10.1007/978-3-030-26534-2_4

Radio and the Postwar, Radio and the Next War: The Radio Imaginary in the 1920s

We think of wars as ending on very specific dates, such as November 11, 1918, at 11:00. But contemporaries knew better: Ernst von Salomon, a German-nationalist writer of the late 1920s and early 1930s, coined the term "postwar" to refer to that messy, indefinite period immediately after the war, when conditions were still unsettled, and the path forward was still unclear.[2] The public perception of, and discourse about radio in the post- or interwar period was deeply colored by the war, and the "normality" of the postwar peace was constantly perturbed by ripples and eddies of instability left over from the conflict.

We can distinguish (somewhat arbitrarily) two levels of this discourse. First, there was a practical or immediate level. On a very simple level, the militarization of society, which persisted after the war (and was not caused by the war, only amplified by it) also found expression in radio. Thus, the first Great German Radio Exposition of 1924 featured the reproduction of the radio room of not only the contemporary "Amerika-Zepplin", but also of a U-Boot of the Great War.[3] And of course, members of postwar radio clubs were often treated to public lectures by veterans on radio or telegraphy during the Great War.[4] Also, on a very practical level, one characteristic problem of the 1920s was how to cope with the millions of crippled and wounded left by the war.[5] Among them were thousands of blind men, joining a much smaller pre-war population with the same handicap. Radio, as an acoustic medium, seemed a natural response to

[2] The term "Postwar" (*Nachkrieg*) was used by the contemporary nationalist writer Ernst von Salomon in several works as an analytical category. See: *Die Geächteten*, (Reinbek bei Hamburg: Rowohlt Verlag, 1962 (1930)) and *Nahe Geschichte. Ein Überblick* (Berlin: Rowohlt, 1936).

[3] Postrat H. Thurn, "Ausstellungsberichte" "Bilder von der Ausstellung der Deutschen Reichspost auf der Großen Deutschen Funkausstellung", *Der Radio-Amateur* 3, No. 6 (February 6, 1925): 154–158. The article included six large photographs, including one photo of both radio rooms.

[4] "Die Vortragsfolgen" Beiheft, "Mitteilungen", *Der Radio-Amateur* 3, No. 47 (November 20, 1925): 24–29. In this case, a "Herr Bieleit" gave a talk on "Feldfunkerei" to the Lichtenberg chapter of the Deutscher Funktechnischer Verband (DFTV). Similar talks became increasingly common in the later 1920s and early 1930s.

[5] See, for example, Deborah Cohen, *The War Come Home: Disabled Veterans in Britain and Germany, 1914 to 1939* (Berkeley: University of California Press, 2001).

their plight, and radio clubs worked with governments and charities to provide radios to the army of the blind and sight-impaired.[6]

On a more abstract level, other forms of war-related loss, those involving the entire nation, were also projected onto radio, which in some cases even became an indirect way of addressing the national trauma of the lost war, or of speaking of conflicts which were over, but not yet resolved. So, for example, a 1923 article about radio and new ways of providing pilots with meteorological information suddenly referred to the Versailles Treaty and its proscription of equipping German planes with radio telephony, whereas planes of all other lands did not face the same restriction.[7] Thus, the loss and injustice of the Versailles Treaty came to dominate even a minor technical piece about radio. We will see below that the theme of loss of status or strength as compared with other countries will occur again and again in the discourse around radio.

Clearly, discourses around radio often served as ways of addressing larger conflicts relating to the recent war. One recurrent theme involved the former German territories "lost" after the war or under Allied occupation. In the early 1920s, radio was often used as a proxy with which to criticize the Allied occupation of the Rhineland and Saarland. After all, even if possession of an ordinary broadcast receiver was rather restricted in Germany proper, the Allies forbade their very possession in the occupied Rheinish territories for several years, giving occasion to a series of complaints and solemn pronouncements against this injustice in the radio press.[8] In many cases, this critique of the postwar settlement and Allied policy went well beyond the actual situation in the occupied territories themselves. For example, a report on the founding conference of the

[6] H. Koslik, "Helft unsern Schwerkriegsbeschädigten!", *Der Radio-Amateur* 3, No. 33 (August 14, 1925): 821; "Die Versorgung der Kriegsblinden Unterfrankens mit Rundfunkgeräten", *Der Radio-Amateur* 4, No. 3 (January 15, 1926): 63.

[7] Dr. K. Keil, "Flugwetter", *Der Radio-Amateur* 1. No. 5 (December 1923): 119–120.

[8] See, for example, on the Allied occupied Rhineland: Die Schriftleitung, "Rechtsprechung. Vorläufig noch keine Freigabe des Rundfunks im besetzten Gebiet", *Der Radio-Amateur* 3, No. 8 (February 20, 1925): 208, or: "Brief aus dem Linksrheinischen Gebiet", signed "Mit treudeutschem Gruß, Ein Radio-Amateur-für Viele", under the regular rubric "Mailbox" ("Briefkasten"), *Der Radio-Amateur* 3, No. 12 (March 20, 1925): 302, or again, under the regular rubric "Mailbox", "Der Rundfunkskandal im besetzten Gebiet" (by "an anonymous German amateur"), *Der Radio-Amateur* 3, No. 19 (May 8, 1925): 483–484. For a larger perspective, see: Daniel Siemens, "Politik ist die Fortsetzung des Krieges mit anderen Mittel: Deutungskämpfe um den Ersten Weltkrieg in der politischen Kultur der Zwischenkriegszeit", *Geschichte für heute* 2, No. 1(2009): 40–58.

"Deutscher Funktechnischer Verband" in 1925 (formed out of the fusion of the "Deutsches Funkkartell" and the "Deutscher Funktechnischer Verein", two large radio organizations),[9] contains this remarkable passage:

> The German Radio-Technical Association remembers the occupied German lands cut off from Germany by the Versailles Treaty, and sends these territories, and especially the radio hobbyists (*Funkfreunde*) in them its warmest and most faithful greetings. It expresses the expectation, that the occupational authorities will no longer withhold from the occupied territories the cultural asset (*Kulturgut*) which is broadcast radio. The German Radio-Technical Association expects that the amateur radio enthusiasts (*Amateuren*) of the entire civilized world (*Kulturwelt*), in the interest of their own reputations, will leave no means unused in order to help the German radio hobbyists come into the full exercise of their rights.[10]

It is important here to note the use of the words "*Kulturwelt*" and "*Kulturgut*", with their echoes of the wartime discourse about the superiority of German *Kultur* over Western "civilization"—these were still loaded words in the 1920s. The text indirectly asserts that Germany (still) belongs to the "civilized world" of major industrial powers, as opposed to the uncivilized savages of the rest of the world, and that radio is one of the major signs of this belonging to a higher sphere of cultured nations. (More subtle is the use of the foreign loan-word "*Amateur*" for radio hobbyists outside of Germany, whereas German radio enthusiasts are denoted with the German term "*Funkfreund*"—more on this below.)

So, for example, an article in the radio hobby press explicitly comparing the state of radio in Germany to that in foreign countries ended with the exclamation: *"We have a right to that which other countries have unfortunately been given before us! Germany cannot and must not lag behind in*

[9] See pages 52–53 above.

[10] "*Der Deutsche Funktechnische Verband gedenkt der durch den Versailler Vertrag abgetrennten und besetzten deutschen Landesteile, er sendet ihnen und besonders den dortigen Funkfreunden seine herzlichsten Treugrüße und spricht die Erwartung aus, daß die Besatzungsmächte den besetzten Gebieten nicht länger das Kulturgut des Rundfunks vorenthalten. Der Deutsche Funktechnische Verband erwartet von den Amateuren der Kulturwelt, daß sie im Interesse auch ihres eigenen Ansehens kein Mittel unversucht lassen, den deutschen Funkfreunden zu ihrem Recht zu verhelfen*". "Tagung des Deutschen Funkkartells in München am 28 Juli 1925", *Der Radio-Amateur* 3, No. 31 (July 31, 1925), special insert on the annual meeting of the Deutscher Funkkartell on July 28, 1925: 1.

this!"[11] Another article, this time focused specifically on ham radio, stated that radio was "a first rate means of international communication".[12] "The person transmitting, who takes a Morse code key in his hands, speaks to people as far as his transmission energy can reach. He speaks for Germany, when he sends code. He represents Germany. His transmitter must therefore absolutely correspond to the rules of operation which have already become international, otherwise, he will look bad and will soon no longer receive reception reports. The Operations Office[13] therefore has the serious duty to teach this knowledge as quickly as possible, and to make sure that the rules are respected. Only in this manner can the new amateur transmission live up to its duty of being a means of understanding between the technicians of the world".[14]

Radio was clearly seen as something important, a criterion by which a country's degree of civilization was to be measured. The degree to which this was echoed in the radio hobby press is perhaps understandable, since these were true believers in the radio gospel "preaching to the choir". Radio enthusiasts naturally felt that radio was important, and middle-class hobbyists tended to define that importance in cultural terms. But beyond

[11] "*Wir haben ein Recht auf das, was andere Länder leider vor uns erhalten haben! Deutschland kann und darf herein nicht zurückstehen!*" (italics in the original). A. Klein "'Drahtlose' Reiseeindrücke", (Fortsetzung von S. 90), *Der Radio-Amateur* 1, No. 5 (December 1923): 114–116.

[12] "*ein internationales Verkehrsmittel ersten Ranges*". L(udwig). von Stockmayer, Vorsitzender des Oberdeutschen Funkverbands, "Die Tätigkeit und Organisation der Amateursender", *Der Radio-Amateur* 3, No. 34 (August 21, 1925): 844–845.

[13] The "operations office" of the Upper German Radio Association. This was a committee dedicated to amateur transmission or "ham" radio. It was later transferred to the Deutsche Amateur Sende- und Empfangsdienst (DASD), the national organization of German hams, when it was founded in 1926. See below, pp. 127–131.

[14] "*Dieser Telegraphieverkehr hat neben der Leistungssetigerung der Geräte und der Funker im sportlichen Sinne dadurch eine ganz große Bedeutung, daß er ein internationales Verkehrsmittel ersten Ranges ist. Der Sender, der die Taste in die hand nimmt, spricht zu Menschen, soweit seine Sendeenergie reicht. Und Kurzwellen gehen weit. Er spricht für Deutschland, wenn er gibt. Er vertritt Deutschland. Sein Sender muß sich deshalb den bereits international gewordenen Verkehrsregeln unbedingt anschließen, sonst fällt er peinlich auf und wird bald keine Empfangsmeldungen mehr bekommen. Die Verkehrsleitung hat deshalb die ernste Aufgabe, diese Kenntnis so bald als möglich zu vermitteln, und darüber zu wachen, daß die Regeln befestigt werden. Nur so kann das neue Senden seiner Aufgabe gerecht werden, ein Verständigungsmittel der Techniker der Welt zu werden*". L(udwig) von Stockmayer, Vorsitzender des Oberdeutschen Funkverbands, "Die Tätigkeit und Organisation der Amateursender", *Der Radio-Amateur* 3, No. 34 (August 21, 1925): 844–845. Note that the article also praised the contribution of ham radio in the US to the war effort.

this, what is striking here is the extent to which the conviction that radio was an important sign of civilization was accompanied by a fear bordering on panic that Germany was in danger of losing its status as a civilized nation because of its backwardness when it came to radio, a fear clearly echoing greater worries of national decline and resentment at losing the war.

So, for example, in 1925, the Deutscher Radio-Klub e.V., then the most important umbrella association of radio hobby clubs, officially declared in the preamble to its governing statutes that the technology of radio was so important that those countries which do not understand it will lag behind ("*zurückstehen*") in the competition among nations ("*Wettstreit der Völker*").[15] The implication that Germany ran this risk if it did not do more to support radio was obvious. It appears so often in other contexts that it is clearly much more than a rhetorical tool to gain support for the clubs, it expressed a real fear. Another article concerning ham radio in particular and hoping that the German government would soon grant the authorization to licensed private individuals to transmit over the radio expresses the conviction that ham transmission in Germany "will develop as quickly and as well as the other amateur activities have, and that the time when we will catch up to the other civilized European nations in this area is no longer a distant one".[16] And conversely, many discussions of radio show German fear that Germany might be falling behind other nations; note how often the term "*Kulturvölker*" was applied to countries involved in radio.[17]

Radio after the Great War was clearly seen as a highly strategic technology, and one which was an important matter of government—indeed, possession

[15] L. von Weiher, "Mitteilungen des Deutschen Radio-Klubs e.V.", *Der Radio-Amateur* 2, No. 4, (April 2, 924): 122–125. The club soon changed the official spelling of its name from "Klub" to "Club".

[16] "*Wir sind überzeugt, daß sich die Sendertätigkeit in Deutschland ebenso schnell und gut entwickeln wird, wie sich das Amateurwesen überhaupt entwickelt hat, und, daß der Zeitpunkt nicht mehr allzu fern ist, wo wir den übrigen europäischen Kulturstaaten auch auf diesem Gebiet auf die Ferse rücken werden*". Note the term "*Kulturstaaten*". "Beginn des Privatverkehrs in Deutschland", part of the announcements from the "Hamburger Radio-Klub e.V." (successor to the former "Funkverband Niederdeutschland e.V."), under the rubric "Mitteilungen befreundeter Vereine", *Der Radio-Amateur* 3, No. 28 (July 10, 1925): 28–30.

[17] Hugo Koslik, "Tagung des Funkkartells", *Der Radio-Amateur* 3, No. 21 (May 22. Mai, 1925): 528. At the end, the Kartell had a thought for the Germans in the occupied territories who are prevented from listening to radio. "*Mit einem Aufruf wendet sich das Kartell an die Regierungen aller Kulturstaaten, an alle Sendegesellschaften und Amateurvereinigungen der ganzen Welt und fordert die Aufhebung des unwürdigen Verbotes des Rundfunkempfangs im besetzen Gebiet.*

Möge dieser Appell nicht ungehört verhallen!"

of a robust radio culture was seen as a sign of being a major state and a "*Kulturnation*". It was a measure of national greatness. Radio was both an abstract measure of national strength, and also a very concrete one: radio was a key strategic weapon in the struggle between nations.[18] In a 1924 article encouraging radio listeners to learn Morse code, the scientist Manfred von Ardenne openly stated that knowledge of Morse code would surely be important in the next war.[19] And just as radio was an important tool of war, it was also immediately seen to be an important tool in the war for people's minds: propaganda. Again, this was admitted quite candidly in the radio press.[20]

The introduction of radio as a mass technology and occupation was accompanied by a series of shocks in Germany and took place in the aftermath of the huge national trauma of the lost war. As was mentioned earlier, at the beginning of the twentieth century, Germany had been the leading nation in the world when it came to both basic science in the field of electricity (and therefore, radio) and to its powerful electrical industry, and occupied a similar leadership position in the chemical industry and in basic physics research. The shock to discover that Germany by the early 1920s was trailing badly behind other major industrial nations like the US, but also Britain and France in the practical development of radio was therefore all the greater. If radio was a symbol of national greatness in Germany, one of the panoply of cultural tools, which defined a civilized nation (in the German term, "*Kulturstaat*" ("civilized state") or even more fundamentally, "*Kulturvolk*" ("civilized people")), as all agreed it was, then the shock of being outpaced by foreign nations (particularly those which had just been its opponents in the Great War) became a panic about German decline. The new medium of radio was never innocent, never free of a

[18] For example, Dr.-Ing. Max M. Hausdorff, "Nur Amateur!", *Der Radio-Amateur* 1, No. 1 (August 23, 1924): 15–16—citing a French general!

[19] Manfred von Ardenne, "Radio-Unterricht. Wie lerne ich morsen?", *Der Radio-Amateur* 2, No. 21, (September 12, 1924): 550–551. He began by stating that it can be important to know Morse code to calibrate your radio, know more about world around you, and so on, but also: "The ability to use Morse code will certainly play an important role in the transmission of information during future wars." ("*Die Fähigkeit des Morsens wird bei der Nachrichtenübermittlung in kommenden Kriegen sicher eine bedeutende Rolle spielen.*")—a footnote from the journal's editors says this was already the case in the First World War—to Germany's detriment. Ironically, von Ardenne later became a socialist, and chose to live in the German Democratic Republic after the Second World War.

[20] For example, "Die kulturelle Bedeutung des Rundfunks", *Der Radio-Amateur* 2, No. 33 "1. Messeheft" (December 5, 1924): 876–880.

relationship to the hopes and fears of the wider German society. Indeed, it quickly became a screen upon which wishes and fears were projected.[21]

Another area in which radio was used as a proxy for larger issues concerns language, the very terms used to describe the components of the new medium and the people who used it. The question was, "whose words" would describe (and thus structure) the new technology and social practice of radio? Would the new medium be described by foreign loan words (in keeping with the practice in most of the rest of the world), or would it reside safely in words of Germanic consonance—even if they had to be newly invented? "*Radio*" or "*Funk*"? "*Amateur*" or "*Funkfreund*"? Would the new medium belong to foreigners, or would it be German? The debate was intimately related to the reception and social valance of the new technology of radio, which needed new words and concepts. Naming conveys power, and in this case, was directly linked to the social reception and, therefore, social place of the new technology. But equally as clearly, the issues involved went far beyond radio itself, and were directly connected to the war, and to the place of Germany in the world. Let me cite two manifestations of this debate within the context of radio.

In 1924, at the very beginning of radio as a mass phenomenon, scientists and experts rushed to write books explaining and popularizing radio technology and giving practical advice on building and operating a receiver. One of the first and most popular was written by Dr. Eugen Nesper, the editor of the journal *Der Radio-Amateur*, and a leading radio activist.[22] The book in question was titled *Der Radio-Amateur. "Broadcasting"*,[23] and sought to explain to a German public just what the new American invention of broadcast radio was all about, how radios functioned, and could be built.[24] It specifically justified an engagement

[21] The topic of cultural pessimism in Germany is too vast to discuss here. See the classic study: Fritz Stern, *The Politics of Cultural Dispair: A Study in the Rise of the Germanic Ideology*. Garden City, NY: Anchor, 1965.

[22] Nesper (1879–1961) worked with Slaby in Berlin before even graduating from high school. He earned a graduate degree (*Diplom*) in machine- and electrical engineering at the Berlin-Charlottenburg Technical University (*Technische Hochschule*) in 1902, and two years later received his doctorate from the University of Rostock. He worked for Telefunken and Lorenz, and after the First World War had a career as a writer and publicist, scientist, and inventor. See: Oskar Blumtritt, "Nesper, Eugen Heinrich Josef", *Neue Deutsche Biographie* 19 (1998): 70–71, http://www.deutsche-biographie.de/pnd129294594.html, Accessed July 16, 2014.

[23] Berlin: Julius Springer, 1923.

[24] Note that the book appeared before broadcast radio began in Germany, therefore, at a time when all civilian-owned radios were principally illegal.

with radio by stating that Germany had suffered during the war due to a lack of technically trained people, particularly in the art of radio and Morse telegraphy. Radio as a hobby was a prime means of educating the German people: "What the Germans really need, is proper technical thinking".[25] The book was thus colored by the kind of German-nationalist thought common among educated Germans, and was in no way an apology for Germany's enemies.[26]

Thus, it must have been a surprise to many when a letter from Senior Engineer (*Oberingenieur*) Ernst Neumann, Frankfurt a./M. appeared in the "Mailbox" rubric of the January 1924 issue of *Der Radio-Amateur* magazine, harshly criticizing Nesper for his use of non-German language. He attacked the use of words like "*amateur*", stating:

> *Amateur* is really too French. There may have been a time, and in several decades there may well come a time again, when one will not need to resent Herr Nesper's use of the French word. But today, particularly at this point in time, a good German should not sin against the German cause, even in little things. And the German language is the German cause. And after the lost war more than ever![27]

[25] "*Was den Deutschen not tut, ist richtiges technisches Denken*" (p. VII).

[26] To be fair to Nesper, he also ended the introduction with the wish: "May radio-broadcasting, among other means, be called upon to do away with the poisoning of peoples which has been spreading continuously ever since 1914, may it bring peoples of all races and clans closer to each other and may it speed the exchange of important cultural capital." ("*Möchte das Radio-Broadcasting mit dazu berufen sein, die sich seit 1914 immer noch weiter ausbreitende Völkervergiftung zu beseitigen, die Völker aller Rassen und Stämme einander wieder näher zu bringen und den Austausch wichtiger Kulturgüter zu vermitteln*".) (Nesper, *Broadcasting*, VIII). In 1923, a radio idealist like Nesper could still be a German nationalist and believe that radio could bring international understanding, all the while preparing the German public for the next war. Nesper was not alone in his beliefs in Germany in the 1920s. Within ten years, the contradiction in this position would be resolved by simply shifting the commitment to international understanding in the discourse around radio to using radio to explain the German side of world issues.

[27] "*Amateur ist wirklich zu französisch. Es mag eine Zeit gegeben haben und vielleicht nach vielen Jahrzehnten einmal wieder geben, wo man Herrn Nesper das französische Wort weniger übel zu nehmen brauchte. Aber jetzt, gerade im gegenwärtigen Zeitpunkt, sollte sich ein guter Deutscher auch im Kleinen nicht an der deutschen Sache versündigen. Und die deutsche Sprache ist deutsche Sache. Nach dem verlorenen Kriege erst recht!*" Letter from Ernst Neumann, Oberingenieur, Frankfurt a.M.-West, in the regular rubric "Briefkasten", "Zu Dr. Nespers erstem deutschen Radio-Amateurbuch "Der Radio=Amateur=Broadcasting"", *Der Radio-Amateur* 2, No. 1 (January 1924): 28–30.

Nesper replied that one shouldn't use foreign words without a good reason, but that words like "*Amateur*" were already in widespread use in Germany, and that it sounded funny to use German expressions such as "*Liebhaber*" or "*Freund*". Moreover, he stated that radio was an international movement, and that it was good for Germany to use the same terminology—like "broadcasting"—as the rest of the world. Finally, he cited Goethe to say that only small-minded people insist on purity of the language, and that if Goethe wasn't a German enough witness, then one should note that the Prussian Army uses all sorts of French words, and surely Neumann wouldn't accuse it of not being "German-national" enough.[28] Nesper could get the final word in this exchange not least because he was the editor of the journal in which it appeared. But the conflict over language itself would not and did not go away, and lasted into the 1940s and the Third Reich. The influence of the (lost) war (and of pre-war trends) continued to affect the nascent technology and social acceptance of radio, even down to the very words used to describe it.

The discussion over language was not limited to hobby circles; even state instances became involved. For example, in April 1925, the Prussian Minister for Science, Art, and Popular Education circulated a bulletin to all schools concerning radio and its place in education.[29] The bulletin originated with the Reich Postal Ministry, and was designed to make university students and high and middle school pupils aware of laws governing radio, mainly in an effort to combat unlicensed radio possession and use.[30] But the bulletin also listed the proper German terms for radio technology, and specifically condemned the use of foreign terms, stating that the Postal Ministry was working with other government authorities and hobby organizations to end their use.

These are but two examples in a much larger nationalist conflict over language which continued through the end of the Second World War. The point here is not to give an exhaustive study of the language conflict, but

[28] Ibid.

[29] The "School Radio" branch of the Funktechnische Vereinigung (FTV) discussed the bulletin (*Merkblatt*) in the regular rubric: "Mitteilungen", *Der Radio-Amateur* 3, No. 24 (June 12, 1924): 27–30. Specifically, this was the "Erlaß of 6.April.1925" from the Prussian Minister für Wissenschaft, Kunst und Volksbildung (i.A. Schellberg), which had just appeared in the *Zentralblatt fur die gesamte Unterrichtsverwaltung*.

[30] This was important not least because broadcast radio in Germany was a state monopoly and was funded by a monthly subscription fee payable for each radio receiver an individual or collective entity possessed.

rather to show how its presence in the discourse around the new technology of radio is one of the many ways in which the First World War colored subsequent discourse and social practice.

For most German nationalists, one of the worst losses caused by the Versailles Treaty was the "loss" or separation of large numbers of ethnic Germans from Germany proper. Large "German" minorities were now part of Poland, France, Belgium, Denmark, Lithuania, Hungary, and Romania. Moreover, the Rhineland and Saarland were under French and Belgian occupation and risked cultural alienation as well. To these "stolen" Germans came the large numbers of Germans who had immigrated abroad since the nineteenth century.[31] They, too, might be "lost" to Germany if steps were not taken to cement their ties to their "homeland" and native language and culture. The potential loss of this "good German blood" was a common fear already before the Great War, and loomed larger afterward.[32] This theme is thus no anomaly in the discourse around radio in the 1920s and early 1930s. For example, the German Postal Ministry set up a transmitter specifically directed at the occupied territories as soon as the Allies permitted ownership of private receivers there. One article describing the plans states: "It is noted that a large portion of a highly cultured people is being closed out from participation in the reception of valuable cultural assets (*Kulturgüter*) and is not in a position to follow the development [of radio]".[33] Again, actions of the Allies are cast as a cultural crime. The German government's plan in the mid-1920s to build a broadcast transmitter specifically to serve the German popula-

[31] I am speaking of perception, rather than reality. Naturally, the labeling of these people as "German" (particularly in an exclusive and inherent way) was also an ideological construct.

[32] The subject of German ethnic minorities abroad is a large one. It finds echo in German citizenship law, which favors blood relation (*ius sanguinis*) as the determinant of belonging. It colored the Party Program of the Nazi Party. And it was a major focus of academic research during the Weimar Republic. As just one example of a voluminius literature, see Gerhard Weidenfelder, *VDA. Verein für das Deutschtum im Ausland. Allgemeiner Deutscher Schulverein (1881—1918). Ein Beitrag zur Geschichte des deutschen Nationalismus und Imperialismus im Kaiserreich,* Europäische Hochschulschriften, III/ Vol. 66, Bern-Frankfurt a/M.: Herbert Lang Verlag, 1976. Note that the "Association for Germans Abroad" (*Verein für das Deutschtum im Ausland*, VDA) was given radio broadcast time.

[33] "*Es wird darauf hingewiesen, daß ein großer Teil eines hochkultivierten Volkes durch die Besatzungsbehörde von der Teilnahme am Empfang wertvoller Kulturgüter ausgeschlossen wird und nicht imstande sei, die Entwicklung zu folgen*". Franz Kirkam, "Das kommende deutsche Rundfunknetz", in the rubric "Neustes", *Der Radio-Amateur* 3, No. 21 (May 22, 1925): 530–531.

tion under Allied occupation in the Rhineland demonstrates that radio was seen as a tool to mitigate, if not undo the Versailles settlement. This extended to all the "orphaned" German populations and indeed, via broadcasts from the powerful shortwave transmitter in Zeesen (Königs Wüsterhausen), even to the still larger Germanic diaspora worldwide.[34] There was great optimism that radio would be able to allow German communities, which had been separated from Germany (or Austria-Hungary) by the post-First World War settlement or who had left the Motherland to seek economic opportunity elsewhere to keep their German identity and culture, and therefore to prevent their "loss" to Germany. This purpose was expressed quite openly in the radio press. Speaking of plans for a powerful new transmitter, an article from 1924 stated it was directed at "the many German men and women who had been torn away from the Motherland by the unfair Peace of Versailles" with the goal "of keeping the idea and the feeling of the German cultural community in the separated territories awake and alive".[35] Radio before the First World War was seen as the key to breaking the information "blockade" imposed on Germany and its colonial possessions by the British ownership and control of the undersea telegraph cables. After the war, the same transcendent nature of radio waves was imagined as a cure for the alienation of German settlers and emigrants from the mother country. In both cases, radio technology is imagined as the solution to political and even racial problems of the German body politic.

[34] Deutsche Welle, *Mit 8 KW rund um die Welt.: Deutscher Weltrundfunk in der Weimarer Zeit. Geschichte des Kurzwellenrundfunks in Deutschland 1929–1932* Berlin: Haude & Spener, 1969. Regular but still experimental shortwave transmissions began in 1926 in Königs Wusterhausen, while the first regular German shortwave transmissions began in 1929 from the new transmitter in Zeesen.

[35] "*die zahlreichen durch das Versailler Diktat vom Mutterlande losgerissenen deutschen Männer und Frauen wendet, mit dem Ziele, den Gedanken und das Gefühl deutscher Kulturgemeinschaft in den abgetrennten Gebieten wach und lebendig zu halten*". "Auslandsdeutschtum und Auslandskunde im Rundfunk", in the regular rubric "Verschiedenes", *Der Radio-Amateur* 2, No. 36 (December 1924): 996. The article announced the beginning of a new program transmitted by the Süddeutsher Rundfunk in Stuttgart about the ethnic Germans abroad, produced by the "Institute for Germans Abroad" (*Deutscher Auslands-Institut*), and announced plans for an entirely new transmitter aimed at Germans outside Germany. Note that ever since the German Empire, nationalists, and demographers in Germany had been concerned by the loss of valuable German blood through emigration, and the efforts via radio can be seen in continuity with this preoccupation.

The Continued Fear of Radio

Of course, there was another side to the understanding that radio could be an important tool for the German state; it could equally well become a tool of those who might seek to subvert it. In the 1920s and 1930s, radio was widely perceived as an extremely dangerous threat, particularly in the hands of socialists or communists. As Brecht pointed out, there is nothing in the technology of radio that per se destines it only to follow the broadcast model. Radio is inherently a two-way medium; transmission and reception are two sides of the same technology.[36] Particularly in the socially divided and political polarized Germany of the Weimar era, there was great fear that radio could aid and abet Germany's enemies, particularly its' internal enemies. These fears were a major factor shaping German broadcast radio as a state monopoly with strictly de-politicized, yet government-supporting content focusing on high-culture and the intellectual and moral uplifting of the general public. It played a direct role in the virtual ban on legal amateur transmitters, and it fed a constant paranoia that ordinary citizens might get out of hand and misuse radio.[37] Some of these fears might have been justified due to the state of radio technology in the early 1920s, such as the possibility of widespread unintended interference caused by improper use of regenerative radios. Others had far deeper roots in the deep social divide between workers and the middle class, which was a constant factor in the German Empire. After all, working-class enthusiasm for radio matched that of the middle class and, thus made it dangerous. Still, wartime fears of espionage and the chaotic experience of the German Revolution immediately after the war were the most immediate wellsprings of this fear.

We have already seen that the Postal Ministry claimed exclusive authority over any and all civilian uses of radio. As we have seen, this was justified by the *Reichstelegraphengesetz* (Imperial Telegraph Law) of 1892 and restated in the Postal Ministry's "*Verordnung zum Schutze des Funkverkehrs*" (Directive for the Protection of Radio Traffic) of March 8, 1924. This was similar to the state of affairs in most European countries in the 1920s where governments claimed far-reaching control over broadcast radio, but expressed in much more extreme form. It stood in marked contrast to

[36] Bertolt Brecht, "Der Rundfunk als Kommunikationsapparat", *Blätter des Hessischen Landestheaters Darmstadt* July 16, 1932.

[37] See below.

developments in the US, though even there, the US Navy had attempted to claim exclusive authority over all uses of radio immediately after the war, only to be rebuffed by a congress under great pressure from civilian hobbyists and business interests.[38] One motivation for this assertion of authority by the Postal Ministry in Germany was simple fear of competition, but there was also a more visceral fear that the sacred privacy of commercial and individual (telegraphic) communications could be violated by those possessing radio receivers.[39] This was sometimes cited as a proxy for other fears (see below), but it is echoed so often by representatives of the Postal Ministry that it must be taken seriously.[40]

But the larger fear was even more basic and lay in the fundamental distrust of an unregulated population. This basic mistrust on the part of the largely former-Imperial civil servants in charge of the Postal Ministry, and seconded by the Ministries of Interior and Defense, was only heightened by what was seen as the real and immediate threat of communist and socialist revolution. Hadn't the Bolsheviks used radio as a tool of subversion as early as 1918?[41] In the view of the German authorities, Soviet Russia was working to bring about world revolution, and was eagerly assisted by organized socialists and communists in Germany, who had already attempted to seize power by force during the German revolution and in a series of putsch attempts thereafter. Radio, if it were not properly regulated and controlled, threatened to bring the infection of socialism into every kitchen and living room, and allowed subversives and spies to speak directly with their handlers in Moscow. The ministries charged with radio matters sought to limit

[38] See Desoto, *200 Meters.*

[39] Postal control over radio was continually justified by arguments that the principle of the secrecy of telegraph communications (the *Telegraphengeheimnis*, the law or principle of privacy of telegraph transmissions, not the *Postgeheimnis* or the principle of the privacy of communications by letter) would be violated by unregulated possession of radio receivers. See, for example, Rechtsanwalt Franz Landsbergin, "Der deutsche Rundfunk", *Der Radio-Amateur* 1, No. 4 (November 1923): 83–88, or Ministerialrat Gieß, "Die Versuchssender im neuen Weltfunkvertrag", *Funk-Bastler Fachblatt des Deutschen Funktechnischen Verbandes e.V.* No. 30 (July 20, 1928): 465–467.

[40] Penalties for violation of the "*Telegraphengeheimnis*" were even increased in the new "Gesetz über Fernmeldeanlagen" passed by the Reichstag without discussion on November 24, 1927. (Radios were included by the Postal Ministry in the definition of "*Fernmeldeanlagen*", "long-distance communications installations"). "Das Neue Telegraphengesetz", *Funk-Bastler*, No. 1 (January 1, 1928): 1–2.

[41] See Alexander Rabinowitch, *The Bolsheviks in Power: The First Year of Soviet Rule in Petrograd* (Bloomington: Indiana University Press, 2007): 164–165.

public access to the technology as a result. For example, the "*Merkblatt für die Genehmigung von Rundfunkempfangsanlagen*" (Information Sheet for the Licensing of Radio Receivers) from the *Reichstelegraphen Verwaltung* (Imperial Telegraph Administration, within the Postal Ministry) in 1923 in the early days of German radio stated that radio receivers would only be allowed for German citizens, but that even then, permission could be refused to any person for whom there was a "well-founded suspicion" that they would misuse the radio. Furthermore, only approved radios which had been inspected and then sealed could be allowed.[42] The law was soon loosened a bit (1925), but the tone of the original legislation is telling. Internal ministerial documents disclose that there was near consensus on the potential dangers of radio, even if the Postal Ministry felt that it would be worthwhile to introduce broadcast radio.[43]

It didn't help that the Left was itself fascinated with the prospects of radio as a tool for both democracy and revolution, or that a specifically socialist radio movement called the Worker's Radio Union (*Arbeiter-Radio*-Bund, or ARB) immediately developed in 1924 as a pendant to the middle-class one. Later, after an internal split in 1929, an openly communist off-shoot, the *Freier-Radio-Bund* (FRB) was created, only increasing middle-class fears.[44] The existence of a nationwide network of left-wing

[42] The "Merkblatt für die Genehmigung von Rundfunkempfangsanlagen" was reprinted in the article "Der deutsche Rundfunk" by Rechtsanwalt Franz Landsberg in: *Der Radio-Amateur* 1, No. 4 (November 1923): 83–88. The strict regulations were justified publicly by the threat to the *Telegraphengeheimnis*. Some exceptions for non-citizens could be permitted. Note that police and officials of the Postal Ministry were allowed to search private property for illegal radios without a warrant.

[43] For example, see: "Niederschrift über die 21. Sitzung der Reichsfunkkommission am Freitag, dem 9. Juni 1922" in: BArch. R/4701/8673 Reichspostministerium, Sitzungsberichte R.F.K. Geh. Reistratur Z, Band 1 1919–1925, pp. 365–381, or "Niederschrift über die 22. Sitzung der Reichsfunkkommission am Montag, dem 25. Juni 1923" in: BArch R/4701/8673 Reichspostministerium, Sitzungsberichte R.F.K. Geh. Reistratur Z, Band 1 1919–1925, pp. 475–482.

[44] See, for example, Peter Dahl, *Arbeitersender und Volksempfänger. Proletarische Radio-Bewegung und bürgerlicher Rundfunk bis 1945* (Frankfurt/M.: Syndikat, 1978). The FRB was closely monitored by the Police of the German states. Detailed evidence of this monitoring and the concern, even fear provoked by the FRB may be found in the Bundesarchiv (BArch), record groups: R/1501/20400 St10 RMdI KPD-Radio-Propaganda Bd. 2 Mai1931–Juli1933; R/1501/20401 St10 RMdI KPD-Radio-Propaganda Bd. 3 Dez. 1929–Febr.1931 St10/62, Bd. 3; R/1501/20402 St10 RMdI KPD-Radio-Propaganda Bd. 4a March 1931–February 1933; R/1501/20403 St10 RMdI KPD-Radio-Propaganda Bd. 4b March 1931–February 1933; R/1501/20404 St10 RMdI KPD-Radio-Propaganda Bd. 5 March 1933–March 1934.

radio clubs, the ability of individuals to both receive direct propaganda from the Soviet Union via radio, and to build illegal transmitters, which might be used for espionage or domestic propaganda haunted the authorities. The development of radio on a broad scale in the Soviet Union was watched with a wary eye.[45] There were constant warnings about the dangers of a politicization of broadcast radio, which was designed in Germany to be politically neutral, but which in fact favored conservative and moderate parties.[46] The fundamental opposition to licensing private transmitters in Germany was another consequence of the fear of the left and of an uncontrolled access to technology in general.[47] The willingness of both right- and left-wing radio amateurs to build and use illegal transmitters only reinforced this fear. Clearly, radio reflected (and often perpetuated) fundamental continuities in German society, such as the deep divide between Right and Left. But these continuities were also exacerbated by the War and its immediate aftermath. The postwar view of radio as a powerful potential tool for both the German state and its enemies owes much to the experience of the war.

The First World War was the "ur-catastrophe" of the twentieth century.[48] It did not create anything lasting, but did serve as a catalyst to accelerate

[45] See, for example, "Rundfunk in Russland", in the regular rubric "Kleine Funkmitteilungen", *Der Radio-Amateur* 4, No. 23 (June 4, 1926): 21–32. The article listed seven Soviet transmitters which could be heard in Germany—together with their frequencies! Even more telling, see the series of radio monitoring reports by the "Office for Monitoring Private German Radio Transmissions at the Police Main Radio Office" ("Überwachungsstelle für den privaten deutschen Kurzwellenverkehr bei der Pol. Hauptfunkstelle Berlin") in BArch, R/1501/20404 St10 RMdI KPD-Radio-Propaganda Bd. 5 March 1933–March 1934. This office monitored both legal and legal shortwave radio transmissions by amateurs, and the broadcasts of Radio Moscow. The Reichswehr was also concerned about Radio Moscow. See: Der Reichswehrminister 45/32 geh. Abw. II (Geheim) "Rundfunkpropaganda der Sowjetunion in Deutschland und ihre Auswirkungen" of 19. March 1932, in: R/1501/20402 St10 RMdI KPD-Radio-Propaganda Bd. 4a March 1931–February 1933, pp. 284–294.

[46] See, for example, "Kleine Funkmitteilungen", *Der Radio-Amateur* 4, No. 15 (April 9, 1926): 30–31, which warned of such politicization, specifically citing Leon Trotsky.

[47] See, for example, Ministerialrat Gieß, "Die Versuchssender im neuen Weltfunkvertrag", *Funk-Bastler* No. 30 (July 20, 1928): 465–467. Ironically, this opposition was ended by the Nazis in 1933, who finally acceded to the wishes of German hams—but only in the context of a regulated and regimented amateur radio association intended to be both a propaganda tool abroad and a major agent for rearmament.

[48] The idea comes from George F. Kennan, who called the war the "great, seminal catastrophy of the 20th. Century". George F. Kennan, *The Decline of Bismarck's European Order. Franco-Russian Relations, 1875–1890* (Princeton: Princeton UP, 1979): 3. The term is trans-

many trends, which were already visible before, and did prepare the ground for future developments. This was certainly the case with radio. While the basic technology was well in place before 1914, radio became a mass phenomenon after the war. The experience of the war prepared the ground for this development. It made people more conscious of, and more comfortable with the technology in all forms. It spread the basic technical skills, which were the foundation of the "radio revolution". At the same time, the war cast a long shadow, and the invention of broadcast radio in the following two decades (at least) was strongly colored by the experience of the war and the resentments accompanying its loss in Germany. This would play a major role in the subsequent hardening of the political debate nationally, and thus also within the radio hobby community. The result, the Nazification of the radio hobby, was not inevitable, but did stem in part from the long "hangover" after the First World War.

Evolution of Radio Technology and Radio Hobby Clubs

Broadcast radio was a solid presence across Europe by late 1920s. But the proliferation of broadcast transmitters brought problems: there was competition for radio spectrum, which led to increasing round of international and European radio conferences designed to allocate spectrum to the various interests involved.[49] The ever more crowded airwaves forced listeners to invest in ever more sophisticated receivers and better antennas, providing motivation to both industry and the do-it-yourself movement. Selectivity (the ability to receive a desired frequency without wavering, and without receiving neighboring frequencies) and sensitivity (the ability to receive weak signals) became increasingly important in radio design. Greater pressure on spectrum and increasing numbers of radio owners also made radio-frequency interference from machines, street-cars, medical

lated into German as "*Urkatastrophe*", which is here re-translated back into English. See also: Ernst Schulin, "Die Urkatastrophe des zwanzigsten Jahrhunderts", in *Der Erste Weltkrieg. Wirkung, Wahrnehmung, Analyse*, edited by Wolfgang Michalka (Munich: 1994): 3–27.

[49] W.M., "Und das Wellenchaos gesteht weiter?", *Funk*, No. 17 (April 26, 1929): 70; Postrat Münch, "Internationale Verständigung im Rundfunk", *Funk*, No. 18 (May 3, 1929): 77 f.

apparatus, and other sources a growing irritant.[50] Among other things, this was a sign that radio listening had really become an important activity, and one which was seen as a right, rather than a technical wonder. In Germany, the fact that owners of receivers paid a monthly fee to the postal authorities also caused them to act like consumers: if I pay for service, I expect to receive it without interference. The result was a growing number of consumer complaints, and a corresponding attempt to seek solutions at all levels, from the national government down to individual listeners. In Germany, 1929 saw a national campaign against radio-frequency interference and the founding of the "Committee on Radio Interference" ("*Ausschuß für Rundfunkstörungen*"), chaired by—who else—von Bredow, and loosely tied to the *Reichsrundfunkgesellschaft*.[51] This was important for two reasons: first, the Committee on Radio Interference brought government, industry, commerce, technical associations, and radio clubs together in a common endeavor. Thus, both the consumer power and the technical competence of the radio clubs were recognized through their inclusion. (Interestingly, the list of participating clubs included even the socialist ARB, a sign of the importance of the problem.) The second reason this is important is more complicated to explain. Radio-frequency interference is a problem of physics and engineering, and in that sense, relatively straightforward, even if the search for the source of any given individual interference could be technically quite difficult. But what is much less straightforward is the fact that the problem of radio-frequency interference quickly became cast in political and national terms. We see by 1929 a general hardening of political divisions in Germany, and a marked rise in the popularity of extreme right- and left-wing political parties. Crowded radio spectrum was increasingly cast as a matter of national defense and even cultural survival. Commentators began to speak in the

[50] Many electrical devices give off radio-frequency radiation, which can interfere with or block radio waves, particularly if they are poorly grounded or insulated. This is still a problem today but was even more troubling in the 1920s and 1930s.

[51] "Kampf gegen Rundfunkstörungen", *Funk*, No. 45 (November 8, 1929): 208. The initiative was announced at a meeting of the "National Association of Radio Dealers" (Reichsverband der Funkhändler) held during the Great German Radio Exposition in 1929. The Committee included: the Postal Ministry, the Reichspostzentralamt, the Prussian Ministry of the Interior, the Reichs-Rundfunk-Gesellschaft, the DFTV, the Arbeiter-Radio-Klub Deutschlands, e.V., the Reichsverband Deutscher Funkhändler e.V., the Verband Deutscher Elektro-Installations-Firmen e.V., the Verband der Funkindustrie e.V., and the Zentralverband der Deutschen Elektrotechnischen Industrie e.V.

radio journals of Germany being "surrounded" by (unfairly) powerful foreign transmitters, against which it had to reply in order to avoid cultural alienation.[52] An inherently technical problem was thus cast as a political one which explicitly involved the rejection of a (culturally and politically) foreign presence. Extraneous signal rejection became xenophobia and paranoia. This is part of the "background noise" to the hardening of the political debate in Germany, even before the effects of the Great Depression began to be felt. It is certainly part of the background to the rise of National Socialism and the later "coordination" of the radio hobby, which we will cover in the chapter after this.

When broadcasting was introduced in Germany in 1923, the design of radio receivers (and all other parts of the system, such as speakers, microphones, transmitters, and even transformers to convert household current into voltages useable within a receiver) was in its infancy. Typically, new designs and components came from science, industry, and "super-users" (advanced hobbyists), and then trickled down to individuals via both radio journals and the radio clubs. The transmission of knowledge was both top-down (from scientists or industry to users) and bottom-up (from hobbyists to industry). In Germany, hobbyists were often better than industry at adopting ideas from abroad, and at designing practical, efficient radio circuits. Radio specialists within Germany by the end of the 1920s were already able to look back at three distinct stages of technical evolution of radio.[53]

In the first period of broadcasting (1923–1925), both do-it-yourselfers and industry searched for affordable, practical designs for consumer receivers. Vacuum tubes were both expensive and hard to obtain, and builders looked for ways to incorporate them into receivers efficiently. The most common type of receiver was the simple crystal receiver without amplification.

[52] "Neuordnung des deutschen Sendewesens. Auch Deutschland baut Großsender—Erweiterung des Gleichwellenrundfunks", *Funk* No. 49 (Dezember 6, 1929): 221. The article speaks of Germany's foreign neighbors "rearming", and warning that Germany now surrounded by a ring of high-power foreign transmitters, which present a direct and present danger to German radioreception. ("*Inzwischen ist Deutschland bereits von einem Ring fertiggestellter oder im Bau befindlicher Großsender umschlossen, die eine unmittelbare Gefahr für die Empfangsverhältnisse bei uns bedeuten*"). See also: Hans Bodenstedt, "Funkarbeit in der deutschen Geltung", *Funk*, No. 52 (Dezember 27, 1929): 235. The article announces a new program on German broadcast radio designed to teach Germans about the importance of Germany's technical contributions. The tone is very nationalistic.

[53] Prof. Gustav Leithäuser, "Fünf Jahre "Funk-Bastler"—fünf Jahre Funktechnik", *Funk-Bastler* No. 18 (May 3, 1929): 273–274.

The focus was on simple reception, with not much thought about fidelity. In nearly all cases, users had to use headphones to listen, for very little was known about designing amplifiers or speakers. It was a period of much experimentation.

In the second period from roughly 1925 to 1927, more emphasis was placed on fidelity or reproduction, mainly through the design of good amplifiers. Reasonable fidelity reproduction became possible through headphones, but speaker design still lagged behind. The development of vacuum tubes for radio reception and ancillary tasks (amplification and transformation of current) made great progress, and receiver design settled down to a handful of standard forms, which had proved themselves through experimentation. These were simple crystal receivers (with or without amplification via a vacuum tube), simple regenerative receivers (with one or more vacuum tubes), and a handful of more complex designs using multiple tubes. Tube design made great progress, with the development of vacuum tubes for more and more tasks (reception, high frequency amplification, current transformation, etc.). Receiving tubes became longer lived and more robust.

The year 1927 introduced the third period of early receiver design. It became easier and more common to run receivers off household current (as opposed to the earlier dependence on batteries), not least because of continued improvement in tube design. During this period, the design of loudspeakers also advanced. The resulting trend was toward more sensitivity and selectivity, better sound, and greater ease of use. In short, radios became more and more consumer devices requiring less and less technical knowledge to operate—though ever greater technical knowledge to build.

The technical progress of radio receivers and amplified speakers naturally had an impact on the clubs. While many first came to the radio clubs to learn how to build a simple and inexpensive crystal receiver, the ability to continue to learn and perfect both one's knowledge and one's building skills led many to stay. The moving target of up-to-date radio receiver design helped many to remain interested in radio and committed to their club. Under the impact of developments in both receivers and speakers (and microphones), a subgroup of the radio hobby developed, which was interested in "high-fidelity" sound reproduction. Many radio hobbyists were also phonograph enthusiasts as well.[54] Some radio hobbyists also got

[54] See note 56 below. In the 1920s and 1930s, electronics manufacturers offered very high-end radios, which paired them with phonographs in the same cabinet and using the same speakers.

involved in amateur television, and even for average radio club members, talks, and demonstrations of television were always a big draw. TV was still in its infancy. Most radio hobbyists saw radio and TV on a technological continuum, and not as separate and rival technologies[55]: for similar reasons, radio enthusiasts were also interested in sound film, which became commonplace from 1927 on in Germany.[56]

It is well that the radio clubs were able to remain interesting and relevant to radio hobbyists, for the abandonment of the *Audionversuchserlaubnis* in 1925 struck a heavy blow to the clubs.[57] Many who were only interested in broadcast listening and were only in the club to gain a radio permit left immediately.[58] The journals *Funk* and *Der Radio-Amateur* were forced to combine in the Fall of 1926. The Great Depression from 1929 into the mid-1930s also hurt the clubs, as fewer and fewer Germans could afford club dues or radio parts. Later, state-mandated economies of scale in the introduction of the various models of the *Volksempfänger* finally issued in an era of affordable manufactured radios in 1933, further hurting the clubs. And yet, from 1925 to 1933, the clubs continued to exist and thrive, albeit with fewer members than during the radio fever of the early 1920s.

If the clubs were no longer a necessary step for those seeking to own a receiver, they remained an important aid for those who wished to save money by building a receiver themselves. By the late 1920s, the major

[55]For example: Ludwig Kapeller, "Sollen wir bildfunken?" *Funk* No. 29 (July 13, 1928) (Hauptteil): 213–214 (note that this article also refers to sending text or "fax"), or Lothar Band, "Fünf Jahre "Funk". Verheißung—Erfülung—Hoffnungen". *Funk*, No. 18 (May 3, 1929): 73–74. The article predicts that TV will soon be as big as radio listening.

[56]Dr. P. Gehne, "Fachzeitschrift und Funkvereine. Eine Fünfjahreserinnerung", *Funk-Bastler* No. 18 (May 3, 1929): 275–276; Prof. Gustav Leithäuser, "Fünf Jahre "Funk-Bastler"—fünf Jahre Funktechnik", *Funk-Bastler* No. 18 (May 3, 1929):273–274. Both articles give lists of technological interests of radio club members.

[57]Remember that this was the "Radio Tube Experimental Permit", which was required in Germany from 1923 to 1925 before an individual could build or use a radio containing one or more vacuum tubes. They were issued by the radio clubs under Postal Ministry authority. Club membership was a prerequisite for taking the exam leading to the issue of the permit, a requirement which swelled club ranks. See pp. 37–41 above.

[58]For example, the local chapter "Wannseebahn" of the DRC (SW Berlin) went from 700 members in March 1925 to around 500 members in March 1926. Losses in other places were likely greater. E. Bley, 1. Chairman of the Chapter, "Das Clubleben in der Ortsgruppe Wannseebahn (Sitz Steglitz) des Deutschen Radio-Clubs e.V.", *Der Radio-Amateur* 4, No. 12 (March 19, 1926): 235–238.

change for the DIY hobbyists was that now they could build much more advanced and sophisticated radios, often of better quality than what was available (or at least, affordable) on the market. The social role of the clubs also remained important, both in the sense of providing a "safe" space within which to interact with modern technology, and in the sense of providing a place for social interaction: the parties continued, though in a more subdued form during the Great Depression. Moreover, the advance of radio and allied technologies meant that the hobby clubs also gained new foci of interest, as just described. If the heady days of 1924 and 1925 were gone, radio hobby clubs remained socially and technologically relevant into the early 1930s.

Shortwave

One of the most important technical developments in radio, and the one which did more than anything else to keep the clubs relevant, was the growing importance of the "shortwave" part of the radio spectrum.[59] By the mid-1920s, many radio hobbyists began to get interested in listening to parts of the radio spectrum not used by broadcast radio, specifically, those in the "shortwave" portion of the spectrum. In the early days of radio, propagation was poorly understood, and scientists assumed that there was a direct relationship between wave length and distance of propagation. Thus, early radio use tended to avoid the "useless" shortwave part of the spectrum, which was left for the transmitting hobbyists of ham radio. Experiments and practical use made by hams rapidly revealed that, in fact, shortwave signals could travel quite literally around the world, though it took longer to discover just how this worked.[60]

[59] The part of the radio spectrum with a frequency between roughly 3 and 30 MHz. (meaning a wave length of between 10 and 100 meters) is commonly called "shortwave". "Medium wave" (used for most European broadcast radio in the 1920s and 1930s) is commonly defined as frequencies between 300 and 3000 KHz. (or 0.3 to 3 MHz.; wavelengths between 100 and 1000 meters). These are working definitions, not absolute ones. See International Telecommunication Union, *Radio Regulations* (Geneva, ITU, 2016), Article 2, Section I, p. 27, http://www.itu.int/pub/R-REG-RR-2016. Accessed January 17, 2019.

[60] It was later discovered that the relatively short waves on this part of the spectrum were being reflected by both the ground and certain layers of the atmosphere, allowing them to bounce for long distances, and sometimes completely around the globe. On the science of radio propagation, see: Henry L. Bertoni, *Radio Propagation for Modern Wireless Systems* (Upper Saddle River, NJ: Prentice Hall, 2000), or Eric P. Nichols, *Propagation and Radio Science* (Newington, CT: American Radio Relay League, 2015).

Gradually, use of the shortwave frequencies expanded. Transmitting amateurs rapidly colonized the shortwave part of the spectrum even before the First World War, so that in Germany in the 1920s and 1930s, the term used by the ham radio community to refer to itself was "shortwave amateurs".[61] From these hobbyists, interest in this part of the spectrum spread, first within the radio clubs, and then among the general public.

By the mid-1920s, while parts of the shortwave spectrum were reserved for hams by international agreement, other parts were beginning to be used by scientific expeditions, shipping, the military, and even long-range broadcasts. That meant that there were interesting and even exciting things to be heard on this part of the spectrum.

A key factor here is that by listening to shortwave transmissions, ordinary listeners—provided they had an appropriate receiver and antenna—could hear foreign countries and very far-off events in real time. Quite literally, the most exotic parts of the world and most exciting world events were brought into the living rooms and kitchens of ordinary listeners. Even in the early days of broadcast radio in the long-wave and medium-wave portions of the spectrum, there was quite a thrill attached to reception of faraway broadcasts. Shortwave transmissions, with their potential for much longer distance propagation, only intensified the thrill. This helped feed the "expedition fever", which was characteristic of the 1920s and 1930s in industrialized countries. Scientific expeditions and other notable geographic events, such as some early transatlantic flights or even regular ship crossings were able to transmit live, using first medium wave, and then by the mid-1920s, shortwave frequencies. A recent article lists 33 major scientific expeditions to polar regions between 1924 and 1937, which were equipped with radios, and which were listened to by radio enthusiasts, and widely reported in the popular press.[62] There were many others. For Germany, in particular, we must also add the travels of prestige vessels such as the cruise ship Bremen, and airships such as the Graf. Zeppelin, which broadcast regularly on short- and medium waves.[63] The importance of this in the history of radio cannot be overemphasized.

[61] "Ham radio" as a term comes originally from the US. It was sometimes used by German hams but was not the preferred term until after the Second World War. See below.

[62] Michael W. Marinaro WN1M, "Polar Exploration", *QST* 98, No. 6 (June 2014): 63–65.

[63] Gillaume de Son, *Zeppelin! Germany and the Airship, 1900–1939* (Baltimore, Maryland: Johns Hopkins University Press, 2005); H(ans) Busch and Otto Reuter, "Erfahrungen im Luftschiff-Funkdienst", *Arbeitskreis Seefunk—Erfahrungen—Probleme—Berichte* 4 (1967), http://www.seefunknetz.de/lzzep.htm, Accessed August 9, 2018; "Die Jungfernfahrt der

While the middle-wave frequencies used for European broadcasting in the 1920s and 1930s could often travel surprising distances at night, this was nothing compared to the ability of shortwave frequency transmissions to quite literally go around the globe. If the (nighttime) search for broadcast transmissions from (mostly neighboring) countries gave European radio listeners a taste for distance, shortwave brought information from even the most remote parts of the earth into domestic spaces in real time. This was real magic.

The series of Steven Spielberg films featuring the archaeologist and adventurer Indiana Jones give a good feeling for the mixture of romantic adventure and science, which characterized both "expedition fever" and shortwave radio in the 1920s and 1930s. The big difference was that while tropes of scientist-adventurers in exotic foreign places were theatrical and exaggerated (campy) in the 1980s, 1990s, and 2000s, they were dramatic reality for radio listeners in the 1920s and 1930s.[64] Foreign music, foreign news, foreign adventure: it was all brought home by the radio, and listeners were fascinated. The ability of radio (chiefly shortwave) to bring the far and foreign into the home was made even all the more magical because ordinary citizens could participate in the scientific research into shortwave propagation by submitting signal reception reports.[65] Moreover, ordinary individuals could even hope to get directly involved in scientific expedi-

"Europa" im Rundfunk", *Funk* No. 48 (November 29, 1929): 218. The Europa was the sister-ship of the Bremen.

[64] The first of these iconic films was "Raiders of the Lost Ark" from 1981. Several others followed, all directed by Stephen Spielberg. A new film in the series may be in the works as of the writing of this book (2018). See the list at: https://www.imdb.com/list/ls053788854/ (last accessed January 17, 2019).

[65] The sending of reception reports was quite important in the early days of radio, as a means of investigating propagation of radio waves. In the early days of radio, hearing a transmission, any transmission, was something of an event, and transmitting stations sought feedback from listeners in order to try to understand how far their signal was going, and why. From the early days of radio into the 1920s and sometimes beyond, anyone who heard a radio transmission could send the transmitter a report documenting that the station was heard. These soon came to be called "QSL cards", from the telegraphic abbreviation "QSL", meaning "I acknowledge receipt of your transmission/can you acknowledge receipt of my transmission?" Often, a thank-you letter for the report was sent from the station back to the listener. Within the amateur radio (ham) community, these reception reports are used to document a mutual contact, either for fun or as part of a contest or award program. Either way, it was a positive proof of an amateur's skills. This quickly grew in the 1920s, to the point that most national ham organizations instituted centralized QSL card offices, which received the reports and forwarded them to the appropriate address.

tions and other major world events by becoming the first to hear breaking news or even a faint distress signal.[66] This was both "citizen science", and something more: shortwave listeners could potentially (rarely) get involved in real adventure, from the safety of their own living rooms.

In Germany in the 1920s and early 1930s, the radio clubs helped fuel this fascination. As I have already mentioned, the abandonment of the *Audionversuchserlaubnis* in 1925 by the postal authorities and the consequent drop in the number of club members left the shortwave amateurs in a relatively more prominent position within the clubs. The shortwave amateurs knew the practical ins and outs of the shortwave spectrum better than anyone, and they had more practical experience in the design and building of shortwave receivers (and, of course, transmitters) than the electronics industry. From the ham radio enthusiasts in the radio clubs, other club members learned of the promise of shortwave, and more importantly, also learned how to build or modify a radio to receive shortwave transmissions.[67] Commercial shortwave receivers became widely available by the late 1920s and early 1930s, but remained quite expensive, compared to simpler long- and medium-wave receivers designed primarily to catch local broadcast stations.[68] The shortwave amateurs pretty much

[66] For example, a police monitoring report of August 1932 states that German amateurs heard transmissions from the airplane of Wolfgang von Gronau, which was then making a round-the-world flight. They were even able to have short conversations with it. "Bericht 21 der Überwachungsstelle des privaten deutschen Kurzwellenverkehrs bei der Pol. Hauptfunkstelle Berlin für die Zeit vom 1–31. VII 1932" of 15. August 1932 (Geheim!), in: BArch R/1501/20062 Reichsministerium des Innern KPD-Radio-Propaganda, 1930–1934: 235. On von Gronau, see Roger Connor, "Wolfgang von Gronau and his "Greenland Whales"", August 27, 2013, https://airandspace.si.edu/stories/editorial/wolfgang-von-gronau-and-his-greenland-%E2%80%9Cwhales%E2%80%9D, Accessed January 18, 2019. As a second example, a January 1933 police monitoring report states that an S.O.S. transmission from a German ship near Maderia was heard in Germany, both by the police and by several amateurs. "Bericht 27 der Überwachungsstelle des privaten deutschen Kurzwellenverkehrs bei der Pol. Hauptfunkstelle Berlin für die Zeit vom 1–31. January 1933" (n.d.) (Geheim!), in: BArch R/1501/20062 Reichsministerium des Innern KPD-Radio-Propaganda, 1930–1934: 254–255.

[67] Shortwave reception required changes in design; a receiver designed specifically to receive middle-wave and long-wave transmissions cannot receive the shortwave transmissions (at least not well) without certain changes, both to the internal design of the receiver and to the antenna.

[68] The "Radio Bauer" catalog (a large Berlin radio dealer) from 1927 lists only one receiver, which is specifically listed as being able to receive parts of the shortwave spectrum, and even then, the word "shortwave" is not used in the description. The radio in question was the

kept up with industry when it came to the latest shortwave radio technology, helped by the fact that they closely followed developments overseas, particularly in the US and Great Briton. The result was that shortwave, along with other technical advances mentioned above, helped keep the radio hobby clubs relevant, at least down to the Great Depression.

The Development of Amateur Transmitting

Some radio enthusiasts were not satisfied with just listening, they wanted to transmit their own messages. These enthusiasts were a small but very important part of the German radio hobby in the first half of the twentieth century, and are today the strongest part of what remains of the hobby. Though numerically only a tiny part of the much larger radio hobby in the 1920s and early 1930s in Germany, amateur transmitters played an important role in the clubs and in spreading the cause of radio in general. As we have seen, they were by the nature of things among the technically most competent members of the clubs, since they had to build nearly all of the own equipment, and their use of shortwave frequencies to make contacts with other hams over long distances put them at the forefront of equipment design. Their powerful engagement with radio as a technology meant that they were often among the most active members of any club. Because of the emphasis on long-distance contacts, but also because ham radio had a strong international organization and strong national organizations in nearly all countries, German hams were also the most plugged in to what was going on in radio outside of Germany, and served as conduits of knowledge from the outside world into the German radio hobby: they were part of an international community of users and makers. As the shortwave frequencies began to catch the attention of the larger radio hobby by 1927, the position of the hams was only strengthened. Though

"Radiofrequenz" two-tube receiver. The price, with a single tuning-spool, was 200 marks. Radio-Bauer, *Katalog Nr. 6 April 1927(Berlin: Buchdruckerei Max Lichtwitz, 1927)*. p. 9. The 1928–1929 Radio Diehr catalog, on the other hand, included both an article on shortwave radio, and listed a handful of commercial radios (ostensibly) capable of shortwave reception for sale. They ranged in price from a 3-tube Saba KE for RM 135.50 (p. 33), to a 6-tube Reico RF 366 receiver for RM 789.00—without tubes! (p. 48). Oberingenieur E. Reichel, "Ueber Kurzwellen-Empfang und Kurzwellen-Empfänger" in: Radio-Diehr, *Neuester Illustrierter Radio-Katalog 1928–1929* (Berlin: Kreditgenossenschaft des Reichsverbandes Deutscher Funkhändler, 1928): 9–10.

often considered a bit obsessive, they also were most clearly touched by the glamor and excitement of being able to hear transmissions from far away, and participate directly in the exploration of the science of radio. And as we will see in later chapters, their acknowledged abilities and skill with radio technology and Morse code, as well as their practical experience using shortwave frequencies made them so useful to the German state and military that the shortwave amateurs were actually encouraged by the Nazis, even as they shut down the rest of the radio hobby.[69] Not least, the history of the shortwave amateurs (as they then called themselves) is a good bellwether for the strength and importance of the German radio hobby in general in the 1920s, for the simple reason that their main preoccupation, transmitting, was strongly resisted by the representatives of the State, particularly the Postal Ministry, Police and Interior Ministry, and possibly the Reichswehr Ministry, which all worked to make legal transmissions by individuals next to impossible. Yet hundreds of German hams existed and somehow still managed to transmit and be active members of the international ham community, despite the law and the best efforts of the authorities. A certain illegality and willingness to flout the rules in the interest of their hobby was thus always present in the German shortwave amateur community until 1949, when the two postwar German states finally legalized the ownership of private transmitters, albeit still under greater or lesser restrictions. The story of these German hams is important when we think of the role of hobbies in structuring people's engagement with technology; what better way to measure the importance of hobbies than to look at those who were willing to flout the law, risk prison and possible even death in order to continue to use radio?

Who were the very first Germans to attempt to transmit their own messages?[70] Some ex-military radiomen certainly must have tried transmitting immediately after the First World War, though we have little or no

[69] See Chaps. 5, 6 and 7 below.

[70] Because of the legal limbo (or outright illegality) surrounding private transmitters in Germany, it is hard to document the earliest origins of German ham radio. We largely have to rely on the memories of the early hams themselves, with all the difficulties that poses for professional historians. The best book on the subject from within the ham community is the history of ham radio by Felix Körner. Though Körner was not a professional historian by any means, he was himself at the center of many of the most important developments of the post-Second World War period, and knew first hand many of the leading German hams of the founding generation from the 1920s and 1930s. Körner, W(olfgang) F(elix), DL1CU, *Geschichte des Amateurfunks 1909–1963* (Hamburg: FT-Verlag Rojan & Kraft, 1963).

reliable record. Engineers and scientists working on radio were also among the first to attempt transmission in the very early days after the war.[71] And in at least one case, we can document short-range transmissions between schoolboys—among other things, to collaborate on their homework.[72]At the very least, scientists and former communications officers became the early leaders of the movement, and worked hard to build it and organize it.[73] Nevertheless, amateur transmission came relatively late to Germany; in Britain and the US, it had existed legally and on a fairly large scale well before the First World War.

Highly important for the development of amateur transmission in Germany was the presence of an already well organized and highly dynamic ham radio community outside of Germany. The US, with its huge size and population, was the world leader under the American Radio Relay League (ARRL), but England, Belgium, the Netherlands, and France also had strong ham communities by the early 1920s. Their presence on the air-waves could not be missed, and German hams worked hard to live up to the example set by these foreign amateurs. Given the bleak shadow still cast by the Great War in the early 1920s, the importance of contacts with foreign hams—all of the countries listed above, save the Netherlands, had been on the Allied side and opposed Germany in the First World War—was not self-evident. But from the very beginning, the international ham movement had worked very hard to be inclusive and stand above political considerations, even as national ham organizations also shared patriotic values and were even often in close cooperation with national military establishments. This open helpfulness is sometimes called "ham spirit"

[71] "Die technisch-wissenschaftlichen Sitzungen während der Haupttagung des Deutschen Funktechnischen Verbandes in Köln", *Der Radio-Amateur* 4, No. 34 (August 20, 1926): 24. The article reports that von Stockmayer told an anecdote about students at the Technical University in Stuttgart transmitting music in 1923.

[72] See the case of Richard Dargatz, below. Adolescents in Germany, particularly those attending a *Gymnasium* or *Realschule*, were often radio enthusiasts at this time. We see a similar pattern in the US, as well. They usually had good backgrounds in math and science, and also had a teenager's lack of respect for the adult world and its rules. This may well be a very important combination when it comes to innovation.

[73] Examples are Lt. Col. (ret.) Ludwig v. Stockmayer and Col. (ret.) Otto Fulda who were officers, or Prof. Esau and Prof. Gustav Leithäuser, who were both scientists. Of course, there were also laymen who were important in the hobby, like Rolf Formis and Richard Dargatz. Amateur radio required both technical knowledge and practical experience, and so laypersons could make real contributions.

and is an important part of the background to the development of amateur radio transmission in Germany.[74]

Aside from scientists and former military communications specialists, there were also some who just caught the radio bug and built transmitters simply because they could. As early as 1919, a young high school student and radio enthusiast in Berlin, Richard Dargatz, built a radio receiver and even asked the postal authorities in writing for permission to use a transmitter. He is one of the first transmitting amateurs we can document in Germany with any certainty, though he is quite certainly not the first German amateur to transmit. In any case, his request was immediately refused by the postal authorities. Dargatz would not be deterred, and within less than a year, he was joined by a school classmate Werner Slawyk, and then gradually by others, many of whom also built transmitters in defiance of the law.[75] These very early transmitters were quite primitive, and still operated on the spark-gap principle, but these very early amateurs soon made contact with Prof. Gustav Leithäuser in Berlin, who taught them about vacuum tubes, leading quickly to better receivers and transmitters, and opening up greater abilities to contact other early transmitters both within and without Germany. Though early German transmitting amateurs were geographically isolated, they had the immense advantage of being able to meet on the air. The Berlin group around Dargatz and Slawyk were soon in contact with Viktor Gramich, an early transmitter in Bavaria, and with Dr. Eugen Nesper, a radio scientist, author, and leading radio hobbyist in Berlin, and then with other budding hams in the Ruhr and elsewhere.[76] Soon, the illegal community of early German transmitting amateurs constituted an informal national network, even though most never met face to face at this time. As others became interested in transmitting, they were soon heard on the air, and integrated into the small but growing community. All of this occurred against the opposition

[74] The term "ham spirit", in English, was often used by German shortwave amateurs from the 1920s onward. It is an aspirational set of values calling on hams everywhere to be helpful and open with their fellow hams regardless of what might tend to separate them. It is still commonly used by hams in Germany and other European countries. For modern definitions from various countries see: Germany: https://www.darc.de/der-club/distrikte/a/ortsverbaende/37/ham-spirit-ehrencodex/; Austria: https://www.oevsv.at/funkbetrieb/funkbetrieb/; France: http://www.f1org.org/articles.php?lng=fr&pg=61&tconfig=0; and finally, the US (where the term "ham spirit" doesn't actually appear as such): http://www.arrl.org/amateur-code

[75] Körner, *Amateurfunk*, 13–15.

[76] Körner, *Amateurfunk*, 13–18.

of the postal authorities and police, who then had very little technical means to discover and identify illegal transmitters,[77] but who were viscerally afraid that radio could get out of control, and particularly opposed to individuals owning private transmitters.[78] Some early hams were nevertheless caught (noisy neighbors or large outdoor antennas were often the reason), though only a few.[79] One or two others, quickly disgusted with the situation, moved abroad, where they were able to legally transmit.[80] Still, illegality was and remained a huge problem. Paradoxically, it led German shortwave amateurs to rapidly develop a formal organization, not least to be able to lobby in favor of opening up legal transmission.

Southwestern Germany quickly became a locus of early transmitting activity. Early transmitters in Stuttgart found a home in the Upper German Radio Association (*Oberdeutscher Funkverband*, OFV). It was one of the very first radio clubs founded in Germany after the beginning of broadcasting, and it became a hotbed of transmitting activity. Using the OFV as a base, early leaders of the transmitting movement gradually created a formal national organization for German hams.

The first legalization of private transmitters came in the March 1924 "Decree on the Protection of Radio Traffic".[81] It specifically allowed pri-

[77] The authorities could easily monitor all shortwave transmissions but could not reliably locate an illegal transmitter (until roughly 1934). See: Polizeihauptmann von Asmuth, "Über die Möglichkeited der Kurzwellenverwendung bei Unruhen durch Staatsfeindliche Elemente, Kontrollmöglichkeit und Vorbereitende Maßnahmen der Behörden" (N.D., but ca. 1930) in: BArch R/1501/20401 St10 Reichsministerium des Innern KPD-Radio-Propaganda, Bd. 3 Dez. 1929–Febr.1931 St10/62, Bd. 3, pp. 146–153.

[78] For examples of the fear surrounding radio, see: "Niederschrift über die 22. Sitzung der Reichsfunkkommission am Montag, dem 25. Juni1923" in: BArch R/4701/8673, Reichspostministerium, Sitzungsberichte R.F.K. Geh. Reistratur Z, Band 1 1919–1925, pp. 475–482; or Reichspostministerium, VZ "(Entwurf) An Woelff's Telegrafischesbüro of 19.March.1924 Sofort!" in: BArch R/4701/10858 Reichspostministerium Z 1/17936, Akten Betr. Funkgesetznovelle 1922, pp. 372–373.

[79] The first German ham to be punished for illegal transmission was Diplom Ingenier Rudolf Horkheimer in Rottenburg am Neckar (K-Y8, later K4YAE), who was sentence in August 1923 for illegal transmissions beginning as early as in 1921. Körner, *Amateurfunk*, 18–21. Horkheimer was Jewish, and emigrated in 1938 to Bolivia, before returning to Germany in 1958.

[80] For example, Felix Cremers, D4xvf, another early ham, moved to France for a time in 1923/24 and worked with French amateurs. Felix Cremers, "Mit 40 Watt über den Atlantischen Ozean", *Der Radio-Amateur* 3, No. 25 (June 19, 1925): 626–629.

[81] Der Reichspräsident [Ebert], "Verordnung zum Schutze des Funkverkehrs. Vom 8. März 1924", *Deutscher Reichsanzeiger* und *Preußischer Staatsanzeiger*, 1924, No. 66 of March 8, 1924.

vate "research" transmitters for scientists, industry, for radio clubs, and for private individuals—in exceptional circumstances.[82] This was something the clubs had fought hard to obtain,[83] but in any case, they were not the focus of this part of the law, but rather accidental benefactors. The main purpose of legalizing private transmitters was to help industry and science. Both needed their own transmitters for tests, experiments, and even quality control. Radio clubs were included under the assumption that they would be participating in "legitimate" science, and not the "sport" of amateur radio.[84] As a result, the Postal Ministry was slow to actually issue licenses to clubs and individuals, and was never willing to do so on a large scale. The radio clubs had to constantly fight to get them, and even then, only a handful were ever issued before 1933. Nevertheless, under great pressure from the radio clubs, in 1924–1925, the Postal Ministry issued licenses for a little over 100 amateur transmitters. Most of these were issued to industry or universities, a handful to radio clubs for club stations, and only a very few licenses were also given to private individuals, mainly to scientists and a few very early shortwave pioneers.[85] German radio

[82] Clearly, neither research nor the production of radios by industry could take place without transmission (even manufactured radios had to be tested against a known signal frequency). Thus, when it came to legalizing private transmission, the "Degree on the Protection of Radio Traffic" represented more of a post hoc legalization of what was already taking place.

[83] Amtsblatt des Reichspostministeriums Nr. 46, included in: *Der Radio-Amateur* 2 No. 8 (May 28, 1924) and Obertelegrapheninspektor Ernst Schulze, "Die neuen Rundfunkbestimmungen", *Der Radio-Amateur* 2, No. 11 (July 4, 1924): 289–292 and Justizrat Dr. Felix Szkolny, "Rechtsprechung. Radio-rechtliche Zeit- und Streitfragen" (part 1), *Der Radio-Amateur* 2, No. 34, 2. Messeheft (December 12, 1924): 935ff and Justizrat Dr. Felix Szkolny, "Rechtsprechung. Radio-rechtliche Zeit- und Streitfragen" (part 2), *Der Radio-Amateur* 3, No. 4 (January 23, 1925): 97–99. See also: "Mitteilungen", *Der Radio-Amateur* 2, No. 29 (November 7, 1924): 778–782.

[84] Reports of police monitoring of private transmissions often complained that few of the radio club stations were being used for the intended scientific purposes. See: Polizeihauptfunkstelle, "Zusammenfassender Bericht der Überwachungsstelle des privaten deutschen Kurzwellenverkehrs bei der Pol. Hauptfunkstelle Berlin für die Zeit vom 15.10.1930–15.10.1931" of 20 October 1931 (Geheim!), in: BArch R/1501/20062 Reichsministerium des Innern KPD-Radio-Propaganda, 1930–1934, pp. 202–214.

[85] Hanns Günter & C. Culatti, *Wer Gibt? Die Funkstationen der Welt, ihre Rufzeichen, Reichweiten u. Wellenlängen einschließl. d. europäischen Amateursender* (Stuttgart: Franckh'sche Verlagshandlung, 1925) lists only 31 such stations, only three of which are listed as belonging to individuals. By law, the club transmitters were to be used only to test radio propagation and make scientific measurements and were even officially called Research Transmitters (*Versuchssender*). Ham contacts were prohibited, though in practice, this is

hobbyists, including not only those who wanted to transmit themselves, but also many radio enthusiasts of all kinds, were furious and confused. Try as they might, the Postal Ministry refused to significantly raise the number of transmitting licenses it issued.

It is clear in retrospect that the refusal to issue more private transmitting licenses was due primarily to political considerations specific to the German situation in the 1920s, but on a secondary level, it was also due to a failure to really understand the new media and the fear of change which this engendered. Powerful interests within the German state were simply too afraid of what ordinary citizens might do with the ability to own transmitters. The main fear was that leftists might gain access to transmitters and use them in some future putsch. Remember that all of Europe was still under the shadow of the 1917 Bolshevik revolution in Russia, and that Germany had gone through a revolution and small-scale civil war in 1918–1919, and a series of politically motivated putsches and murders in 1921–1923. Strong (legal) socialist and communist movements existed in Germany. The German organs of state in the Weimar Republic were still full of the representatives of the old imperial order, and were deeply afraid of the power of the Left and of the prospect of some future left-wing revolution or putsch. In closed-door meetings, notably of the interministerial "Imperial Radio Commission" (*Reichsfunkkommission*), high-level government representatives openly expressed fear that the Left would use radio to subvert the German state.[86] Early meetings of the group confirm that while transmitters were seen as a particular danger, even the simple possession of radio receivers by the public was considered a great risk, meaning really that it was the entire medium of radio which was feared.[87]

what most club stations were used for. See Polizeihauptfunkstelle, "Zusammenfassender Bericht der Überwachungsstelle des privaten deutschen Kurzwellenverkehrs bei der Pol. Hauptfunkstelle Berlin für die Zeit vom 15.10.1930–15.10.1931" of 20.October.1931 (Geheim!), in BArch R/1501/20062 Reichsministerium des Innern KPD Radio Propaganda, 1930–1934, 202–214.

[86] See the various protocols of the meetings of the Imperial Radio Commission (Reichsfunkkommission) in BArch R/4701/8673 Reichspostministerium, Sitzungsberichte R.F.K. Geh. Reistratur Z, Band 1 1919–1925; R/4701/8666 Reichspostministerium, Reichsfunkkommission, Geh. Reistratur Z, Band 2, 1925–2927 and BArch R4701/8994 Reichs=Postamt, Akten betreffend Reichsfunkkommission 1918–1924, Geh. Registratur Z, Band 2.

[87] For example, see: "Niederschrift über die 22. Sitzung der Reichsfunkkommission am Montag, dem 25. Juni 1923" in BArch R/4701/8673 Reichspostministerium, Sitzungsberichte R.F.K. Geh. Reistratur Z, Band 1 1919–1925, 475–482, or "Niederschrift

The radio clubs, spurred on both by a desire to spread engagement with radio and by the pressure of the shortwave enthusiasts in their ranks, constantly kept the issue alive, and pressured the Postal Ministry to issue more licenses. The DFTV, in particular, contained a strong element, which was interested in transmitting. Their presence forced the DFTV to keep the issue of private transmitter licenses at the forefront of its political lobbying. It had a special subcommittee for transmitting, and eventually became the official German member organization in the International Amateur Radio Union (IARU). As the clubs and shortwave amateurs continued to get nowhere with their demands, there was a great deal of speculation as to who was really to blame. As the ministry most closely involved with the issue, the Postal Ministry got most of the blame, to the point where some postal officials privately tried to direct the attention of the angry shortwave enthusiasts elsewhere.[88] A look at the documents shows that resistance to radio in general and private transmitters in particular was much more complex, and much more generalized. All the ministries represented in the Imperial Radio Commission feared that radio could hurt the security of the state, but the degree of opposition varied.[89] Rivalry between ministries and fears of losing control over the new medium to the people also played a major role, particularly for the Postal Ministry, which clung tightly to its monopoly control over radio. In the closed meetings of the Imperial Radio Commission, the Interior Ministry and specifically the Police were usually the most vocal in opposing the issue of transmitting licenses. The Postal Ministry seems to have accepted that a few private transmitters were necessary for research and testing purposes, but definitely wanted to limit their number.[90] The Reichswehr (military) tended to remain fairly silent and avoid direct pronouncement about the

über die 21. Sitzung der Reichsfunkkommission am Freitag, dem 9. Juni 1922" in BArch R/4701/8673 Reichspostministerium, Sitzungsberichte R.F.K. Geh. Reistratur Z, Band 1 1919–1925, 365–381.

[88] Körner himself blamed the military. Körner, *Amateurfunk*, p. 56.

[89] For example, see: "Niederschrift über die 22. Sitzung der Reichsfunkkommission am Montag, dem 25. Juni 1923" in BArch R/4701/8673 Reichspostministerium, Sitzungsberichte R.F.K. Geh. Reistratur Z, Band 1 1919–1925, 475–482, or "Niederschrift über die 21. Sitzung der Reichsfunkkommission am Freitag, dem 9. Juni 1922" in BArch R/4701/8673 Reichspostministerium, Sitzungsberichte R.F.K. Geh. Reistratur Z, Band 1 1919–1925, 365–381.

[90] See, for example: "Niederschrift über die 26. Sitzung der Reichsfunkkommission am Donnerstag, den 2. Juli.1925", in BArch R/4701/8673 Reichspostministerium, Sitzungsberichte R.F.K. Geh. Reistratur Z, Band 1 1919–1925, 510–534.

issue of transmitters, though was certainly also quite concerned about the misuse of radio by spies or revolutionaries, and tended to vote with the majority when it came to limiting the number of private transmitters. On the other hand, the Reichswehr (specifically, the Navy) was secretly supporting the German shortwave enthusiasts, and had very close ties to their organization, not least because the leaders of the shortwave enthusiasts included a large number of former officers.[91] The military thus played a double game in negotiations about radio and the issue of private licenses to transmit. The other ministries involved in the discussions all, in the end, supported the strictest of limitations when it came to private transmitting licenses, and followed the lead of the Post, Interior Ministries, and Military. Thus, a coalition of interests at the highest levels of state together blocked the issue of more private transmitting licenses due to their collective fears of left-wing subversion. Fear of the Left was of course a fundamental problem in the Weimar Republic, and poisoned the entire political debate. Though their worries were not completely unfounded—as we have seen, radio clubs directly tied to both the Social Democrats and Communists existed in Germany[92]—the authorities had very limited technical means to track down illegal transmitters until the early 1930s,[93] and so in the end, by their restrictive policies, the German authorities merely succeeded in angering the radio enthusiasts who wanted to transmit, without really being able to stop any illegal transmission.

If the postal authorities thought that German amateurs would be satisfied with the bare handful of transmitting licenses on offer, they were sadly mistaken. The few legal transmitters only whet appetites for more and made it very difficult to determine which transmitters heard on the airwaves were legal and which were not. Moreover, the presence of transmitters in the major clubs created spaces where practical shortwave

[91] On Navy covert subsidies to the DASD, see: BAMA, RM20/1978 In7 No 20330/34 g. Kdos. In 7 Ia of 8. November 1934 (Abschrift) an A.H.A. Betr.: Regelung aus dem x = Fonds, I.A. gez. Fellgiebel. Remember that DASD Presidents Fulda and Stockmayer were both former professional officers, as were most DASD Presidents between 1933 and 1945.

[92] See pp. 103–106 above and pp. 234–241 below.

[93] Polizeihauptmann von Asmuth, "Über die Möglichkeited der Kurzwellenverwendung bei Unruhen durch Staatsfeindliche Elemente, Kontrollmöglichkeit und Vorbereitende Maßnahmen der Behörden" (N.D., but ca. 1928 or 1929) in BArch R/1501/20401, St10 RMdI KPD-Radio-Propaganda, Bd. 3 Dez. 1929–Febr. 1931 St10/62, Bd. 3, 146–153.

construction and use could be practiced and taught.[94] As a result of the interest in shortwave in general and transmitting in particular, most clubs by 1927 had a special shortwave section, which concerned itself with both shortwave listening and transmitting.

As early as 1926, the OFV began to carry out nationwide tests of propagation, and collated reports of reception of special broadcasts, which were announced in the radio hobby press. This helped to spread the desire to transmit and began to bring shortwave listeners and potential transmitters into contact with the Stuttgart club.[95] Many shortwave enthusiasts first learned of the possibility of transmitting via one of these club stations and the amateurs involved in them. Moreover, German shortwave transmitting enthusiasts were spurred on and excited by events abroad. In the mid-1920s, the international organization of amateur radio coalesced, not least under the sponsorship of the powerful ARRL and its President Hiram Maxim in the US. By 1923 or 1924, there were enough hams worldwide to make some form of international organization necessary. In 1924, at a preliminary meeting in Paris, representatives from the ARRL and a handful of prominent radio amateurs from other countries decided to hold a larger, international amateur radio meeting in Paris in 1925 in order to found an international organization of ham radio enthusiasts. This led to the founding of the International Amateur Radio Union (IARU), which still exists. While German representatives did not attend the 1924 pre-conference (there was no single German organization capable of representing German amateurs abroad at the time, and very few Germans on the air), a number of German representatives were invited to the April 14–18, 1925 conference. The German Radio Cartel (*Deutsches Funkkartell*) sent a delegation, which included a number of prominent individuals representing the radio clubs, together with a few leading shortwave enthusiasts.[96] The Paris meeting proved to be a real turning point for the shortwave amateurs in Germany, for several reasons.

[94] Early Club stations were often originally licensed for middle- and long-wave frequencies and were later converted to shortwave use. For example, "Mitteilungen befreundeter Radio-Klubs", "Oberdeutscher Funkverband e.V.", *Der Radio-Amateur* 3, No. 7 (February 13, 1924): 29.

[95] For example, "Mitteilungen anderer Vereine", *Der Radio-Amateur* 4, No. 3 (January 15, 1926): 28–29, announcing a special report of reception card (QSL card—see note 65 above) in exchange for reports of reception of the club transmitter.

[96] On the participation of the Funkkartell, see: "Arbeit, Aufgaben und Ziele der Deutschen Vereine von Funkfreunden (= Tätigkeits bericht of Generalsekretär des Deutschen Funk-

First, the German shortwave enthusiasts who attended were profoundly shocked for two reasons: to begin with, they were confronted by the backwardness of German amateur radio. Germany had once led the world in electronics, but coming together with representatives and fellow hams from powerful amateur radio countries like the US, Britain, and even Switzerland, where well-organized and self-confident national amateur radio organizations already existed, and where amateur transmission was allowed, showed the Germans just how far behind the curve they were. It motivated the German shortwave enthusiasts to catch up to the world standard and gave them important models as to how to proceed. If this weren't enough, the German shortwave enthusiasts were just as deeply shocked by the way they were treated by the other international shortwave enthusiasts gathered in Paris. Here were the Germans, international pariahs after the loss of the Great War, attending an international congress in Paris, amongst delegates from the US, Britain, Belgium, and France, and they found themselves completely accepted, even warmly received, not as Germans, but as fellow radio amateurs. Immediately after the opening session, the shortwave hams, Germans included, all went off to their own reception and dinner. Delegates wore formal nametags, which also included their radio call signs. The French organizers also handed out a small trumpet or whistle to each delegate at the reception, and soon all were madly tooting out "on air" contacts with each other in Morse code. Formal names were soon dropped in favor of call signs, and the party soon got into high gear. The liberal provision of alcohol certainly helped. What began as a stiff and formal international conference turned into something closer to a fraternity party. This was international "ham spirit", and the Germans were just some of the boys.[97] Just as important, in the next few days, the delegates broke up into working groups to design and implement the future shape of the new IARU. The Germans participated on the same basis as all the other delegates, and were deeply impressed by the open, businesslike, and practical discussions which followed. This was

Kartell, Friedrich Schmidt, auf 28.Juli.1925 meeting in Mü)", *Der Radio-Amateur* 3, No. 32 (September, 1925): 795–797. Körner states that the following German delegates represented the shortwave faction: Oberstltn. von Stockmayer, Rolf Formis, Dr. Jäger, Prof. Dr. Esau, and Felix Cremers. He also states that Hermann Kraus, formally representing the FTV, was converted to transmitting while in Paris. Körner, *Amateurfunk*, 36–37.

[97] Felix Cremers, "Internationaler Radio-Amateurkongreß in Paris", *Der Radio-Amateur* 3, No. 19 (May 8, 1925): 468–470 and Hermann Kraus, "Über Amateur-Senden", *Der Radio-Amateur* 3, No. 48 (November 27, 1925): 1125–1127.

unexpected, shocking, and delightful for the Germans, who expected to be treated as pariahs. The international fraternity of amateur radio operators had opened its doors to Germany and shown itself to be both welcoming and highly organized. The attraction was immediate, and powerful. This was far different from the discourse about radio going on in Germany at the time, which was still mired in issues of class and politics.

The German delegates to the founding congress of the IARU came away with one more important lesson as well. They had gone to Paris together with other German delegates from the radio clubs, men who were mainly interested in broadcast listening or in DIY radio building and saw themselves as representing those interests. Immediately, a deep gulf developed in Paris between the German shortwave (transmitting) enthusiasts, who were immediately received into the international community of fellow hams, and the others, who were (aside from a few on-the-spot converts to the transmitting community) lukewarm or even opposed to promoting the transmitting side of the hobby. This made the need for a national German organization exclusively devoted to ham radio apparent and gave impetus to the founding of a nationwide organization in Germany for shortwave (transmitting) enthusiasts.

After the Paris meeting, German shortwave enthusiasts moved to create their own national organization and organize their hobby such that it could take part in the international practice of ham radio. Just as in the period immediately after the Second World War,[98] the Stuttgart-based amateurs led by Lt. Colonel (ret.) Ludwig Friedrich von Stockmayer took the lead. Stockmayer and Rolf Formis, another leading member of the OFV, began to lay the groundwork. They had both been to Paris and were very much interested in seeing Germany participate in the international ham movement. For this, there needed to be a strong German organization for shortwave enthusiasts.

The very first step was to allow German shortwave enthusiasts to begin to meet face to face. Rolf Formis took the initiative and helped bring about a fairly informal first face-to-face meeting of early transmitters at the house of "MARS" (Schmitz) in Mühlheim/Rhein. There were only ten men at this first physical get-together, but it was a beginning: the German transmitters began to meet each other physically and talk in ways they could not yet do on the air.[99]

[98] See Chap. 6 below.

[99] Körner, *Amateurfunk*, 16–18. There is a group picture of the meeting on page 17.

One early initiative by German shortwave hobbyists who helped consolidate the hobby was to begin large-scale systematic tests of radio reception in Germany. Radio hobbyists were asked to send in reports if they were able to receive a given transmission on a given date or dates, reporting on the quality of the signal at their location and relevant atmospheric conditions. This followed similar initiatives in the US and elsewhere.[100] While the scientific benefits of such tests were important, they had much more far-reaching consequences. Stressing the scientific contribution, amateurs could make a way of both attracting new people to shortwave and putting amateurs into a good light with the authorities. Getting them to send in their observations was also a way of learning who was active and getting together a list of names. Systematic observations also required a greater degree of organization on a national level than had existed before, spurring further developments. The OFV from its inception worked to facilitate the exchange of QSL cards and helped organize the first national German reception test in February 1926.[101]

Out of this engagement, the OFV founded the "German Reception Service" (*Deutscher Empfangsdienst*, DED) in the Spring of 1925.[102] The OFV began in June 1925 to issue "Radio Monitoring Numbers" ("*DE-Nummern*") to any German radio enthusiast who wanted one and could fulfill the technical conditions. This was administered by Rolf Formis, who also set up an office to distribute QSL cards. This was done outwardly in connection with the club's propagation experiments, but actually to fill the void left by the Postal Ministry's lack of support for amateur radio.[103] At the beginning, this was an independent initiative by

[100] See "Mitteilungen des Deutschen Funktechnischen Verbandes e.V., Berlin", Beilage, "Mitteilungen befreundeter Radio-Vereinigungen und Verbände", *Der Radio-Amateur* 4, No. 8 (February 19, 1926): 31.

[101] "CQ des O.F.V.", *Der Radio-Amateur* 3, No. 31 (July 31, 1925), p. 29.

[102] Körner, *Amateurfunk*, 50–52. This was legal, because it was ostensibly only for those interested in listening to shortwave transmissions.

[103] "DE" stood for "*Deutscher Empfangsamateur*" or "German Amateur Monitor", and they were quickly called "DE-Numbers". They consisted of the letters DE and then an individual number. Leo H. Jung, DH4IAB, "QSLs erzählen deutsche Amateurfunkgeschichte", part 3, " 'Amateure' auf Sendung (unlis) und Empfang (DEs)", *Funk-Telegramm* No. 6 (2003): 28–29. (available at: http://dokufunk.org/upload/folge_03.pdf, accessed October 16, 2018). See the list of DE-Number holders now in the Dokufunk Archive: DL6KQ, "Funkamateure mit DE-Nr. Aus den Jahren 1926 bis 1945, von 00001 bis 07400. Sammlung von DL6KQ" M.S., 1 July 1998, with later additions. Dokufunk Archive, DE Rufzeichenliste, dok_DASD_RZL_05.

the OFV with no national or international legal standing and only involved monitoring (listening to) shortwave transmissions, but it quickly became the basis for something greater.

The decision to issue DE-Numbers turned out to be a stroke of genius. By international law, every radio transmitter (or its owner), be it commercial or amateur, required a unique call sign. These are a short means of identifying the station and its whereabouts, and must by international law be repeated on-air at intervals for identification of the station. In principle, these call signs are issued by a national organization, usually governmental, in conformation with an international system today supervised by the International Telecommunications Union (ITU), a UN agency. Though the use of call signs goes back to the earliest days of radio, they began to be systematized in the early 1920s, and were given their modern form in 1927 at the Washington International Radiotelegraph Conference.[104] Any radio listener today is still familiar with the system. "WHRO" is the call sign of a Norfolk, VA, public radio station, "D4xvf" was the call sign in the 1920s of the German amateur Felix Cremers, cited above. Call signs are important; hams exchange them as proof of a mutual contact, and use them not only as names, but also as addresses for the sending of QSL cards, which document a contact. Owning a legal call sign is thus as important as owning a legal name, they are proof that a person exists. Moreover, amateurs are not supposed to make contacts with illegal transmitters. But what to do if legal transmissions are banned or severely restricted? The German shortwave amateurs were willing to transmit illegally, but they needed some sort of a call sign in order to be acknowledged on the air as members of the international ham movement. The DE-Numbers, though not legal call signs themselves, were a step in that direction. Significantly, they were not issued by the Postal Ministry or other governmental organization in Germany but were rather a private initiative by the hobbyists themselves. They helped cement a common identity for the shortwave enthusiasts, and served as the legal, public face for a subsequent illegal call sign system.

The formation of a formal national organization to represent all the German shortwave enthusiasts gained wide support. Not least, Germany needed an official organization in order to be represented in the IARU. The first German "Shortwave Conference" (*Kurzwellentagung*) was called in

[104] See https://www.itu.int/en/history/Pages/RadioConferences.aspx?conf=4.39, accessed October 31, 2018.

Jena on January 16–17, 1926, by Formis and Stockmayer. It was the first national meeting for all German shortwave transmitting enthusiasts. Some 40 men, most already members of the DFTV, attended. Prof. Dr. Abraham Esau, a leading physicist and advocate of amateur transmission, provided rooms at the Jena Technical University where he taught. After hearing a report on the DED, the delegates decided to form a national organization for shortwave transmitting enthusiasts and brought the "German Transmission Service" (Deutscher Sendedienst, DSD) to life as a subgroup within the DFTV. Prof. Esau became President, von Stockmayer the Manager (*Geschäftsführer*) and Rolf Formis the Secretary. A telegram was sent to the IARU announcing the formation of the DSD.[105] Immediately after the Jena Conference, the OFV held the first "German Transmission Day" (*Deutscher Sendetag*) competition, a large-scale contest to report reception of transmissions from the OFV club transmitter. In total, 164 German shortwave enthusiasts took part (a mix of DEs and those with transmitters), who sent in 840 observations. By mid-1926, the DSD had over 400 members.[106]

A second Shortwave Conference was held in Berlin on September 4, 1926, coinciding with the German Radio Exposition. The fledgling DSD was threatened by tensions between Stuttgart and Berlin. Berlin enthusiasts argued that the seat of the DSD should be in the capital and lobbied for this to happen. Formis felt personally hurt and resigned his function as "Chief Traffic Manager" ("*Hauptverkehrsleiter*").[107]

A schism between north and south threatened to split the new organization. In the end, the division was barely avoided.

The Third Shortwave Conference was held in Kassel from March 19–20, 1927. The meeting was again largely organized by Rolf Formis. Some 300 voting members attended. At this conference, the DSD and

[105] "Kurzwellentagung Jena", *Der Radio-Amateur* 4, No. 4 (January 22, 1926): 88. Körner, *Amateurfunk*, 53. Officially, the DFTV became the member organization of the IARU for Germany, but the men who represented the DFTV with the IARU were always from the DSD/DASD.

[106] "die Vortragsfolgen", "Mitteilungen", Der Gruppenverkehrsleiter des DRC, "Erster allgemeiner Deutscher Amateursendetag!", in: *Der Radio-Amateur* 4, No. 8 (February 19, 1926): 28–31.

The test was announced, among other places, in: "Mitteilungen des Deutschen Funktechnischen Verbandes e.V., Berlin", *Der Radio-Amateur* 4, No. 6 (February 5, 1926): 32–33. Körner, *Amateurfunk*, 54.

[107] Körner, *Amateurfunk*, 54–55.

DED were combined under the new name of "German Amateur Transmission and Reception Service" ("Deutscher Amateur- Sende- und Empfangs-Dienst", DASD). Though the QSL office was temporarily left in Stuttgart, the Berlin faction succeeded in moving the headquarters of the DASD to Berlin. A new President and Manager, Colonel (ret.) Otto Fulda, was elected, and A. Kofes (EK 4 ade) became Traffic Manager. Formis cut all ties with the organization, whereas v. Stockmayer remained a public voice for amateur transmission. The DFTV (of which the DASD was a part) guaranteed the DASD a seat on its governing board.[108] At the 1928 German Shortwave conference, the DASD voted to accept a formal set of by-laws, completing its organizational structure.[109] Already by the Fall of 1927, the DASD was organized into 13 different club regional districts spread over all of Germany. Each regional district was led by a district Traffic Manager, and was responsible for administering the membership test, which, if passed, led to the award of a DE-Number.[110] In May 1927, the DASD also got its own journal, *CQ*, which was both sent to members and also appeared as a supplement to *Funk*.

The ability of the young DASD to organize was remarkable, but ultimately dictated by the amplitude of the tasks at hand. The example of amateurs in other countries was a strong motivating factor, and contacts with foreign amateurs brought ideas and examples. Moreover, the fledgling DASD could count on the support of the DFTV and the other radio clubs to which its members belonged. Nevertheless, the amount of progress made both organizationally and technically bears witness to the dedication and commitment of hobbyists who helped build the organization.

Interestingly, the shortwave community in Germany remained firmly anchored within the larger radio club universe even after an organization for German hams was created, despite the often-real divergence of opinion between the different groups. This was a very smart decision. Until 1933, the shortwave community in Germany remained under the larger umbrella of the radio hobby clubs by its own choice. Ostensibly, the ham organization

[108] Körner, *Amateurfunk*, 63–67.

[109] (Otto) Fulda, "Die Kurzwellentagung 1928", *Funk* No. 24 (June 8, 1928): 188.

On the by-laws, see: (Otto) Fulda, "Mitteilungen der H.V.L.", "Der Deutsche Amateur-Sendedienst", *CQ Mitteilungsblatt des Deutschen Amateursendedienstes (D.A.S.D.)* 1, No. 7 (November 1927):127 and "Die neuen Satzungen des D.A.S.D.", *CQ* 1, No. 13 (Mai 1928): 263–266.

[110] (Otto) Fulda, "Mitteilungen der H.V.L.": "Der Deutsche Amateur-Sendedienst", *CQ* 1, No. 7 (November 1927): 127ff.

(first called the "German Transmission Service", and later the "German Amateur Transmission and Reception Service"[111]) was only a subordinate branch of the larger club universe, specifically, a sub-group of the DFTV. This brought the hams the advantages of greater numbers—there were a lot more radio listeners and DIY enthusiasts than there were shortwave transmitters. It gave the shortwave hams access to the club assets, such as the meeting rooms, journals, workshops, libraries, and club transmitters. And it kept them in contact with the larger world of the radio hobby, from which they drew nearly all of their members. Disagreements over policy were fought out as internal matters and often on the pages of the club journals, but there was no radical break-up.[112] On the whole, this was good for the clubs, too. Despite sometimes differing interests, the shortwave amateurs in Germany remained connected with the other radio hobbyists, to their mutual benefit.

The creation of a national organization of German shortwave enthusiasts was only half the battle. Before German hams could become a full part of the international world of ham radio, they needed to solve the difficult question of licensing and call signs. Without call signs, German shortwave amateurs could not make contact with other hams internationally and could not normally receive QSL cards—both of these requirements were part of the international practice of the ham radio community. What German hams needed was not just call signs, there had to be a centralized system to issue them in an orderly manner, and a means of enforcing the respecting of international standards as far as procedure on the air and general knowledge of radio. There thus had to be uniform training and some sort of a license test in order to get a call sign. In most countries, these were all done by some arm of government. In Germany, the Postal Ministry was legally responsible for issuing radio licenses of all kinds. But despite repeated efforts by amateurs and their organizations to pressure the Postal Ministry to expand the issue of permits for amateur transmitters, there was no real movement on the question until the Nazis came to power in 1933. Legal transmission was limited to the handful of club stations and even smaller number of individuals with Postal Ministry licenses. The vast majority of shortwave enthusiasts in Germany were shut out. Unless, of course, they broke the law. Many did.

[111] "*Deutscher Amateur-Sende-Dienst*" DSD, "*Deutscher Amateur-Sende- und Empfangs-Dienst*", DASD. See below.

[112] See, for example, L. von Stockmayer, "Zur Frage des deutschen Amateur-Sendedienstes", *Der Radio-Amateur* 4, No. 30 (July 23, 1926): 599–600.

Illegal transmitters in Germany first used all sorts of fantasy call signs—after all, they couldn't give their real names and addresses. By 1927, most illegal transmitters patterned their made-up call signs on the legal model as enshrined in international law. That way, it was hard for an outsider to know if they were legal or illegal, and thus, other hams were willing to make contact with them and send them QSL cards. At that time, all (legal) German amateur call signs began with the letter "K" or "EK", followed by a number and two letters. In 1928, the international rule changed, and the "K" was replaced with a "D" for Germany. Most of the illegal transmitters followed suit. This looked good, and allowed contacts with illegal transmitters to take place, but it was still problematic. How would the illegal stations receive the coveted QSL cards to document their contacts? Some sort of central system had to be put into place, yet it had to be kept from the prying eyes of the Postal Ministry and Police. The whole issue was tricky. German hams needed call signs which appeared legal in order to make contacts with foreign hams and legal domestic ones. They also needed a way to match the call sign with a real name and address, so that they could receive QSL cards sent after making those contacts. But at the same time, the German authorities had to be unable to do the same thing, or else the illegal German hams would quickly get into deep trouble. The DSD and later DASD took care of the QSL card problem by setting up their own QSL card distribution office. Foreign hams sent their cards to the central office, addressed simply by call sign. The QSL office then took charge of forwarding them (discretely) to the right person. A practice which developed internationally as a way to save time and money, became, in the German case, a way of preserving secrecy as well.

There remained the problem of uniformization. The illegal call signs had to be unique, and there had to be a consistent pattern to them if they were to successfully fool outsiders. Moreover, there had to be some guarantee that those transmitting knew what they were doing, or else they might interfere with the others. To address these problems, and to demonstrate that they were capable of imposing order on German hobbyists, by 1928, the DSD/DASD began to secretly develop a licensing testing system and to issue completely illegal transmitting call signs on a national basis.[113] The illegal system built on the existing DE-Number process,

[113] As an illegal system, there is naturally a shortage of official sources to back this up. Ernst Fendler (DL1JK) and Günther Noack (DL7AY), *Amateurfunk im Wandel der Zeit* (Baunatal: DARC Verlag, 1986): 30 and Körner, *Amateurfunk*, 77–80. This is borne out by an exami-

which was legal and above board. It also followed the international model for structuring call signs used by the Postal Ministry for the handful of legal transmitters it allowed. Again, adhering to international standards made the DASD look disciplined, and also made it difficult to tell their illegal call signs from the legal ones. As a further twist to distinguish between the legal and illegal call signs, one of the DASD members involved had the idea of issuing the "official" illegal call signs with three letters instead of two: D+number+three letters, instead of the legal form D+number+two letters.[114] In keeping with international law, the DASD also soon added two exams of basic radio theory and practice, first to become a member of the DASD at all, and then another as a prerequisite to gaining an "official" illegal call sign. Ostensibly, the DASD administered the exam for the "*Funkzeugnis C*", a legal German radio permit for commercial transmitters. Candidates who passed the exam would find attached to the announcement a blank piece of paper with three letters written in pencil, providing them with their "official" illegal call sign.[115] The DASD kept a record of these "official" illegal call signs in order to distribute QSL cards. In order to prevent it from falling into the hands of the authorities, it was kept in a code, which was regularly changed.[116]

All of this was designed to hide the true nature of what was going on. In actual practice, both the German authorities and the international IARU had to look the other way if it were to succeed. Given that most German amateur transmitters were illegal before 1933 (and even many well thereafter), the IARU had to stretch its own rules to accept the "official" illegal call signs in such widespread use, or at least pretend not to notice that they were illegal. And the German authorities were no dupes, either. Documents prove that the Postal Ministry,[117] Police/Interior

nation of QSL cards collected in the Dokufunk Archive. The collection of historic QSL cards in the *Dokufunk* archive offers reasonable proof that Körner and other contemporary memoirs are correct. A series of historic QSL cards sent by or to Körner himself and now found in the Dokufunk Archive document the evolution of both legal and illegal call signs in Germany over the course of the twentieth century.

[114] Körner, *Amateurfunk*, 77–80.

[115] Körner, *Amateurfunk*, 77–80.

[116] See: The code key for the period: September 1, 1930 to April 3, 1933, in the Dokufunk Archive. See also Fendler & Noack, 30, which contains a photo of another of the code keys.

[117] The Postal Ministry conducted extensive monitoring of both foreign and domestic transmissions. It also received the secret police reports on monitoring cited below in the following note.

Ministry,[118] and Reichswehr[119] all were well aware that widespread illegal transmissions were being made by German hams, but all chose, officially, to do nothing about it on any large scale—as long as the illegal transmitters were not communists, and as long as the DASD was able to enforce some sort of discipline. This (unspoken) policy was partly out of calculation, and partly out of necessity: no one was able to reliably and quickly locate any illegal transmitter, which took reasonable precautions before 1934 or 1935,[120] and at least the DASD brought some order and discipline to the on-air activities of its members.[121] Keeping most transmission illegal meant they could be easily stopped if they got out of hand, and allowed the authorities to gather intelligence about illegal transmitters quietly. Some illegal transmitters were caught and punished, but they were few in number. The vast majority of German hams transmitted illegally and thrived. By the mid-1930s, the German police estimated that there were some 600 illegal transmitters regularly on the air, most of whom were held to belong to the DASD, in addition to the roughly 140 legal transmitters.[122] The illegality, which became deeply ingrained into German ham culture, later came back to haunt the DASD after the Nazis took power.

The German shortwave amateurs remained firmly anchored within the larger landscape of radio hobby clubs, but also built a well-organized

[118] See the regular monthly reports by the "Office for Monitoring Private German Radio Transmissions at the Police Main Radio Office" ("Überwachungsstelle für den privaten deutschen Kurzwellenverkehr bei der Pol. Hauptfunkstelle Berlin") in BArch R/1501/20404 St10 RMdI KPD-Radio-Propaganda Bd. 5 March 1933–March 1934.

[119] See, for example, Der Reichswehrminister 45/32 geh. Abw. II (Geheim) "Rundfunkpropaganda der Sowjetunion in Deutschland und ihre Auswirkungen" of March 19, 1932, BArch R/1501/20402 St10 RMdI KPD-Radio-Propaganda Bd. 4a March 1931–February 1933, 284–294.

[120] Polizeihauptmann von Asmuth, "Über die Möglichkeited der Kurzwellenverwendung bei Unruhen durch Staatsfeindliche Elemente, Kontrollmöglichkeit und Vorbereitende Maßnahmen der Behörden" (N.D., but ca. 1930), BArch R/1501/20401, St10 RMdI KPD-Radio-Propaganda, Bd. 3 Dez. 1929–Febr. 1931 St10/62, Bd. 3, 146–153.

[121] Ironically, the DASD always argued that it was better to allow legal amateur transmitters and depend on the amateurs themselves to maintain order and fight illegal transmission. In effect, this is exactly what the Postal Ministry (RPM) and other agencies did, without conceding legal status to the amateurs.

[122] Polizeihauptfunkstelle, "Zusammenfassender Bericht der Überwachungsstelle des privaten deutschen Kurzwellenverkehrs bei der Pol. Hauptfunkstelle Berlin für die Zeit vom 15.10.1930–15.10.1931" of 20.October.1931 (Geheim!), BArch R/1501/20062 Reichsministerium des Innern KPD-Radio-Propaganda, 1930–1934, 202–214.

parallel organization, the DASD. This was, after all, partly in the nature of the hobby: it didn't take much organization to simply listen to a radio broadcast, but amateur transmission was another matter. Two-way transmissions had to follow a set of rules, some of which were international, some national, and some simply to create order for the proper consummation of a contact. All this required good organization and discipline. German shortwave amateurs never got the hotly wished-for large-scale issue of licenses for private transmitters before 1949. Because of that, they were forced into widespread illegality, even as they continued to push the Postal Ministry to allow legal licenses. Yet this was a strange form of illegality; international standards had to be respected, and the DASD itself had to set up a working training and testing procedure to unofficially "license" new transmitters. Both international and national authorities alike knew of the widespread illegality, but each for its own reasons did little to stop it. We will have occasion to speak at great length about the DASD in the following two chapters.

Women and the Radio Hobby

What about women in the radio hobby? The 1920s are generally thought of as a time when women became "modern",[123] and began to seek and experience emancipation on a large scale. How do these "flappers" and "new women" fit into the radio hobby? Kate Lacey states very plainly that "the history of radio cannot properly be understood if gender is absent as a category of analysis", and goes on to point out that women emerged as "citizens with a political voice" at exactly the same time as radio became a mass medium, "with its potential for reinvigorating the public sphere".[124] Though Lacey is mainly interested in broadcast radio, what she says applies equally well to the somewhat different examination of the radio hobby undertaken here. Women were a central part of the story. The problem is, they are largely absent from the written record, and so the process of reconstructing just what their roles were is rather difficult. On the one

[123] For basic histories of (German) women in the 1920s and 1930s, see: Ute Frevert, *Frauen-Geschichte. Zwischen Bürgerlicher Verbesserung und Neuer Weiblichkeit* (Frankfurt: Suhrkamp, 1986); Frauengruppe Faschismusforschung, *Arbeitsbuch und Mutterkreuz. Zur Geschichte der Frauen in der Weimarer Republik und im Nationalsozialismus* (Frankfurt: Fischer, 1981); Gisela Bock, *Frauen in der europäischen Geschichte: Vom Mittelalter bis zur Gegenwart* (Munich: C.H. Beck, 2005).

[124] Lacey, *Feminine Frequencies*, pp. 221–223.

hand, we can provide a rather cursory list of the handful of women who do appear—mainly in the radio journals and their accounts of club activities. At the very least, there is documentation (albeit sparse) that women were present in the clubs, and that they did play a role in the radio hobby. Their numbers, and the importance of their roles, however, has to be read between the lines. Within the radio clubs, some women occupied the "traditional" female roles behind the scenes: mothers, wives, girlfriends, and siblings. This is not trivial, for their household labor (and forbearance) provided men the space to engage in a hobby in the first place.[125] But beyond this, these women would also become radio listeners, once a radio was present in the household, no matter whose name was on the postal permit. As such, they certainly were a part of the wider radio world, and presumably had at least some influence as consumers.[126] The broadcast authorities recognized the important role of women as consumers (listeners) and as citizens by aiming programming specifically at women.[127] In fact, women were likely to spend more time in the home in direct contact with the radio than men, since so many women were either housewives or worked from the home (as seamstresses, washers, etc.); the nature of female labor in Germany in the 1920s and 1930s practically guaranteed women to have more exposure to radio broadcasts than men.

Women certainly played an indirect role in the radio clubs. Again, they mostly appear as wives, siblings, and girlfriends. And once again, these roles are not trivial at all. As we have already demonstrated when discussing the social function of clubs, the association of women with largely male groups was extremely important in attracting new members. The fact that one might well meet a future spouse within the context of club activities was not negligible, nor was the ability to turn club activities into family activities by including wives and children. Moreover, both men and women needed to come to terms with the new mass technologies of the 1920s for them to succeed. Even those women who were not club members still felt the influence of the club's relationship to technology.

[125] Haring, *Ham Radio*, chapter 6.

[126] Manufactured radios were heavily marketed to women, even though their high cost meant that men would likely have the final decision at this time. Some of the marketing involving women is certainly because "sex sells", but women clearly influenced consumer decisions at the time, particularly those concerning the home. Lacey, *Feminine Frequencies*, 176. She cites, among other sources, Daniel Miller, "Consumption as the Vanguard of History", in *Acknowledging Consumption: A Review of New Studies* edited by Daniel Miller, 1–57 (London: Routledge, 1995).

[127] Lacey., *Feminine Frequencies*, passim.

As I have already argued, radio hobby clubs were important vectors for spreading positive awareness of radio and often other new technologies. They were hugely influential in "domesticating" technology and science. Clearly, it was mainly men who were in a position to devote much time to hobby pursuits and be active in the clubs directly. Nevertheless, it would be a mistake to think that the influence of the clubs did not extend outside the clubhouse walls. Women participated in club activities even as family members or friends of hobbyists. They knew of the positive engagement with radio of the people around them, discussed radio with family members, heard the radios the men built and brought home, and sometimes built their own.[128]

Despite all of the impediments to their direct engagement with radio, some women became full members of the clubs, not just for the social activities, but because they were just as interested in the technology as the men.[129] They tend not to appear often in the public record, but we know they were there. As Haring has pointed out in her study of US ham radio clubs, women who were club members very often took on important (but largely unsung) administrative duties such as secretary, treasurer, and organizer of various social functions.[130] This may have been less the case with German clubs in the 1920s, 1930s, and 1940s, given that the secretary and treasurer typically belonged to the club's governing board, and thus held public leadership positions, but we can assume that women probably took on many of the behind-the-scene duties. We know, too,

[128] Note that this "optimistic" view of women and radio was challenged by some contemporary commentators, who feared that radio might compete with women for male attention, and thus hurt relationships. See Mary Peacock for example, as cited in Lacey, p. 39.

[129] When they were clearly present in the clubs as members. A group photo taken at the occasion of the fusion of the formation of the DFTV included at least four, and probably five women, and 67 men. See: "Die Gründung des Deutschen Funktechnischen Verbandes", *Der Radio-Amateur* 3, No. 32 (August 7, 1925): 792–795. The photo is on p. 793. In another example, the Lichtenberg branch of the German Radio Club reported that "Miss Michaelis" had resigned her club membership in writing, as of March 31, 1926. See "Die Vortragsfolge", "Mitteilungen", *Der Radio-Amateur* 4, No. 14 (April 2, 1926): 28–31. There was no explanation why she left. Nevertheless, she had been a full member. Several clubs report the admission of female members in 1928; for example, both the Funktechnischer Verein e.V., Magdeburg and the Deutsche Funkgesellschaft e.V. See: "Aus den Vereinen und Verbänden", *Funk* 5, No. 7 (February 10, 1928) Programmteil: 5–8.

[130] Haring, *Ham Radio*, 47–48.

that women did get involved in the technical side of the clubs.[131] In published results of the frequent radio building competitions, women occasionally appear as award winners: while building and tinkering are marked as typically male activities, we have proof that some women beat the men at their own game—perhaps not least because neatness and attention to detail (important criteria in build) were seen at the time as typically female attributes.[132] Perhaps the most prominent example of a women who began to build radios, then began to tinker with the technology is the Austrian actress "Heddy Lamar". She learned how to build a simple radio receiver as a young woman. While it can't be documented that she ever participated in a radio club, she did continue to be an active tinkerer with radio technology, and eventually went on to invent (and patent) "spread-spectrum" technology, without which we would have no portable phones, Bluetooth, or Wi-Fi.[133]

Some women even became a part of ham radio (amateur transmission), the most technical part of the hobby of all. A handful of women were members of the DASD, and held DE-Numbers, indicating that they took and passed an exam on radio theory, and were active in club activities.[134]

[131] As one example, in 1928 the chemist Miss Sonnenschein ("Fräulein Chemikerin Sonnenschein") was elected to the club's Technical Commission. "Aus den Vereinen und Verbänden", "Saar-Radio-Club e.V.", *Funk* 5, No. 14 (March 30, 1928): 9.

[132] For example, in 1928, the Funktechnischer Verein Magdeburg held a "Funkausstellung 1927", with prizes for home-made radios; in the competition Helene Hünerbein was one of three people to win third prize. "Aus den Vereinen und Verbänden", *Funk* No. 2 (January 6, 1928), Programmteil, 9–12.

That same year, another woman, Fräulein Anneliese Scharfenberg, won second place in a design competition for a radio certificate to be issued by the DASD: "F", "Preisgekrönte Diplomentwürfe. ein erfolgreiches Preisausschreiben des D.A.S.D.", *Funk-Bastler* No. 25 (June 21, 1929): 399.

[133] The most recent biographies are Richard Rhodes, *Hedy's Folly: The Life and Breakthrough Inventions of Hedy Lamarr, The Most Beautiful Woman in the World* (New York: Doubleday, 2011) and Stephen Michael Shearer, *Beautiful: The Life of Hedy Lamarr* (New York: St. Martin's Press, 2010). See also the recent film biography: Alexandra Dean: "Bombshell: the Hedy Lamarr Story", Reframed Pictures, 2017.

[134] See "Funkamateure mit DE-Nr. Aus den Jahren 1926 bis 1945, von 00001 bis 07400. Sammlung von DL6KQ" in: Dokufunk Archive, DE Rufzeichenliste, dok_DASD_RZL_05. The list is a postwar attempt to identify all those who held a DASD DE-Number. Names exist for over half the DE-Numbers known. Of these, 15 could be identified as women by their names. The actual number of women holding DE-Numbers was certainly greater, though was certainly well under 100. Examples include Hedwig Döring, Stettin (DE01316B), and Elfriede Schäfer, Berlin-Fronau (DE06887F).

An even smaller number of women held transmission licenses,[135] or possibly even transmitted illegally, just as so many male radio hobbyists did.[136] The number of women active in the DASD probably rose slightly during the war, not least because most men were under arms, but also because of the progressive militarization of German society under the Nazis. At least one woman held one of the few special wartime call signs (*Kriegsrufzeichen* or "*K-Nummer*").[137] And one woman, Else Bödigheimer, even officially took over the administration of the DASD Regional Group (*Landesverband*) "T" Hessen from her husband, when he was called up for service.[138]

Some of these women became acquainted with the DASD through male friends or spouses,[139] others almost certainly through the Bund Deutscher Mädel (BDM), the female division of the Hitler Youth (HJ). Beyond the radio hobby, many more women were introduced to radio and radio technology during the war not as hobbyists, but as female auxiliaries in the armed forces or BDM.[140] Few of these women ever found the time

[135] Ibid. See also: "Liste der YL-Stationen 1909–1939", http://www.dokufunk.org/amateur_radio/contributions/index.php?CID=2031 Accessed November 26, 2018. Examples include: Käthe Jacobs, D4BSG (also DE1282G) and Else Müller, D4BDH.

[136] See "Bericht 25 der Überwachungsstelle des privaten deutschen Kurzwellenverkehrs bei der Pol. Hauptfunkstelle Berlin für die Zeit vom 1–3. XI 1932" of 17. December 1932 (Geheim!)", BArch R/1501/20062 Reichsministerium des Innern KPD-Radio-Propaganda, 1930–1934, 248–249. This monitoring report specifically mentions a woman who engaged in long "chats" on the air with other hams. The terms of the report are very unflattering and stereotypical; as in other areas, women were not always welcome in male-dominated activities.

[137] This was Olga Lippert, who was specifically authorized to use the wartime call sign of her husband, Dr. Erich Lippert, D4AZS. See: "Liste der Kriegsfunkgenehmigungsinhaber (Stand 25.August.1944)", Dokufunk Archive.

[138] "Anschriften der Landesverbandsführer, nach dem Stand vom 15.4.43", *CQ. Nachrichtenblatt des DASD* No. 1 (May 1943): 5–6; DASD Landesverband (T) Hessen und bei Rhein, "Rundschreiben an die Kameraden des LV/T an der Front und in der Heimat" (January 15, 1943) (edited by Else Bödigheimer, Deputy Region Leader), Dokufunk Archive.

[139] An example is Annemarie Stork (born Krieberger) of Gießen (DE02772T), whose husband was DE02771T.

[140] Some 500,000 women served as Wehrmacht auxiliaries, not all, of course, as radio technicians or telegraphists. See: Ursula von Gersdorff, *Frauen im Kriegsdienst 1914–1945*, Beiträge zur Militär- und Kriegsgeschichte 11 (Stuttgart: Deutsche Verlagsanstalt, 1969); Kathrin Kompisch, *Täterinnen. Frauen im Nationalsozialismus* (Köln/Weimar/Wien: Böhlau, 2008); Birthe Kundrus, "Nur die halbe Geschichte. Frauen im Umfeld der Wehrmacht zwischen 1939 und 1945" in: *Die Wehrmacht. Mythos und Realität*, edited by Rolf-Dieter Müller and Hans-Erich Volkmann, 719–735 (Munich: Oldenbourg, 1999);

to engage in radio as a hobby, even after the war, yet the numbers of women in technical hobbies, including ham radio, has steadily risen since the 1920s. The numbers of women in the radio hobby in the 1920s and 1930s was small, but a sign of things to come. The wave of new, mass media technologies which washed across the industrialized world in the twentieth century also affected women. They were influenced by radio as a hobby, and in some cases took up the challenge of becoming hobbyists themselves. We have to acknowledge not only that they were there, but that they played a major role in the hobby even if they do not appear more often in published documents.

Technically, the late 1920s and early 1930s were a time of great progress, when it seemed that the radio hobby stood before a bright future. This changed dramatically in 1933, for reasons we will cover in the next chapter. Technological change played a role, but the real problem was politics.

Franka Maubach, *Die Stellung halten: Kriegserfahrungen und Lebensgeschichten von Wehrmachthelferinnen* (Göttingen:Vandenhoeck & Ruprecht, 2009); Franz W. Seidler, *Blitzmädchen. Helferinnen der Wehrmacht* (Augsburg: Bechtermünz, 2003); Gerda Szepansky, "*Blitzmädel", "Heldenmutter", "Kriegerwitwe". Frauenleben im Zweiten Weltkrieg* (Frankfurt/M: Fischer, 1997). See also the memoir by Hilde Kerer, *Ich war ein Blitzmädel. Frauenkamaradschaft in der Wehrmacht. Aufgezeichnet von Thomas Hanifle* (Bozen: Edition Raetie, 2014).

CHAPTER 5

The Nazification of the Radio Clubs, 1929–1935

> It's not as if we only intended to transmit [Nazi] Party-related programs. We want to give the widest of room to entertainment, recreation, games, jokes and music; but everything must have an inner relationship to our times. Everything must help carry our great reconstructive project, or at the very least, it must not stand in contradiction to it. For this to happen, we need above all: the strict centralization of all broadcast policy efforts, the implementation of the [Nazi] Leadership Principle, clarity in all philosophical underpinnings and the broadest elasticity in the translation of these philosophical underpinnings into the practical work of broadcasting.[1] (Joseph Goebbels at the opening of the Great German Radio Exposition, 1933)

[1] "Nicht, als wenn wir die Absicht hätten, lediglich Parteiprogramme zu senden. Wir wollen der Unterhaltung, der leichten Muse, Spiel, Scherz und Musik breitesten Spielraum geben; aber alles soll eine innere Beziehung zur Zeit haben. Alles soll die starke Note unserer großen Aufbauarbeit tragen, oder es soll sich doch mindestens dazu nicht in Widerspruch befinden. Dabei ist vonnöten: eine straffe Zentralisation allen rundfunkpolitischen Schaffens, der Vorrang ihrer geistigen Aufgaben vor den technischen, die Durchsetzung des Führerprinzips, die Eindeutigkeit der weltanchaulichen Tendenzen und die weitherzige Elastizität, mit der diese weltanschaulichen Tendenzen in die praktische Sendung übersetzt werden". Joseph Goebbels, "Der Rundfunk als achte Großmacht. Eröffnungsrede zur Rundfunkausstellung am 18. August 1933", in *Signale der neuen Zeit. 25 ausgewählte Reden von Dr. Joseph Goebbels* (Munich: Eher Verlag, 1934): 205.

B. B. Campbell, *The Radio Hobby, Private Associations, and the Challenge of Modernity in Germany*, Palgrave Studies in the History of Science and Technology, https://doi.org/10.1007/978-3-030-26534-2_5

This chapter comes with a sound track. It is the sound of marching feet,[2] loud voices, and a few screams. It is the sound of a dictatorship establishing itself, of terror rampant. It is the sound of Germany in the Spring and Summer of 1933. Sound and terror are both ubiquitous mediums; when they are present at all, they are everywhere. It is important to keep the sound of marching feet in mind, not least because radio is an aural medium with relatively few written traces.[3] The Nazification of the radio hobby is the story of violence threatened, more than violence carried out. It remained mostly as a threat, which did not need to be carried out in the narrow organizational realm I will describe, for its reality was so manifest in other aspects of daily life. There was some real violence, but a little went a long way in a climate where it, like its sound, was ubiquitous. The result was the Nazification of Germany, and the end of all but one of the radio hobby clubs.

The shoulder between 1929 and 1935 was an important one for radio enthusiasts. It should, and could have been the highpoint of the radio hobby movement: radio technology was maturing rapidly, and economies of scale were beginning to drive prices lower, meaning that more people had greater access to better radios than ever before. Technically, radio receivers were a mature technology by the early 1930s.[4] By this point, the tube radio had clearly triumphed over the more primitive crystal radios (even though thousands of crystal receivers remained in use and even in industrial production, due to their low cost). Audio technology (a function of both improved receivers and loudspeakers) was now able to reproduce sound in a pleasing and accurate manner, so that some radio enthusiasts could by now more accurately be termed "high-fi" enthusiasts.

[2] In the evocative French term, literally "*le bruit des bottes*".

[3] This is a nod to the new field of Sound Studies. For basic works, see: Michele Hilmes, "Is There a Field Called Sound Culture Studies? And Does It Matter?", *American Quarterly* 57, No. 1 (March 2005): 249–259; Trevor Pinch and Karin Bijsterveld, eds., *Oxford Handbook of Sound Studies* (Oxford: Oxford University Press, 2011); Florence Feiereisen and Alexandra Merley Hill, eds., *Germany in the Loud Twentieth Century* (Oxford: Oxford University Press, 2011); Greg Goodale, *Sonic Persuasion: Reading Sound in the Recorded Age*, Studies in Sensory History (Champaign: University of Illinois Press, 2011).

[4] Radio by the 1930s was accessible to a majority of the population. By then it required fairly little technical knowledge (unless one wished to construct one's own equipment). Civilian radio receivers were in serial production by a host of different companies both at home and abroad, and in a wide variety of designs. Perhaps most importantly, radios were a self-evident part of the popular imagination, figuring in both works of fiction (film, novels, even pictorial art), and in political or economic works.

Radio hobbyists had access to more and more sophisticated equipment, not just involving radio receivers (and for some, transmitters) but also including the ability to record records, make not only home movies, but now sound movies, or experiment with the transmission of moving images: television.

And yet, the Great Depression was in full swing, severely restricting the economic and often personal resources available to hobbyists and casual listeners alike. Though radio ownership (and thus, presumably, radio listening) continued to expand in Germany during the Depression,[5] there was a disturbing trend for some to give up their radios, and the radio industry faced declining sales and a shortage of capital for investment: the economic situation certainly slowed the public interest in radio, even though it may, paradoxically, have extended the life of the do-it-yourself part of the radio hobby, which continued to build radio receivers at home because this was the only affordable option. The clubs, however, saw membership stagnate, even decline; many could no longer afford to pay the dues.

But this was not all. In the Depression, in no small part *because* of the Depression, the political debate in Germany grew increasingly polarized and ideological. Radio was not spared this development: the external debate about the content of broadcasting in Germany became increasingly shrill and divisive, and the radio clubs themselves could not remain immune to this influence forever. The discourse about and around radio inevitably became as politicized as the rest of society. This was not good for the radio hobby community. The great break came with the rise of National Socialism to power. Within a year of coming to power, the Nazis had forced the radio hobby clubs (and all other aspects of civil society) into a stranglehold, thereby gutting the spontaneous and diverse radio hobby and essentially killing it. Only "ham radio", that part of the radio hobby, which transmitted as well as received, was preserved in a tightly regulated form, and only because of its utility to rearmament. But we need to be clear about two things:

First, while the Nazis were responsible for the death of the radio hobby clubs by their direct and conscious efforts to Nazify all of society in the Spring and Summer of 1933, the process had begun long before then. The polarization of the political debate from 1928 or 1929 on, and particularly, the founding in 1930 of a politically motivated and exclusively

[5] See Table 2.1 above.

conservative organization of radio listeners, the "*Reichsverband Deutscher Rundfunkteilnehmer*" ("Imperial Association of Radio Participants", RDR), had hardened and politicized the radio world long before the Nazi takeover of German society in 1933. Though originally linked to and founded by mainstream conservatives, the RDR was later taken over by the Nazis in an internal coup in December 1931. The RDR was the vector or tool the Nazis used to Nazify the radio hobby clubs, but RDR members, other conservatives, and Nazi supporters had long been present in the middle-class radio clubs, so that the Nazi takeover in 1933 was often experienced in these clubs more like an internal coup than an outside compulsion. The RDR merely prepared the way for the later Nazification of the radio hobby, not least by coarsening the discourse around radio. And of course, when the Nazis took over, the working-class radio clubs were destroyed through direct compulsion and violence.

Second, while it was the political struggle that destroyed the radio hobby culture in Germany, the decline of the radio hobby clubs was also indirectly a product of the evolution of the electrical and radio industry, and of changes in the technology of radio. The Nazis Nazified the radio hobby clubs (along with all other private associations), and largely dismantled them for political reasons, but the clubs were already at a turning point by the early 1930s and at the beginning of a decline due to other factors. Two prime motivations had been behind the rise of the radio hobby in the 1920s. People had sought to learn something of the exciting and novel, if frightening new technology of radio, not least because it symbolized modernity. Very practically, they had also sought to learn how to build a radio themselves at an affordable price. Both of these reasons were in decline by 1933–1934. By then, radio was no longer quite as novel as it had been, and with 4,307,772 radios registered by the German Postal Authorities in 1932, radio was in fact on its way to becoming ubiquitous.[6] The very ubiquity of radio by the early 1930s, combined by radio's technical maturation (expressed not least by a growing simplicity of use) made the radio clubs less important than they had been as a pathway toward access to radio technology. For related reasons, by 1933–1934, economies

[6] Kurt Wagenführ, ed., *Jahrbuch: Welt-Rundfunk 1937–1938* (Heidelberg-Berlin: Kurt Vohwinckel Verlag, 1938). Table: "Entwicklung der Hörerzahlen in den Europäischen Ländern 1922–1936", 130–131. At the time, Germany had roughly 66 million inhabitants. *Westermanns Deutscher Reichsatlas* (Braunschweig, Berlin, Hamburg: Georg Westermann ca. 1933). Table: "Staatl. Einteilung, Fläche u. Bevölkerung des Deutschen Reiches 1933", XI.

of scale (and pressure on prices during the Great Depression) meant that commercially manufactured radios were increasingly affordable. Ironically, it was again the Nazis who stepped in to tip the scales in favor of commercially manufactured radios, with their affordable "*Volksempfänger*" (People's Receiver) project.[7]

In 1933, Nazi leaders bullied a consortium of virtually all the German companies active in radio manufacture to produce a standardized model radio receiver, the famous *Volksempfänger* or "People's Radio", and bring it to market in large numbers at a standard price, accompanied by generous credit terms. This was a propaganda effort by the Nazis, who sought both to be seen as the defenders of the purchasing power of ordinary citizens, but also to place a radio in every home so that their propaganda message would be omnipresent. The unintended consequence was, however, that there was suddenly much less need for most members of the radio hobby clubs to continue, so that when the clubs were Nazified and largely dismantled, Germans adjusted very quickly.[8] (The only exception

[7] On the Volksempfänger see: Ansgar Diller, "Der Volksempfänger: Propaganda-und Wirtschaftsfaktor", *Mitteilungen des Studienkreises Rundfunk und Geschichte* 9 (1983): 140–157; Wolfgang König, "Der Volksempfänger und die Radioindustrie. Ein Beitrag zum Verhältnis von Wirtschaft und Politik im Nationalsozialismus", *Vierteljahreshefte für Sozial- und Wirtschaftsgeschichte* 90 (2003): 269–289; Wolfgang König, "Mythen um den Volksempfänger. Revisionistische Untersuchungen zur nationalsozialistischen Rundfunkpolitik", *Technikgeschichte* 70 (2003): 73–102; Wolfgang König, *Volkswagen, Volksempfänger, Volksgemeinschaft. "Volksprodukte" im Dritten Reich: Vom Scheitern einer nationalsozialistischen Konsumgesellschaft* (Paderborn: Ferdinand Schöningh, 2004); Uta C. Schmidt, "Der Volksempfänger. Tabernakel moderner Massenkultur" in: *Radiozeiten. Herrschaft, Alltag, Gesellschaft (1924–1960)*, edited by Inge Marßolek and Adelheid von Saldern, 136–159 (Potsdam: Verlag. f. Berlin-Brandenburg, 1999); Kilian J. L. Steiner, *Ortsempfänger, Volksfernseher und Optaphon. Entwicklung der deutschen Radio- und Fernsehindustrie und das Unternehmen Loewe 1923–1962*, (Essen: Klartext Verlag, 2005); Joachim Krausse, "Volksempfänger: Zur Kulturgeschichte der Monopolware", *Kunst und Medien* (1984): 81–112.

[8] The Nazi Volksempfänger is important both for its rationalization of production and for the scale at which it became available. But specialists also point to another inexpensive radio receiver, the Loewe oe333, introduced as early as 1926, which pointed the way to affordable yet useful manufactured radios, and was even called a "people's radio" long before the Nazi Volksempfänger. Its production at a comparatively low price point was made possible by the invention of a special radio tube, which essentially combined three functions—basically an entire receiver—in one unit. The Loewe oe333 cost only RM 39.50, whereas the first Volksempfänger cost RM 76.00 for the version, which ran off household current, or RM 65.00 for the battery-driven version. In 1938, a redesigned Volksempfänger was able to be sold for only RM 35. This was possible because it also used a specially designed multifunction tube. The Loewe family was Jewish, and the company was later Aryanized and production of

was the small but solid community of amateur radio hobbyists, as we will see below and in the next chapter.)

Thus, changes in the world economy surrounding the Great Depression, changes in the political debate within Germany, and the maturation of radio technology itself all worked against the continued thriving of the radio hobby. The rise of National Socialism was, of course, by far the most important element in the equation, and was directly responsible for suppressing and "coordinating" radio clubs and the entire radio hobby. But changes in the technology of radio receivers (becoming internally more complex, yet also easier to use for the uninitiated), and changes in the manufacturing of commercial radios (economies of scale and standardization, leading to drastically lower prices) also led in the same direction: for technical and economic reasons, the home-building of radio receivers became more and more a small niche hobby.

The Political Context of Clubs (*Vereine*) in the Weimar Republic

Some aspects of the culture of private associations in Germany played a particular role in the process of Nazification or *Gleichschaltung*. For example, it was a common tenant of nearly all middle-class private associations in Germany, be they sports clubs, choirs, or garden societies, that politics should ostensibly be excluded so as not to divide the club or distract from its main purpose (be it gymnastics, singing, or growing the best roses). The importance of the principle of political neutrality had historical roots going back to Absolutism and had real value both in avoiding complications with the authorities and in reducing internal conflict within any given club or association.[9] Yet club life was always highly political. How could it not be so, when clubs in Germany were such a fundamental part

the oe333 stopped, in favor of the Volksempfänger. See: Ansgar Diller, "Der Volksempfänger. Propaganda- und Wirtschaftsfaktor", *Mitteilungen des Studienkreises Rundfunk und Geschichte* 9 (1983): 140–157 and Table 2.1 above. The point is that mass production and simplification of design were inherent in the development of radio technology at the time, and were not Nazi inventions. The Nazis merely instrumentalized existing trends.

[9] In the special case of amateur radio (those with private transmitters), the avoidance of political themes was, and is still today, also a part of international law. See International Telecommunications Union, *Radio Regulations: Articles, Edition of 2016* http://handle.itu.int/11.1002/pub/80da2b36-en, accessed February 15, 2019.

of the tissue of the public sphere?[10] There was the obvious issue of internal club politics, which are rarely completely harmonious. Beyond this, the subculture of elective associations in Germany was deeply riven by class and confessional differences. For any given interest, there could be separate clubs for each of the major confessions (protestant, catholic, and sometimes Jewish), plus at least one separate working-class club, which often later split during the Weimar Republic into separate socialist (SPD) and communist (KPD) sponsored groups. We have already seen this in the world of the radio clubs with the founding of the "Worker's Radio Union" (ARB) in 1924.[11]

The principle of political neutrality in the world of private associations was, in fact, often a means to exclude those with differing religious or political beliefs. In particular, the "bourgeois" or middle-class clubs often used it as a way of excluding those with opinions outside of the mainstream of middle-class values, and this is a major reason why there existed in Germany a parallel world of working-class associations with the same purposes and functions as the middle-class ones. In the world of the radio hobby, while there were not open religious divisions (perhaps because of the scientific nature of radio, perhaps simply because radio was a new, modern interest without the historical baggage of older associations), there was a division between the middle-class clubs under the Deutscher Funktechnischer Verband e. V. (DFTV), and the separate working-class clubs forming the *Arbeiter-Radio-Bund* (Worker's Radio Union) and the later communist-dominated *Freier-Radio-Bund* (Free Radio Union), which split off from it in 1929. Individual hobbyists could choose to overlook these differences, the different umbrella-organizations could work together to lobby for favorable legislation, but the differences were of crucial importance in the fate of the clubs during the period of the Nazi takeover: the socialist or working-class clubs were immediately disbanded in February (Freier-Radio-Bund) and July 1933 (Arbeiter-Radio-Bund), and their members persecuted and kept under surveillance or arrested and sent to camps and prison. The middle-class clubs were Nazified and their

Article 25, "Amateur Services": 25.2 § 2 1 "Transmissions between amateur stations of different countries shall be limited to communications incidental to the purposes of the amateur service, as defined in No. 1.56 and to remarks of a personal character".

[10] My understanding of the public sphere draws heavily on Habermas. See: Jürgen Habermas, *Strukturwandel der Öffentlichkeit: Untersuchungen zu einer Kategorie der bürgerlichen Gesellschaft* (Frankfurt: Suhrkamp, 1990).

[11] See Chaps. 1 and 3 above.

umbrella organization (the DFTV) fused with the RDR. They continued to exist on paper in this new, unified form until 1934 or early 1935, but then they disappeared. Their ultimate fate was not much different from that of the working-class radio clubs, but this ignores important differences: the middle-class clubs enjoyed a short shadow existence within the RDR, and they were subject to much less violence and intimidation. Many of their members, though certainly not all, even welcomed the coming of the Third Reich. The members of the ARB were encouraged, even expected to join the RDR after their own club was destroyed, but they were subject to great violence and intimidation in the short term, and heightened surveillance and harassment in the long term.

Though ostensibly seeking to end class divisions, the Nazi-led social revolution in Germany left middle-class interests and the state bureaucracy largely intact, and always treated workers with suspicion. But at least, if workers and socialists conformed, they could potentially become a part of the national community. Jews and other racial minorities were eliminated immediately and entirely from all radio clubs and all other private associations not specifically reserved for Jews.

The views held by radio club members were naturally quite diverse. As a technical hobby, radio enthusiasts tended to be better educated than the average German. Many radio enthusiasts, like technology enthusiasts in general, were persuaded in the 1920s that radio would serve the cause of world peace and bring people together.[12] Some radio enthusiasts were close to the Esperanto movement or other, similar attempts to create a synthetic world language.[13] Many were quite open to the world and to foreign points of view. There were many formal and informal contacts between German radio hobbyists and those of other countries.[14] And foreign laws and developments were often held up by the German radio clubs

[12] For example, L. von Stockmayer, Vorsitzender des Oberdeutschen Funkverbands, "Die Tätigkeit und Organisation der Amateursender", *Der Radio-Amateur. Zeitschrift für Freunde der Drahtlosen Telephonie und Telegraphie. Organ des Deutschen Radio-Clubs* 3, No. 34 (August 21, 1925): 844–845. The importance and ubiquity of the idea that radio would serve to bring greater international understanding is underscored by the fact that von Stockmayer was a conservative, later a Nazi Party member and former Lieutenant-Colonel.

[13] See, for example, "Deutscher Ido-Radio-Klub", in the rubric "Klubmitteilungen", *Der Radio-Amateur* 2, No. 11 (July 4, 1924): 296–298.

[14] For example, see: W. Ehrenhaus, "Radiobrief aus Amerika", *Der Radio-Amateur* 2, No. 5, (April 16, 1925): 139–140, or Robert Wunder, "Über einen neuartigen Amateur-Empfänger", *Der Radio-Amateur* 2, Nr. 5 (28 May 1924): 195–198.

as examples for German lawmakers, particularly when the private possession of transmitters, or the right of private individuals to build their own radios was concerned.[15] Certainly, there was some fear on the part of the authorities (and sometimes shared by conservative radio hobbyists) that the radio hobby was filled with internationalist and leftist elements. And yet, within the middle-class radio clubs, and even among some members of the Arbeiter-Radio-Bund, there was a strong streak of German patriotism and even nationalism, which could not be denied, and which, if anything, grew stronger into the late 1920s and early 1930s. This middle-class German nationalism was not, in itself, necessarily a bad thing, or particularly different from the situation in other countries,[16] yet it was a precondition for the later Nazification of the radio hobby. It had many roots.

The Germans were likely no more and no less patriotic or nationalistic than other European or western peoples, though compared to other major powers, German patriotism could seem more shrill and brittle in its expression. Certainly, patriotism or nationalism were deeply embedded in the German middle classes well before the First World War and continued to be strong into the Weimar era. German engineers (like members of many other skilled professions), tended to be quite nationalistic, and to link their professional understand of themselves to an ethic of service to the German people and/or state.[17] Given that many German radio hobbyists came from the middle class, and given the strong presence of engineers in the radio hobby, it should be no surprise that nationalism was not absent among radio hobbyists. It could and did exist alongside the open and internationalist aspects of the radio hobby. Many radio hobbyists, as we have already seen, were also war veterans, another social group linked to both nationalism and a sense of national service.[18] Moreover, as we have already seen, radio, because of its strategic and symbolic nature, was par-

[15] Cand. Ing. Oswald von Bergen, "Radio in der Schweiz", *Der Radio-Amateur* 2, No. 5 (April 16, 1925): 140–141.

[16] The one major German peculiarity (to be sure, shared with other countries such as Austria and Hungary) was resentment at the post-First World War Versailles settlement. See above, Chap. 3.

[17] On the engineering profession in Germany, see Konrad H. Jarausch, *The Unfree Professions: German Lawyers, Teachers and Engineers* (Oxford: Oxford University Press, 1990) and Kees Gispen, *New Profession, Old Order: Engineers and German Society, 1815–1914* (Cambridge: Cambridge University Press, 2002).

[18] See the discussion of the influence on radio from the First World War in Chaps. 3 and 4 above.

ticularly influenced after 1918 by the trauma of the lost war. Yet none of this made Germany significantly different from any other European or Western nation, which had participated in the First World War. In short, patriotism or nationalism do not, by themselves, lead automatically to National Socialism, fascism, or any other political ideology, though the trauma of losing the First World War certainly made such developments more likely. The question is, rather, how did German nationalism or patriotism express itself within the radio hobby, and how was it manipulated and instrumentalized by political groups such as the Nazis? In the German case, nationalism quite clearly helped make National Socialism acceptable and successful. It was the prime vector for injecting National Socialist ideas into the radio hobby and other parts of civil society.

Naturally, radio hobbyists (like any other hobby group), identified themselves as a part of their hobby. Radio was even labeled a "movement" by its supporters in Germany (and elsewhere). But for most, this hobby identity existed as only one of many possible identities or personas, simultaneously jostling or competing within a given individual with other identifications. Radio hobbyists, by definition, claimed an identification with science and technology. Some radio hobbyists actually were scientists or engineers, and many others claimed to be their indispensable helpers through their engagement with the technology and science of radio. This was naturally both an international and a national form of identification, for science was universal. Yet "science" was not a neutral category in Germany (or elsewhere) and had a particular national significance. In the German Empire, science and scientists were seen as a source of national pride and greatness, and science was seen as having a national mission. German scientific advancements were perceived as quintessentially German and seen as advancing both human well-being and the prestige and power of Germany. Science was a major site for international competition in the nineteenth and twentieth centuries, and one where the Germans could take particular pride. Science was mobilized before and during the First World War for national defense, and in the Interwar period, it was common for thinkers in Germany to link the crisis in science with the crisis of the German state.[19] In short, science in Germany was an immanently

[19] In general, see: Ralf Jessen & Jakob Vogel, eds., *Wissenschaft und Nation in der Europäischen Geschichte* (Frankfurt, New York: Campus Verlag, 2003); Fritz Stern, *Einstein's German World* (Princeton: Princeton University Press, second edition 2016); Sören Flachowsky, *Von der Notgemeinschaft zum Reichsforschungsrat: Wissenschaftspolitik im*

nationalist domain. Moreover, the science of the radio enthusiasts was made social in the form of private associations and clubs, institutions, which were and are very strong and highly developed in Germany. The form is thus quite important, for middle-class private associations arguably played a crucial role in the development of a German-national identity in the early nineteenth century and became even more drenched with nationalism as time wore on. Thus, both the culture of German engineering and science and the culture of (middle-class) private associations in general were strongly and pervasively nationalist by the twentieth century, and this could not help but be reflected in the radio hobby. That the inherent nationalism of many radio hobbyists ultimately clashed with the international modernism of radio, and its strong connections with the outside world could be overlooked in the roaring twenties. The contradiction was harder to ignore by the early 1930s. Paradoxically, the transnational nature of the radio hobby actually fed particularistic German nationalism and German fears.

The Reichsverband Deutscher Rundfunkteilnehmer and the Polarization of the Public Sphere

In line with the growing conservative attack on the Republic, the content of state-controlled broadcast radio, the state *Reichs-Rundfunk-Gesellschaft* (Imperial Broadcast Radio Society) and the semi-private regional organizations, which controlled broadcast radio programming and policies, came under increasing scrutiny and criticism in 1929 and 1930. While the ostensible issue was a lack of suitable "national" content and charges of corruption, the real issue was control over broadcast radio itself. Conservative circles sought to dominate radio just as they sought to dominate the print press. In 1930, the so-called *Reichsverband Deutscher Rundfunkteilnehmer* (Imperial Association of Radio Subscribers; RDR) was formed by a coalition of conservative groups in the orbit of the nationalist media mogul Alfred Hugenberg. It was founded as a Registered

Kontext von Autarkie, Aufrüstung, und Krieg, Studien zur Geschichte der Deutschen Forschungsgemeinschaft 3, Edited by Rüdiger vom Bruch, Ulrich Herbert, and Patrick Wagner. (Stuttgart: Franz Steiner, 2008); Isabel Heinemann (et al), *Wissenschaft, Planung, Vertreibung. Der Generalplan Ost der Nationalsozialisten. Katalog zur Ausstellung der Deutschen Forschungsgemeinschaft* (Deutsche Forschungsgemeinschaft, 2006): 12. http://www.dfg.de/pub/generalplan/downloads/dfg_wissenschaft_planung_vertreibung_katalog.pdf, accessed May 15, 2012.

Association (*eingeschriebener Verein*) in August 1930.[20] The RDR soon published a weekly magazine to propagate its views and serve as a radio program, called *Der Deutsche Sender* (*The German Transmitter*). In the beginning, this was a German-Nationalist organization dominated by representatives from the Hugenberg media, the veteran's organization "Der Stahlhelm", and the German National People's Party. There was originally only a single National Socialist on the Governing Board.

The RDR seemed to respond to genuine concerns held by conservatives. Broadcast radio in Germany was tightly controlled by the state, and extremist views were limited (albeit more on the Left than on the Right). The very high-brow and didactic nature of most programming was a sore point for many radio listeners, who felt alienated by the focus on intellectual topics, and yearned for more light entertainment programming. And very many listeners simply felt alienated from the decision-making process about programming. In the beginning, the journal of the RDR stuck to a high tone and cast itself as the simple mouthpiece of patriotic and right-thinking Germans. In fact, this was a cynical populist ploy from the beginning, though the tone and the cynicism both got much worse as the RDR was taken over by the Nazis.

By the Fall of 1931, the *Nationalsozialistische Deutsche Arbeiterpartei* (NSDAP) decided it would launch an effort to take over the entire organization, and founded a "Club Syndicate of National Socialists" ("*Verbandsgruppe Nationalsozialisten*"), a sub-group of National Socialists within the larger organization, whose goal was to subvert the organization and take over its leadership.[21] It soon published its own monthly magazine, *Deutsch der Rundfunk* (*The Radio is German*). (Remember, this is well before the Nazis came to power.) The Nazi takeover of the RDR occurred at the yearly General Membership Meeting of the club on December 19, 1931. The Nazis by then had managed to obtain a majority of the members in their support and forced a reorganization of the Governing Board, which gave the NSDAP a controlling influence. The new Chairman of the

[20] Vereinsregister, Amtsgericht Charlottenburg, August 12, 1930, Nr. 6270. It was dissolved in 1936.

[21] Heinz von Fehrentheil, *Aus Geschichte und Organisation des RDR*. In: *Rundfunk im Aufbruch. Handbuch des deutschen Rundfunks 1934 mit Funkkalender* (Schauenburg: Lahr 1934): 12. This came in the context of a broader interest in radio and its potential for propaganda within the Nazi Party. For example, within the Political Organization of the Party, regional (*Gau*) and later local (*Kreis*) groups began to institute a new position of "Funkwart" (radio superintendent).

Board, and the two "*Geschäftsführende Vorstandsmitglieder*" (Governing Board members with executive authority for the running of the association) were now all National Socialists. Two members of the conservative veteran's organization, the *Stahlhelm*, were also made members of the Board, as was one member of the hard-line conservative German National People's Party (DNVP).[22] The legal and seemingly democratic by-laws of the RDR (like those of all Registered Associations in Germany) appeared to have been followed; the change in the membership of the Governing Board appeared to merely respect the democratic wish of the majority of RDR members present at the General Meeting. In fact, the RDR—to be sure, already a conservative-nationalist organization—was actually subverted and taken over by a small group of dedicated and disciplined Nazi Party members. This kind of apparently legal subversion and takeover would become a model for the later forced Nazification of all radio hobby clubs after the Nazi takeover of the government two years later. Democratic rules were cynically twisted to validate the Nazi takeover. Opposition was thus doubly difficult, for the appearance of democracy was maintained, while the Nazis claimed to speak in the name of common conservative and nationalist values shared by all RDR members. The preservation of a mixed Governing Board only seemed to prove that the club was still independent and expressed the will of all of its members. This was only a first step.

In March 1932, the Governing Board called a second General Meeting, and engineered a vote replacing the three non-Nazi board members with Nazis. At this same meeting, a new set of by-laws were adopted, which enshrined the Nazi "Leadership Principle", effectively the abandonment of democratic procedures of governance within the RDR. A second General Meeting that year, in October, saw Joseph Goebbels himself elected as the Chairman.[23] After this, there was no more need for the separate "Club Syndicate of National Socialists", and it was dissolved: the entire organization was now essentially Nazi. The RDR henceforth became an important organ of Nazi propaganda against the state-controlled broadcast radio system.

This is very important for the later fate of the radio hobby clubs. Via the ostensibly independent RDR, as well as in its open propaganda organs, the Nazi Party was able to hijack the debate about control over broadcast

[22] M. Weiß, "Warum Reichsverband Deutscher Rundfunkteilnehmer?", *Der Deutsche Sender. Die Nationale Funkzeitschrift* 3, No. 12 (1932): 2–3.

[23] Again, this model was repeated in all private associations in the Spring of 1933.

radio. In a characteristic style, the debate was sharpened, with wild charges of corruption and deliberate censorship of nationalists and patriots. The language of the debate was coarsened, with a rise in violent and military metaphors. And the decent, highbrow, and middle-of-the road popular content of broadcast radio was condemned in wild terms as "Marxist", "degenerate", "foreign", and "Jewish". The sharpening and radicalization of the debate over state-controlled broadcast radio carried over into the radio hobby clubs (whose members were all, by definition, radio listeners). In this fashion, the sharpening of the culture wars in Germany in the state sphere was transposed onto the private sphere of clubs and hobbyists, laying the ground for the later Nazification of the clubs in the Spring of 1933. The Nazi "Seizure of Power" and the Nazification ("*Gleichschaltung*") of German society and institutions would not have been possible without this ideological preparation in the late 1920s and early 1930s. The Nazis were not the initiators of this process—that dubious honor belonged to their erstwhile German-Nationalist allies. The Nazis were simply more ruthless and better able to use what their allies had started. By 1936, the state was fully in Nazi hands, and there was no more need for the RDR. It dissolved itself in 1936, and its print organ, *Der Deutsche Sender*, was taken over by the "Reich Radio Chamber" (*Reichsrundfunkkammer*) within the Reich Chamber of Culture, and renamed *NS-Funk*.[24] By this time, all the private radio hobby clubs had also disappeared.

In many ways, the internal Nazi takeover of the RDR in late 1931 and early 1932 resembles the later Nazification of the radio clubs in 1933. In both instances, Nazis and conservative sympathizers within the association used the governing laws in a cynical fashion to take over first the leadership, then to impose an authoritarian reorganization of the club structure, and finally to purge the general membership. Many of those who participated in this process within the RDR later were at the forefront of the Nazification of the clubs in 1933. By then, they already knew how to subvert a club from within. The model function of the Nazi takeover of the RDR cannot be denied.

The larger domestic political debate in Germany in general was radicalized from about 1929 or 1930 onward. The agitation of the RDR mirrored this national trend and was the agent for introducing the new tone into the world of radio. It gave a voice to many who were unhappy or

[24] Thomas Bauer, *Deutsche Programmpresse 1923 bis 1941. Entstehung, Entwicklung und Kontinuität der Rundfunkzeitschriften*, Rundfunkstudien 6 (Munich: Saur, 1993).

uncomfortable with the modern cultural forces heard in radio, and more importantly, it brought a radical, aggressive, and explicitly political element into the radio hobby. This gradually changed the tenor of how people thought of radio, and prepared the way for the later *Gleichschaltung*, by making many Nazi ideas and slogans seem closer to the mainstream and anchoring them as a discourse within the world of radio, rather than as an outside import.[25]

The passion and division provoked by discussions of broadcast content may seem strange today, but we should not be surprised. Political divisions in the late Weimar Republic were stretched to the breaking point, and anything could become a focus for these larger issues. Moreover, media are always very personal. This is certainly true of radio. Radio not only hit home, it came home. Remember that it bought its content into the most intimate spaces of the home in a way no other media form had done until then. From the beginning, radio also was a symbol of modernity and thus a sounding board for an individual's feelings about the modern world in general. Radio was a very personal medium, and thus, so was its content. There was great reason for the fierceness of the debate beyond any deliberate troublemaking on the part of the RDR. Any echo of our modern "culture wars" in the US or other contemporary societies is also no accident. We live in the same world; media are important and personal.

As the political conflict in Germany heated up in 1930, 1931, and 1932, both Right and Left increased their criticism of existing government-set policy regarding the content of radio broadcasts and demanded that their voices be heard more on the airwaves. This was a political demand, but also one that took up existing issues such as listener complaints about the price of a radio subscription (particularly for the unemployed), criticism of the highbrow nature of much of the programming content on offer, difficulties with interference from foreign stations, etc., in order to attract as much support as possible. These demands were both populist and elitist.

Complaints from the Left that their concerns were ignored in the official broadcast programming were often quite justified, for the principle that radio should remain "apolitical" meant, in fact, that working-class and Social Democratic voices were deliberately and systematically excluded,

[25]What cannot be assessed at this time in the context of the gradual nationalization of the radio hobby and radio discourse in general are the roles, which may have been played behind the scenes by the German military and military-industrial establishment.

in favor of a very mainstream form of state-affirming German-nationalist programming. On the Right, things were different, but dissatisfaction was still very real. Despite the actual conservative tone of much of German broadcasting, many nationalists felt that programming was insufficiently patriotic or too cosmopolitan (usually meaning that it was too "Jewish") or insufficiently regional. Internal divisions within Germany came to the forefront in the radical and violent Nazification of society in 1933.

The Nazi "Seizure of Power" and the Nazification of German Society, 1933–1934

On January 31, 1933, Hitler was named co-Chancellor of a coalition government of conservatives and Nazis. Over the next 18 months, the Nazis swept aside their conservative allies and established a dictatorship. The goals of National Socialism when it came to power in Germany were extremely ambitious, for the Nazis wanted nothing less than the complete Nazification of German society. While their first tasks in early 1933 were to expand their power base and destroy potential challengers to power, they did not wait, and immediately began to make more fundamental changes. Today, we tend to use the Nazis' own term, "*Gleichschaltung*", ("coordination") to describe this process of immediate power expansion combined with a more ambitious Nazification of society. The *Gleichschaltung* of the radio hobby was merely one small part of this larger series of events. It is important to study, not only due to the strategic nature of radio, but also because the hobby culture around radio was an important part of the web of private, civil associations or *Vereine*, which built so much of the public sphere in Germany.[26]

The Nazis wasted no time extending and consolidating their power and influence once they came to power in late January 1933. Their ambitions

[26] The *Gleichschaltung* is unevenly studied. While the seizure of the state apparatus and the forced "coordination" of the federal states are well known, comparatively little research has been done on the very wide range of private civil associations in Germany. The one major exception involves sports clubs, where the investigation is comparatively recent and ongoing. See: Lorenz Peiffer, *Sport im Nationalsozialismus: zum aktuellen Stand der sporthistorischen Forschung; eine kommentierte Bibliografie* (Göttingen: Verlag Die Werkstatt, 2004; Hajo Bernett, *Der Weg des Sports in die nationalsozialistische Diktatur* (Schorndorf: Verlag Karl Hofmann, 1983)); and most recently, Deutscher Alpenverein, Österreichischer Alpenverein and Alpenverein Südtirol, "*Berg Heil!" Alpenverein und Bergsteigen, 1918 bis 1945* (Vienna: Böhlau Verlag, 2011).

were huge, and they did not worry about legal niceties in doing so. Nazi ambitions were so great that it is impossible to write a book on radio or any other subject in Germany at the time without considering Nazi acts and plans. While the most pressing task for the Nazis and their allies was the takeover of state power and the disruption of those social forces, which could potentially mount genuine resistance, such as the socialist parties, Nazi efforts took place on a very broad front, for their ultimate goal was a profound reorientation of the German society. The Nazification of the radio hobby was merely one part of this larger process. Of course, having said that, radio, because of the strategic nature of the very technology, including the entire gamut from state-controlled broadcast radio down to the smallest of hobby clubs, was at the very center of the Nazi project. This was not because radio had been an important tool for the rise of the Nazis—quite the contrary, they had been largely kept off the airways due to the government insistence that radio be "apolitical". Instead, radio was a particularly important target for the Nazis because Goebbels saw its potential as a propaganda weapon and wanted to make very sure that his enemies did not have access to it.

"*Gleichschaltung*"[27] can be translated as "forced nazification of society", though we usually use the shorter term "coordination". It included both the forced absorption of National Socialist ideology, and the elimination of "enemies", both real and imagined. This seems straightforward, but at close inspection, it lacks firm contours. The term was first coined by the then National Minister of Justice, Franz Gürtner, in the Spring of 1933, who was the architect of much of the necessary legislation, even though he was not yet a member of the Nazi Party.[28] We generally use it to describe events from January 30, 1933, when Hitler was named Chancellor, down through the Summer or Fall of 1934. Yet in fact, it was an ongoing process which continued, really, down until the end of the Third Reich. Nazi ambitions were so big that Nazification was always really a moving target, which could never be met in reality.

The *Gleichschaltung* of amateur radio was a part of the much larger "coordination" and Nazification of the German society, which accompa-

[27]The term, like the related "*Machtergreifung*" ("Seizure of Power"), comes from the Nazis themselves, and is not unproblematic. Nevertheless, it is used here, partly for precision, since the English "coordination" is so broad.

[28]See Lothar Gruchmann, *Justiz im Dritten Reich, 1933–1940. Anpassung und Unterwerfung in der Ära Gürtner*, (Munich: Oldenburg, 3rd. improved edition 2001).

nied the Nazi accession to power. What happened with the radio hobby was very much in keeping with the same process in other parts of society. But it was not, for all that, a trivial or simple process. The stakes with radio were particularly high, for radio had both a great strategic importance, and was particularly close to the heart of Joseph Goebbels, who saw radio broadcasting as a key medium for propaganda. In the case of the radio hobby, *Gleichschaltung* included not only Nazification, but also a partial eclipse of the Reich Postal Ministry as the legal body in charge of radio, in favor of the new Ministry for Popular Enlightenment and Propaganda (*Ministerium für Volksaufklärung und Propaganda*, RMVP) as the key institutional governor of radio in all forms.[29] The struggle over ideology and power was also the struggle over bureaucratic control and influence between old and new elites, and between two powerful ministries.

The initial process of the Nazi takeover and Nazification of society was chaotic yet rapid. Though there was guidance from the top, much also was improvised locally and regionally. The Nazis entered national government as the ostensibly junior partner with mainstream conservatives, and only held a few (key) ministries. All parties to this coalition government shared the same basic goal of ending the democratic system; the Nazis were simply more radical and more focused on seizing power than their partners. In the few areas where Nazis gained direct control over important state institutions and ministries in the new coalition government of January 30, 1933, they rapidly consolidated their position and then used it to expand control to other areas, often with the help of emergency decrees.[30] Individual Nazi Party leaders also scrambled at all levels of government to gain government appointments and offices. This happened out of a mix of political and personal calculation, a typical feature of the Nazi system. The force of the *Gleichschaltung* rested on three pillars: the violence of the SA and allied paramilitary organizations,[31] the early control of pieces of the

[29] In fact, the ambitious Propaganda Ministry controlled all forms of communications media and arts: newspapers, books, radio, theater, and music.

[30] For example, Hermann Göring was named Prussian Minister of the Interior, giving him control over the police in the largest German state.

[31] At this time (and until July 1934), the SS was technically a part of the SA, albeit one with great autonomy. The SA was aided and abetted in its terror by members of the Stahlhelm (a right-wing veterans organization and paramilitary group with ties to the DNVP) and sometimes also by members of the Kyffhäuserbund (a veteran's group) and other right-wing paramilitary organizations, until they themselves were absorbed by the SA and disbanded. See: Stefan Hördler, ed., *Der SA-Terror als Herrschaftssicherung. "Köpinecker Blutwoche" und*

state apparatus (such as the Prussian Police) by the Nazis, and the broad-gaged and insidious pressure of thousands of individual Nazis and their sympathizers on the rest of society. We should also never forget that naked terror accompanied the entire process of Nazification. The chief objects of this terror at the beginning were parties and organizations on the Left, but even those who approved of the violence against the Left were themselves gradually intimidated by it. Once the Socialists and Communists were sufficiently weakened, the Nazis then turned on the liberal and conservative mainstream political parties, including, increasingly, their former allies.

Individual Nazis were already members of all sorts of organizations and clubs, and they immediately moved to solidify National Socialist influence over these organizations, but they could not have succeeded alone. Again, there was a mix of political calculation and personal ambition. The Nazi Party in 1933, though by contemporary German standards a huge mass political party the likes of which had never been seen before, was still comparatively small, and always lacked enough skilled and competent leaders and administrators in numbers commensurate with the Party's ambitions to control all of society. This meant that in the beginning, there were usually too few competent Nazi Party members to take over everywhere and everything. Instead, a handful of Nazis (often less competent, but always more ruthless than the alternatives) acted together with existing conservative elements within the radio clubs and organizations who were "acceptable" enough, or pliable enough, to work with the National Socialists in the takeover of their own organizations. Often, these were national conservatives who were persuaded of the necessity of a realignment of society in a direction, which made them Nazi allies.[32] Often, too, they were ambitious men and women who sought to profit from a change in circumstances and elites.

In nearly all cases (outside of the comparative handful of organizations which, for racial or ideological reasons, could not or would not be coopted), *Gleichschaltung* took the form of a combination of threats and positive enticements, the "stick" and the "carrot". And if the "stick" could

Öffentliche Gewalt im frühen Nationalsozialismus (Berlin: Metropol Verlag, 2013); Richard Bessel, *Political Violence and the Rise of Nazism: The Stormtroopers in Eastern Germany, 1925–1934* (New Haven: Yale University Press, 1984). See also: Michael Wildt, *Volksgemeinschaft als Selbstermächtigung. Gewalt gegen Juden in der deutschen Provinz 1919 bis 1939* (Hamburg: Hamburger Edition. 2007).

[32] See: Hermann Beck, *The Fateful Alliance: German Conservatives and Nazis in 1933. The Machtergreifung in a New Light* (New York, Oxford: Berghahn Books, 2008).

be very sharp, it was often the carrot, which was the most important. A little bit of terror went a long way, and was usually applied directly only to the Left, the Jews, and those who genuinely resisted. In most cases, a combination of ideological affinity, the offer of advantages to both the organization and individuals, and simple ambition were the most effective tools in the Nazi *Gleichschaltung* of the radio movement. For the mainstream radio clubs and hobbyists, the positive promise of Nazification lay in the hoped-for end to class conflict and the elimination of organized working-class interests. For many, it was a belief that the Nazis would restore national greatness, or at least put people back to work. Either way, the Nazification of the radio clubs—at least, the middle-class ones—was not simply a matter of coercion, there was plenty cooperation, too.

In many cases, the Nazi members of existing clubs and organizations and their conservative allies did not wait for Party orders, legal justification, or a central plan, they simply acted, particularly once *Gleichschaltung* began to gain momentum. Measures in one area (such as the removal of Jews from the civil service) served as signals and were quickly emulated in other areas. Sometimes these spontaneous actions had to be reined in by the Nazi Party leadership, but in most cases, they were simply allowed to happen, and the Party only worried later about managing the end result. The product of these dynamics was a savage and effective wave of little "seizures of power", which were difficult to predict and hard to resist. Moreover, the *Gleichschaltung* of the radio hobby was the product of both action and reaction: pressures came from both Nazis and their allies within the clubs themselves, and from outside actors. Clubs made preemptory changes that they thought would satisfy the new regime (in German, the term is "*vorauseilender Gehorsam*"), but were also forced to adopt a series of calculated measures demanded by the NSDAP and its supporters to seize control over all institutions of society, including private ones. As a result, it is often hard to determine the boundary of "spontaneous" and Party-mandated changes. The mix of plan and spontaneity, political and personal calculation in the larger process of *Gleichschaltung* should not be forgotten, particularly when it comes to private, non-political institutions.

The Larger Process of Gleichschaltung

In content, the goal of the NSDAP was an ideological, institutional, and organizational revolution of the German society. The Nazis wanted, of course, to seize control over any and all institutions which could be potential sites of resistance. They further wanted to imbue society with Nazi ideology and values, in a manner which would be fundamental and long-lasting. They argued that this was the only way to deal with what they said was a direct and dangerous socialist threat to Germany, one which liberal democracy was too weak, too corrupt, and too complicit to address. This meant removing all traces of electoral democracy, and their replacement with the "*Führerprinzip*" ("leadership principle"). It also meant an end to pluralist, confessional, and class-based organizations, and their replacement whenever possible with a single central umbrella organization, often on a corporatist basis. Put this crudely, it was a bald recipe for dictatorship, and that was exactly the end result. But at the time, it could seem quite attractive and desirable. For one thing, it promised greater efficiency and simplicity, objectives which could appeal to a fairly wide constituency beyond those who simple desired an end to class struggle and division, and who were tired of political conflict. The much evoked and often illusive national or racial community, the *Volksgemeinschaft*, was a very seductive goal and could seem quite desirable for those who were not a priori excluded from it. And for many clubs and organizations, the reorganization of their particular little corner of society could also seem to hold the promise of greater social and government recognition, a clear social role and identity, and a shaking up of old hierarchies which might conceivably bring new opportunities for personal advancement. This was certainly the case in the radio hobby.

Legally, a series of laws and decrees provided a framework for suspending the constitution and seizing control of the individual state and local governments, and is always cited when explaining the Nazi seizure of power. While they generally only concerned fairly limited areas of government and the law directly, they were applied much more broadly. They served as signals about the direction in which the government was moving and were therefore imitated and emulated in nearly all areas of society. The Nazis felt obliged to maintain the illusion of legality as far as possible in order not to risk their legitimacy, but they did not have any commitment to actually abide by the law. Instead, they used legal procedures to subvert the rule of law itself.[33]

[33] The important laws and decrees included:

The "*Verordnung des Reichspräsidenten zum Schutz von Volk und Staat*", also known as the "*Reichstagbrandverordnung*" (February 28, 1933, RGBl I, 83),

The "*Gesetz zur Behebung der Not von Volk und Reich*" (March 24, 1933, RGBl I, 141),

While this drumbeat of legislation left the Nazi Party in control of all important social and governmental institutions by the end of 1934 (with the possible exception of the churches and the military, both of which were also at pains to collaborate with the new regime and draw advantages from their cooperation), it was only the tip of the iceberg, and was, in many ways, more of a byproduct of events, rather than their trigger. Even more important than legislation was the unremitting terror on the street spearheaded by the SA and other paramilitary groups, and the actions of countless individual Nazis and sympathizers, acting individually and in groups. In many cases, changes were forced (or people were persuaded to make changes before force was applied) without legal basis. In the case of private associations like the radio clubs, this was certainly the case. None of the laws mentioned above applied to the radio clubs directly at all.[34] Instead, they served to set the tone and provided an example of the direction in which the regime intended to move. There was thus no specific legal basis for the *Gleichschaltung* of the radio clubs, but in the face of the avalanche of violence and aggression, which broke over German society in the Spring of 1933, that didn't matter at all.

The "*Vorläufiges Gesetz zur Gleichschaltung der Länder mit dem Reich*" (March 31, 1933, RGBl I, 153) and the "*Zweites Gesetz zur Gleichschaltung der Länder mit dem Reich*" (April 7, 1933, RGBl I, 225) (both of which only dealt with state and local administrations and parliaments)

The "*Gesetz zur Wiederherstellung des Berufsbeamtentums*", (April 7, 1933, RGBl. I, 175)

The "*Gesetz gegen die Neubildung von Parteien*" (July 14, 1933, RGBl. I 479) and the subsequent "*Gesetz zur Sicherung der Einheit von Partei und Staat*" (December 1, 1933, RGBl I, 1016), which left the NSDAP as the sole legal party,

and the disbanding of the unions and the creation of the German Labor Front (May 10, 1933, made law January 20, 1934, in the "*Gesetz zur Ordnung der nationalen Arbeit*" RGBL I, 45–56).

[34] Of all of these laws and decrees, the *Reichstagbrandverordnung* was the most important and most directly relevant to the Nazification of the radio clubs, for it suspended habeas corpus, allowing the terror to take place, and specifically allowed the government to suspend the laws of private associations (*Vereinsrecht*) and the freedom of assembly anchored in the constitution. In particular, its preamble was considered to be the legal basis for the dissolution of many clubs and associations during the Gleichschaltung. See: Klaus Vieweg, "Gleichschaltung und Führerprinzip—Zum rechtlichen Instrumentarium der Organisation des Sports im Dritten Reich", in *Recht und Unrecht im Nationalsozialismus*, edited by Peter Salje, 244–271 (Münster: Wissenschaftliche Verlagsgesellschaft Regensburg & Biermann, 1984). Here pp. 254–260.

The Takeover of the Radio Clubs (*Gleichschaltung* in Practice)

The *Gleichschaltung* of the radio hobby seemed to occur, in retrospect, in a systematic fashion, though at the time, the actual process was chaotic and confused: first Socialist and Communist radio clubs were immediately closed, and their assets seized. Next, leadership changes were imposed on the "middle-class" clubs, giving National Socialists and their allies de facto control of the board of directors, while Jews and progressives were eliminated from the leadership. These early changes, often involving only a few key members, were made at both the national and local levels. At this stage, the Nazis were careful to preserve an illusion of legality, so new by-laws were imposed on the clubs and duly voted into force by the members. They enshrined the Nazi "leadership principle" and the exclusion of Jews and other "enemies" in new by-laws, effectively imposing both Nazi control and Nazi ideology on the clubs. The national umbrella organization of radio clubs, the DFTV, was forced to merge with the RDR, giving the Propaganda Ministry complete control over the clubs.

Next, once the Nazis and their allies controlled the leadership of the clubs, they moved to purge the greater rank and file membership of Jews, socialists, democrats, non-conformers, troublemakers, and so on. They also forced the many clubs to merge into a more hierarchical and simplified structure, so that they lost their (by then only nominal) independence and became simply local branches of the RDR. Finally, the RDR became a part first of the official, state-controlled *Reichsrundfunkkammer*, and then later a part of the *Reichskulturkammer* (Reich Culture Chamber), weaving the once independent civil associations of the radio hobby deeply into the web of Nazi corporatist organizations under the larger umbrella of the Propaganda Ministry.

In the specific case of radio, the driving force behind its *Gleichschaltung* was above all the new Ministry of Popular Enlightenment and Propaganda.[35]

[35] The new Reich Ministry for Popular Enlightenment and Propaganda was created by presidential decree on March 13, 1933, led by Josef Goebbels, the Gauleiter of Berlin and NSDAP Party Propaganda Director. ("Erlaß zur Errichtung eines Reichsministeriums für Volksaufklärung und Propaganda", RGBl I, 104; see also "Verordnung über die Aufgaben des Reichsministeriums für Volksaufklärung und Propaganda" of June 30, 1933). In general, see Peter Longerich, *Joseph Goebbels: Biographie* (Berlin: Siedler, 2010) and Ralf Georg Reuth, *Goebbels* (Munich/Zürich: Piper, 1990). See also Elke Fröhlich, ed., *Die Tagebücher von Josef Goebbels* (Im Auftrag des Instituts für Zeitgeschichte und mit Unterstützung des Staatlichen Archivdienstes Rußlands) (18 vols.) (Munich: Sauer, 1998–2004).

Though the *Gleichschaltung* of radio should of course also be seen in the context of the Nazi turn against the mainstream liberal and conservative parties in the late Spring of 1933, its power as a propaganda medium and the depth of Goebbels's interest in it put it in a special category. It would be hard to find another, more ambitious and tenacious agent of Nazification than the Propaganda Ministry and its leadership. The new minister, Joseph Goebbels, understood the power of propaganda and had a particular interest in seizing the new media of radio. He was determined to take control of as much of the cultural apparatus as he possibly could. While broadcast radio was the main target, Goebbels and his assistants were clever enough to understand that the large community of radio hobbyists could also play an important role in the Nazification of radio, and represented a potentially disruptive consumer group if not managed effectively. Moreover, the idea that private individuals could build and maintain transmitters capable of virtually uncontrolled contact with foreign countries was just as frightening to the Nazis as it was for the state institutions of the Weimar Republic, and thus taking control over the shortwave amateurs (the group with the technical skills to do so) was also of great importance.[36] The radio hobby was made a direct target of *Gleichschaltung* and was included in the larger takeover of radio by the Propaganda Ministry.

Aside from the Propaganda Ministry, there were other outside actors in the Nazification of the radio hobby as well. They include the Reich Postal Ministry (declining in influence under the assault by the Propaganda Ministry but still strong), the Interior Ministry (the Police), and, in the particular case of the Deutsche Amateur Sende- und Empfangs-Dienst (DASD), the German armed forces, still a very powerful social force in the 1930s, but one which was compelled to act largely behind the scenes so long as the pretense of adherence to the Versailles Treaty had to be maintained. Both the Postal Ministry and the Armed Forces had their own, long-standing interest in radio, and were usually able to force Goebbels and the RMVP to consider their wishes, even if Goebbels was now able to take the lead in setting policy. Yet if Goebbels did not entirely get his way, this did not mean that the Nazification of radio was any less thorough. Neither the armed forces nor the Interior Ministry nor the Postal Ministry objected to the Nazification of radio per se, they simply wanted to ensure their own interests were respected.

[36] For example, SPD leaders in Mannheim were accused in 1933 of maintaining an illegal "red transmitter" and arrested. Herbert Hoffmann, *In Gleichschritt in die Diktatur? Die Nationalsozialistische "Machtergreifung" in Heidelberg und Mannheim 1930 bis 1935* (Frankfurt, Bern, New York: Peter Lang, 1985): 164.

The concrete process by which the radio clubs were Nazified in 1933 combined the appearance of "business as usual" punctuated with a series of sudden and abrupt changes. It would be highly useful if we had video tapes of the decisive membership meetings of the major clubs in the Spring of 1933, or at least, honest and complete minutes of the process whereby, willingly or unwillingly, the clubs Nazified themselves. Sadly, we do not have such information today, not only for the obvious reasons (there was no video tape in 1933), but due to the more basic fact that the Nazis were eager for the takeover of individual clubs to happen in secret and largely behind the scenes, to avoid disruption and potential resistance. Barring some future fortunate discovery of a cache of appropriate documents, we can only follow the process through the medium of the incomplete and potentially censored reports in the club section of the journal *Funk*. At this remove, the process of *Gleichschaltung* takes on an eerie, zombie-like quality, as great evil happened behind the veil of an outward normality, and the once dynamic radio hobby clubs become more like the living dead.

Into the early Spring of 1933, the news from the radio clubs gave the impression that nothing unusual was happening, but in fact, much was brewing below the surface. The Nazification of *broadcast radio* (as opposed to radio hobby clubs) had begun almost immediately after the Nazis came to power and served as a model for what happened with the radio clubs. As in other parts of civil society, Nazi efforts at the *Gleichschaltung* of the radio clubs began first with working-class and socialist institutions.

First, all the socialist or working-class clubs of any sort were immediately disbanded. First, in February 1933, came those associated with the communist Free Radio Union (Freier-Radio-Bund). Then, in July 1933, the socialist Workers Radio Union (Arbeiter-Radio-Bund) was also formally banned, though by then many of the members were already under arrest or lying low due to Nazi terror and attacks on the Social Democrats and leftists in general. In each case, the members of the working-class radio clubs were persecuted and kept under surveillance. Many, particularly communists, were also either murdered or arrested and sent to concentration camps. Property belonging to the clubs and often the personal radios of their members were seized and given to Nazi Party organizations. Though ostensibly seeking to end class divisions,[37] the NSDAP-led

[37] The members of the working-class radio clubs were encouraged, even expected to join the RDR after their own clubs were destroyed, but were subject to great violence and intimidation in the short term, and heightened surveillance and harassment in the long term.

social revolution in Germany left middle-class interests and the state bureaucracy largely intact, and always treated workers with suspicion. But at least, if workers and socialists conformed, they could become a part of the national community. Jews were eliminated entirely from all radio clubs and all private associations not specifically reserved for Jews. Of course, they were later eliminated from humanity altogether.

The *Gleichschaltung* of the radio clubs began in earnest for the middle-class radio clubs in April 1933. The first clear move in the direction of *Gleichschaltung* was the forced creation of a "working arrangement" ("*Arbeitsgemeinschaft*") between the Deutscher Funktechnischer Verband e. V. (DFTV) and the Reichsverband Deutscher Rundfunkteilnehmer (RDR).[38] This occurred sometime in the first half of April 1933.[39] Ostensibly, this was done in the name of unity and efficiency, with the RDR taking over responsibility for political questions, and the DFTV limited only to guidance on technical and scientific matters. In practice, it meant the takeover of the central umbrella organization of the radio hobby by the RDR, but in a way, which seemed to preserve as much normality and legality as possible. In reality, it was very clear that decisions were henceforth to be made by the RDR leadership, with its close ties to the Propaganda Ministry. As a "sign of the union of the two organizations", Wolf Ziegler, the Secretary (*Schriftführer*) of the RDR, immediately joined the Board of Directors ("*Vorstand*") of the DFTV.[40] Not that the leadership of the DFTV was particularly reluctant to move in a National Socialist direction: by 1933, it was made up of staunch German nationalists, most of whom continued to hold posts of responsibility later in the Third Reich.[41] This was the beginning of the end of the radio clubs as an

[38] Remember that the DFTV was the umbrella organization for the entire radio hobby, and was made up mainly of other associations of individual clubs, although some individual clubs belonged directly to the DFTV. It is easy to confuse the DFTV with the Funktechnischer Verein Berlin E.V. (FTV). The FTV was an umbrella association of smaller radio clubs, and was a member organization in the DFTV. The two also shared offices and had overlapping leadership, making them easy to confuse even for participants in the radio hobby at the time.

[39] "Arbeitsgemeinschaft zwischen "Deutschem Funktechnischen Verband" und "Reichsverband Deutscher Rundfunkteilnehmer"", *Funk. Die Wochenschrift des Funkwesens*, No. 16 (April 14, 1933), Hauptteil, 62.

[40] Ibid.

[41] "Die führenden Männer des Deutschen Funktechnischen Verbandes", *Der Deutsche Sender* 4, No. 16 (April 16–22, 1933): 6. Included (with photos) were: the President, Dr. Esau, Managing Chairman (*Geschäftsführender Vorsitzender*), Dr. Gehne, Treasurer (*Schatzmeister*) Oberstlt.aD. von Stockmayer, Secretary (*Schriftführer*) Prof. Dr. Leithäuser, and Manager (*Geschäftsführer*), Oberleutnant a.D. Zerlett.

independent scientific or political force. Though the new, combined umbrella organization promised to better focus the efforts and represent the interests of those attached to radio, it was, in fact, simply a tool for the expansion of Nazi power, and the radio clubs were soon to become a dead letter. The first step in the Gleichschaltung of the radio hobby was thus the seizure of the "commanding heights" of the main umbrella organization of all radio clubs, the DFTV.

Meanwhile, into April, the news from the individual clubs remained strictly routine. There were the usual periodic lectures on technical subjects, the holding of classes in radio science, and the usual social evenings ("gesellige Abende"), all part of the day-to-day activities of the radio clubs. The FTV in Berlin even announced a new working group on sound film. Yet the internal *Gleichschaltung* of the individual clubs began to get underway in early to mid-April, and lasted through July, though most of the changes happened in the second half of April and the first half of May 1933.

The Nazi desired to maintain the appearance (and only the appearance) of legality when Nazifying private associations and clubs, as in nearly every aspect of the *Gleichschaltung*, meant that it was necessary to force clubs to hold general membership meetings to vote on the new changes. This is due to the way in which private associations were organized in Germany. Radio clubs were organized according to the usual, formal German model for private associations. Most were "legally registered associations" ("*eingeschriebene Vereine*", e.V.), meaning that they were set up as formal, legal clubs with a written set of statutes and a board of directors or officers (*Vorstand*). Generally, the by-laws of registered associations required that the club officers/board of directors be elected yearly at a general membership meeting. This was a highly regulated, legalistic, and relatively democratic structure, and utterly anathema to the Nazis. They sought—and largely achieved—to replace this democratic structure with an appointed, Nazi-dominated board organized according to the Nazi "leadership principle". This, together with the expulsion of racially and politically "undesirable" elements, was at the heart of the process of *Gleichschaltung*. Obtaining the semblance of democratic legitimation for the end of internal democracy and independence for the radio clubs seems, today, to be just a particularly egregious insult. At the time, it played an important part in preventing open opposition to the expansion of Nazi control.[42]

[42] Klaus Vieweg, "Gleichschaltung und Führerprinzip", 256–260.

As a first step, club meetings were manipulated so that new, provisional club officers/boards of directors were appointed.[43] The sudden changes in leadership (and the beginnings of a purge of the membership) were followed slightly later by new by-laws, though the appearance of legality was always maintained to a greater or lesser degree by having the changes approved (usually retroactively), by a general membership meeting. In most cases, the first outward sign that something was happening was that the announcement of the usual yearly general membership meeting of a given club was made with greater than usual urgency.[44] Other clubs began to suddenly call extraordinary general membership meetings. Others simply urged as many members as possible to attend a coming regular meeting, because of "special announcements". Yet despite Nazi efforts to make the process seem normal, some club announcements, out of either enthusiasm or stubbornness, spoke frankly about "*Gleichschaltung*", even using the term.

Perhaps the first radio association to see changes in its officers was the Saar-Radio-Club. The Saarland was of course a politically sensitive place in the early 1930s, since it had been removed from Germany proper and placed under French administration by the Versailles Treaty; this may be why it was first, or it may simply have been because Nazis and Nazi-sympathizers in the club were in a hurry. Whatever the reason, on March 25, 1933, the Saar-Radio-Club held new elections for its governing board. After keeping the same Chairman and Treasurer since 1931 [at least], it suddenly replaced both officers. Only the Secretary (an official in the postal service at the relatively high rank of *Oberpostinspektor*) remained unchanged, but even here, positions for the "Second Secretary" and the "Second Treasurer" were now created.[45] A club member sympathetic with

[43] It is not clear who made these appointments, but if the pattern for the *Gleichschaltung* of other kinds of clubs holds here, then it was very likely local and regional NSDAP Party branches, and in Berlin, probably the RDR/RMVP directly.

[44] Since many (but not all) radio clubs coincidentally held their yearly general membership meetings in the Spring, these regularly scheduled meetings conveniently came at just the right time to provide an outwardly legal occasion for instituting a Nazi-dominated board, just as the *Gleichschaltung* of private institutions was developing steam. In other clubs whose regular meetings came later in the year (or had already occurred before April), however, special, extraordinary meetings had to be engineered in order to give a legal appearance to compulsory Nazification.

[45] "Saar-Radio-Club", "Aus den Vereinen", *Funk*, No. 14, (April 3, 1931), Programmteil, 2–4; "Saar-Radio-Club", "Aus den Vereinen", *Funk*, No. 17 (April 21, 1933), Programmteil, 2–3.

the Nazi Party was placed in charge of the club as First Chairman (1. *Vorsitzender*),[46] and flanked with new officers in an expanded board of directors packed with Nazis and their supporters. This was a model for what later took place in all the other radio clubs.

The Nazification of the FTV Berlin also came early in the process, both because it was in Berlin and thus close to the centers of power, and because of its close relationship with the DFTV. The FTV was the largest single piece of the DFTV, and was a huge umbrella organization of some 46 individual member clubs or local branches.[47] This was a hugely important part of the organized radio hobby, both because of its size, and because of its location in Berlin, the German capital and the most important single site of the German electrical industry. The *Gleichschaltung* of some of the FTV's member clubs and local branches came in early April, and that of the central office followed soon thereafter. The Nazification of the remaining FTV member clubs proceeded in a rolling fashion in April and May.

An April 11 meeting of the full membership of the FTV signaled the *Gleichschaltung* of the central leadership. The meeting was held ostensibly to discuss "cooperation" ("*Zusammenarbeit*") with the RDR. While the new, provisional board of directors was "accepted" at this meeting, it was decided to postpone a formal general membership meeting of the entire FTV until new by-laws could be worked out, probably by mid- to late May. The new, provisional board—in Nazi hands—was meanwhile given the job of running the DFTV "according to the guidelines issued by the National Revolution".[48] Meanwhile, the *Gleichschaltung* of the individual clubs continued, and gained momentum.

On April 6, the FTV "Gruppe Beuth-Schule" at the Berlin engineering school of the same name held a general membership meeting. After a short mention of the decisive meeting in the May 5 issue of *Funk*, the club

[46] It cannot be determined at the present if the new First Chairman, a Saarbrücken dentist named Schiffler, was a Nazi Party member in early 1933. The surviving records of the central Nazi Party membership file do not contain a card for Schiffler, but are incomplete.

[47] See the nearly full list in the May 26, 1933 issue of *Funk*. The FTV was the largest single organized part of the German radio hobby. Roughly half of the FTV was made up of club branches located in the large, Berlin-based electrical producers such as Siemens and Telefunken. Yet the member organizations of the FTV also reached far beyond Berlin as far away as Göttingen, in isolated cases. "Aus den Vereinen", *Funk*, No. 22 (May 26, 1933), Programmteil, 1–4.

[48] "im Sinne der durch die nationale Erhebung gegebenen neuen Richtlinien". FTV, "Aus den Vereinen", *Funk*, No. 19 (May 5, 1933) Programmteil, 1–4.

reported pompously in the May 12 issue that "In accordance with the nationalist currents which predominate within in our Group, the following board of directors, which in its totality stands firmly on a national foundation, was elected for the current semester".[49] This was not a club which seemed lost or unsure about how it should react, it is a joyful announcement that it stood behind the new regime.

In a similar fashion, the new, provisional board of directors of the FTV "Gruppe Siemens" called an extraordinary general membership meeting for April 12, 1933. After a moment of silence for a deceased club member, they spoke to the membership, emphasizing the absolute necessity of the fusion between the FTV and the RDR. The new leadership explained the path which the club would have to follow in the future, and said that all of the individual clubs making up the FTV would also have to place themselves on a "national foundation" ("*nationale Grundlage*"), as the national government had decreed. Ultimately, the meeting confirmed the provisional board of directors in their positions. There must not have been much discussion, however, for the entire meeting lasted only an hour and 15 minutes.[50]

The announcement by the FTV member-club "Gruppe Aronwerke" that changes were being made in its leadership was particularly blunt, possibly because it was one of the first radio clubs to be Nazified, or perhaps simply because it was, after all, in Berlin, which always had a reputation for blunt speech. The April 28, 1933, issue of *Funk* carried the announcement in a separate box on page 2, stating: "Due to the guidelines for the coordination of civic associations and clubs issued by the German National

[49] "*In Verfolg der in unserer Gruppe vorherrschenden nationalen Bestrebung, wurde für das laufende Semester nachstehend aufgeführter Vorstand gewählt, der in seiner Gesamtheit fest auf nationaler Grundlage steht*". FTV, Gruppe Beuth-Schule, "Aus den Vereinen". *Funk*, No. 20 (May 12, 1933), Programmteil, 1–4. The board of directors named in the announcement was larger than most, indicating the enthusiasm the club expressed for the new situation, though the key officers actually changed little. It is not surprising that the entire board "stood firm on a national foundation", since German student organizations had become pro-Nazi long before 1933. See: Geoffrey Giles, *Students and National Socialism in Germany* (Princeton: Princeton University Press, 1985), and Noyan Dinçkal, Christof Dipper and Detlev Mares, eds., *Selbstmobilisierung der Wissenschaft: Technische Hochschulen im "Dritten Reich"* (Darmstadt: Wissenschaftliche Buchgesellschaft, 2010).

[50] FTV, Gruppe Siemens, "Aus den Vereinen", *Funk*, No. 19 (May 5, 1933) Programmteil, 2–4. The reference to a decree by the national government was probably the "Reichstagsbrandverordnung" (Feb. 28) or the "Vorläufiges Gesetz zur Gleichschaltung der Länder mit dem Reich" (March 31).

Government, the entire board of directors of the club have resigned. Herr Muche has been named provisional First Chairman. Herrn Zillmann, Franke and Rex have been confirmed in their former positions on the board".[51] While this is unusual in its frank reference to government guidelines, it was a fair description of what was going on, and might well have come from any radio club in April or May of 1933. Note in particular that the main change at first was simply the replacement of the club Chairman (*Erster Vorsitzender*) with a pro-Nazi successor.[52] Other members of the board were kept in place, at least for the beginning. This was not at all unusual; in many cases, some or all existing board members were already either Nazis or German-Nationalists. In other cases, as many of the officers of the former board as possible were kept simply as a fig leaf. It was enough to have a committed, forceful National Socialist in charge, who could ram through a handful of strategic changes (usually together with a few German-Nationalist or National Socialist allies in the club). Other parts of the FTV followed the same pattern, and were simply a little more circumspect. Two other member clubs (the "Gruppe Lorenz" and the "Gruppe Berlin Lichtenberg") simply announced "special meetings" in the same April 28 issue. One added ominously that the meeting was necessary "since announcements of decisive importance will be made on this day".[53]

[51] "*Auf Grund der von der deutschen Reichsregierung herausgegebenen Richtlinien zwecks Gleichschaltung der Vereine und Verbände ist der Gesamtvorstand der Gruppe zurückgetreten. Zum kommissarischen 1. Vorsitzenden wurde Herr Muche ernannt. Die Herrn Zillmann, Franke und Rex sind zur Weiterführung ihrer bisherigen Ämter bestätigt worden.*" "Aus den Vereinen und Verbänden", *Funk*, No. 18 (April 28, 1933), Programmteil, 1–4 (here: p. 2). It is not known exactly what the term "*Richtlinien*" (guidelines) used in this announcement refers to. Similar but slightly different terms appear in other announcements, and may refer either to some, now missing formal set of guidelines, or, more likely, simply to the "*Reichstagsbrandverordnung*" (Feb. 28) or the "*Vorläufiges Gesetz zur Gleichschaltung der Länder mit dem Reich*" (March 31). Note that these last two laws did not technically apply to private clubs, but were nevertheless clearly taken as signals of what the regime intended.

[52] The fact that he was termed a "provisional chairman" ("*kommissarischer 1. Vorsitzender*") in the announcement means that he was appointed without a vote of the entire membership at a general membership meeting. Presumably, the appointment was forced by outside pressure, either by the local NSDAP, the company, or even directly from the RMVP.

[53] FTV local branches "Gruppe Lorenz" and "Gruppe Berlin Lichtenberg, which also announced a change in the location of its office (suggesting a change in personnel on the board of directors), and said that the meeting was necessary because of a special announcement: "*da an diesem Tage Mitteilungen von entscheidender Bedeutung bekanntgegeben werden.*" Aus den Vereinen und", *Funk*, No. 18 (April 28, 1933), Programmteil, 1–4.

Events outside of Berlin followed the same pattern. In Halle, the "Funkvereinigung Halle e.V." was combined with the local chapter of the RDR by early May, and a new board of directors was appointed.[54] Similarly, the board of directors of the "Triersche Funkgesellschaft e.V., Trier", resigned in May as a group "in order to make a renewal or re-organization possible". An extraordinary membership meeting was scheduled for May 31.[55] Again, we see a pattern of little resistance to the decisive changes in leadership. Instead, some clubs tend to be simply at a loss as to how to respond in the face of the determined takeover by the National Socialists, while others enthusiastically support it—at least in public.

The now Nazi-dominated combined DFTV/RDR, as the national umbrella group for all radio clubs, apparently told its local member clubs that it would issue a new set of model by-laws which could be expected in early to mid-May.[56] But since the coordination of the individual local and regional clubs continued before they were ready, there was some initial confusion, and a great deal of disruption. The "Gesellschaft der Funkfreunde e.V. Hannover", in particular, seemed to drift for several weeks. It was clearly waiting to learn what was expected by the new by-laws, announcing to its members: "Until the further shape of the DFTV is cleared up, there will be no further regular meetings. To orient our members, there is an informal meeting every Wednesday evening in the clubhouse in the Weberstrasse. In early May, the guidelines under which we will have to work will be issued by the RDR on the one hand, and the DFTV on the other hand. We ask our passive members to come to the clubhouse on one of the coming Wednesday evenings, so that they can be appraised of the new orientation ("*Gleichschaltung*") of our club".[57] The

[54] Ortsgruppe Halle, "Ortsgruppen Nachrichten", *Funk und Bewegung* 1, No. 4 (May 1933): 8.

[55] "Triersche Funkgesellschaft e.V., Trier", "Aus den Vereinen", *Funk*, No. 21 (May 19, 1933) Programmteil, 1–4.

[56] FTV, "Aus den Vereinen", *Funk*, No. 19 (May 5, 1933) Programmteil, 1–4.

[57] "*Bis zur Klärung der weiteren Gestaltung des Deutschen Funktechnischen Verbandes finden keine regelmäßigen Veranstaltungen statt. Zur Orientierung der Mitglieder ist jeden Mittwoch eine zwangslose Zusammenkunft im Klubheim, Weberstraße. Anfang Mai werden die neuen Richtlinien, unter denen wir in Zukunft zu arbeiten haben, vom RDR. einerseits und vom DFTV. andererseits bekanntgegeben. Wir bitten auch unsere passive Mitglieder, an einem der nächsten Mittwochabende im Klubheim zu erscheinen, um sich mit der Neuorientierung unserer Gesellschaft (Gleichschaltung) vertraut zu machen*". "Aus den Vereinen", *Funk*, No. 19 (May 5, 1933) Programmteil, 2–4.

invitation to an informal orientation meeting in the clubhouse continued to appear in *Funk* for the next several weeks,[58] though in the following week, the announcement contained the bizarre addition: "A discussion among the active members of our club, the [sic] unanimously expressed its joy that finally a consortium of all true friends of the radio sciences is coming together in Hannover under decisive leadership, in a way to which our new board of directors has aspired from the very beginning".[59] The original German is just as tortured and painful as this English translation, and indicates some mix of submission, coercion, resistance, or distress. A week later, the club was still drifting, this time announcing that, since the preparations for the fusion of the FTV and the RDR had been delayed, the holding of new radio classes for beginners would be suspended for the summer.[60] Another week later, and the club announced somewhat plaintively that the regular general membership meeting (a legal requirement in the by-laws), which should have been held in the Spring, had been postponed until the question of the club's *Gleichschaltung* was been resolved. It said that the new guidelines were now expected to arrive at the end of the month (May), and that the meeting could be expected in mid-July.[61] Nor was this the only club placed in disarray by the lack of firm guidelines from above. The FTV "Gruppe Siemens" was forced to announce in late May that the next regular monthly meeting and the election of a permanent board of directors could not be held until the new by-laws arrived. The club announced that it wouldn't be able to hold any membership

[58] Gesellschaft der Funkfreunde e.V. Hannover, "Aus den Vereinen", *Funk*, No. 20 (May 12, 1933), Programmteil, 1–4; Gesellschaft der Funkfreunde e.V. Hannover, "Aus den Vereinen", *Funk*, No. 21 (May 19, 1933), Programmteil, 1–4; Gesellschaft der Funkfreunde e.V. Hannover, "Aus den Vereinen", *Funk*, No. 22 (May 26, 1933), Programmteil, 1–4; Gesellschaft der Funkfreunde e.V. Hannover, "Aus den Vereinen", *Funk*, No. 23 (June 2, 1933), Programmteil, 1–4; Gesellschaft der Funkfreunde e.V. Hannover, "Aus den Vereinen", *Funk*, No. 24 (June 9, 1933), Programmteil, 1–4.

[59] Gesellschaft der Funkfreunde e.V. Hannover, "Aus den Vereinen", *Funk*, No. 20 (May 12, 1933), Programmteil, 1–4. "*In einer Besprechung der aktiven Mitglieder unserer Gesellschaft wurde einstimmig der Freude Ausdruck gegeben, daß endlich in Hannover unter planvoller Führung eine Arbeitsgemeinschaft aller wahren Freunde der funktechnischen Wissenschaften, wie sie von unserem neuen Vorstande von Anfang an erstrebt wurde, zustande kommt*".

[60] Gesellschaft der Funkfreunde e.V. Hannover, "Aus den Vereinen", *Funk*, No. 21 (May 19, 1933), Programmteil, 1–4.

[61] Gesellschaft der Funkfreunde e.V. Hannover, "Aus den Vereinen", *Funk*, No. 22 (May 26, 1933), Programmteil, 1–4.

meetings or public talks until then, either.[62] But only a week later, it saw itself forced to announce that, contrary to the earlier message, the club would hold "yet another" general membership meeting on June 14. It further stated that, "should the new by-laws from the main club arrive by then", they would be announced at the meeting.[63] The frustration of even the new, "national-minded" club officers was apparent.

The poor "Gesellschaft der Funkfreunde e.V. Hannover" must have finally received their new by-laws in June, for it finally announced late in the month that the long-postponed general membership meeting would be held on July 8. The agenda would be the "*Gleichschaltung*" of the board of directors.[64] Several other clubs followed suit at roughly the same time, and announced meetings to elect new boards of directors and approve the new by-laws.[65] The Verein Ostdeuscher Funkfreunde could even announce in the same issue that it was able to resume teaching the Morse code classes, which had been suspended earlier "due unclear conditions".[66]

A terse news item in the *Berliner Börsen-Zeitung* (*Berlin Stock market News*) gives the rest of the story.[67] In late June, the "National Socialist Radio Supervisors (*NS-Funkwarte*)" occupied and took over the offices of the DFTV and all its member clubs in a concerted national action. Ostensibly, this was because the DFTV had worked together with the now banned ARB, but this was just an excuse. The radio clubs were now all firmly under Nazi control. The Nazification of the clubs could now accelerate.

With the arrival of new by-laws, and with the governing boards of all the clubs now safely in "national" (Nazi) hands, a blanket of false normality

[62] FTV Gruppe Siemens, "Aus den Vereinen", *Funk*, No. 22 (May 26, 1933), Programmteil, 1–4.

[63] FTV Gruppe Siemens, "Aus den Vereinen", *Funk*, No. 23 (June 2, 1933), Programmteil, 1–4.

[64] The term "Gleichschaltung" was used publicly. Gesellschaft der Funkfreunde e.V. Hannover, "Aus den Vereinen", *Funk*, No. 27 (June 30, 1933), Programmteil, 1–3.

[65] For example, the local branch of the Saar-Radio-Club, in Neuenkirchen, or the Radio-Vereinigung Leipzig e.V. of the Mitteldeutscher Funkverband. See their entries in: "Aus den Vereinen", *Funk*, No. 27 (June 30, 1933), Programmteil, 1–3.

[66] Verein Ostdeuscher Funkfreunde, "Aus den Vereinen", *Funk*, No. 27 (June 30, 1933), Programmteil, 1–3.

[67] "Besetzung der Geschäftsstellen des Funktechnischen Verbandes", clipping from the *Berliner Börsen-Zeitung* Nr. 290 of June 27, 1933, in: BArch R/1501/20404 St10 RMdI KPD-Radio-Propaganda Bd. 5 March 1933–March 1934, 79.

descended over the clubs in public in the late Spring of 1933. Clubs continued to announce regular activities for a time. For example, in the same April 28 issue of *Funk*, which presaged the Nazification of the clubs, the "Funkverein Wannsee" within the FTV announced it would hold a "*Funkfahrt ins Blaue*" (a "radio trip into the blue") to celebrate its ninth anniversary.[68] It was one of the last announcements made of the club's activities. Increasingly in the Spring and into the Summer of 1933, only their contact addresses and regularly scheduled meeting times are given,[69] and gradually, there is no more news of any sort. Most clubs didn't even bother to announce the final adoption of new by-laws. Silence became the rule. The middle-class normality of the club world suddenly became a jarring fig leaf for the Nazification of German society. Once the dirty work of Nazification was begun and the old democratic structure had been replaced, the outward shell of the club structure was maintained through the end of the summer of 1933, and everything seemed to be just like before. But by the early summer, the radio clubs were zombies, robbed of their former independence and free will, and simply going through the motions.

By the end of summer 1933, the radio clubs as independent, apolitical organizations part of the larger German public sphere of private associations were dead, even though the DFTV continued to exist as an appendix to the RDR.[70] As of September 1, 1933, the RDR announce that, henceforth, all local chapters of the combined RDR/DFTV would simply be called "*Kreisgruppen des RDR und des DFTV*" ("County Groups of the RDR and the DFTV"). Moreover, all such "*Kreisgruppen*" ("county groups") required confirmation by the RDR leadership, meaning that those which had not been fully "coordinated" could be dissolved. This was done in the name of "simplification of our organization" ("*Vereinheitlichung unserer Organisation*"), but it is clear that this step simply tightened the RMVP's control over the radio clubs, and robbed them of any independence outside of the structure it had created.[71] Typical for the process of *Gleichschaltung*, it also simply made uniform a process

[68] "FTV Funkverein Wannseebahn", in the rubric "Aus den Vereinen", *Funk*, No. 18 (April 28, 1933), Programmteil, 1–4.

[69] *Funk* (1933).

[70] As we will see below, the one exception to this was the fate of the "Deutsche Amateur Sende- und Empfangs-Dienst" (DASD), the small part of the radio hobby interested in actual transmission on the shortwave bands (what we in the US call "ham radio").

[71] "Kreisgruppen-Nachrichten des RDR und DFTV", *Funk und Bewegung* 1, No. 10 (October 1933): 11–12.

which was already underway locally for several months, whereby multiple local and regional radio clubs were combined into a single one. The final step in the concentration and hierarchization of the radio clubs came in November 1933. At this time, the *Reichskulturkammer* (Reich Culture Chamber) was created within the RMVP, with the *Reichsrundfunkkammer* (Reich Radio Chamber) as one subdivision of it. The RDR/DFTV became a formal part of the Reichsrundfunkkammer, completing the absorption of the radio hobby into the State/Party media landscape.[72]

By early 1934, a monolithic radio organization was in place, which was completely, and largely behind the scenes, in Nazi hands. Gradually, the radio clubs were all absorbed into the RDR/DFTV directly and quietly closed, save for the DASD, with its strong protectors in the military and indispensable technical abilities (see below). The final step in the *Gleichschaltung* of the radio hobby came in the summer and early Fall of 1933 with a through purge of the membership of the "Kreisgruppen". As of May 1, 1934, the *Reichspropagandaleitung* announced a membership moratorium for all *Kreisgruppen* of the RDR and DFTV.[73] It was in effect until November 1, 1934. The announcement ominously states that the moratorium will allow the clubs to better school their members,[74] but a later report of a meeting of the Nazi-Party "Funkwarte" (Nazi Party Radio Supervisors) in Gau Berlin speaks even more frighteningly of the moratorium as a time for the "verification and thorough reorganization of the membership".[75] This was a final opportunity to thoroughly weed out "unreliable elements" such as democrats and internationalists, and the few remaining socialists and Jews. The DFTV was later formally disbanded. The *Gleichschaltung* of the radio hobby was now complete.

In the change of year from 1933 to 1934, all separate news and announcement of DFTV member clubs in the major radio journal *Funk* disappear, though the journal continued to be published. Behind the scenes, they were all absorbed into the RDR/DFTV directly and quietly closed, save for the DASD. Though Lothar Brand remained titular editor

[72] "In der Werkstatt der Politik", *Funk und Bewegung* 1, No. 12 (December 1933): 7.

[73] "Mitgliedersperre ab 2. Mai", *Funk und Bewegung* 4, No. 5 (May 1934): 1.

[74] *Funk*, No. 19 (May 4, 1934).

[75] *Funk*, No. 25 (June 15, 1934): 4; "Nachprüfung und Durchorganisierung des Mitgliederbestandes", in Gaufunkwart and Intendant Kriegier, Breslau, "Die Verbandsorganisation. Rede im Schulungslager Berlin", *Funk und Bewegung* 4. No. 10 (October 1934): 5–7.

of *Funk*, the journal became a mouthpiece for the radio interests of the Reichs Propaganda Ministry. The program section and the weekly "do-it-yourself" supplement were dropped as of January 1, 1934. In 1934, the redesigned journal contained two parts: a front section increasingly full of NS propaganda in the form of official announcements, speeches, and reports, and a longer technical section, replacing the "do-it-yourself" supplement. The contract between pure propaganda and pure technology is striking, the face of radio becomes a bipolar monster of rabid NS propaganda and robotic technological normality. Only the separate DASD supplement remained to testify to the once heterogeneous and effervescent radio club landscape.[76]

What brought the diverse and vibrant radio hobby to this point? How could this happen, given that in most cases, it can be assumed that actual Nazi party members must have been in the minority? The middle-class patriotism and conservative political beliefs (often including a diffuse anti-Semitism) held by many educated, middle-class Germans undoubtedly motivated some to go along with the new national government and its program, sometimes enthusiastically, sometimes reluctantly. The wave of expulsions and reorganization brought new men to the leadership of the clubs, and many must have believed the promise that the Nazi-led reorganization would genuinely make the representation of hobby interests more effective. Many others, of course, acted out of opportunism. The elimination of the left-wing Arbeiter-Radio-Bund, and even more, the purge of Jewish members of the other radio clubs, was both a crime and a calamity, and must not be minimized. But it came with a double dose of perceived opportunity for the remaining, non-Jewish, non-leftist, non-oppositional, non-democratic and patriotic radio hobbyists. Opportunity for advancement within any given club was suddenly present, as was the opportunity to change and shake-up existing structures. In this sense, the Gleichschaltung brought a potential for modernization of what were often very Wilhelmine club structures, albeit in a populist and anti-democratic form. But more importantly, opportunity was created for personal, professional, and organizational advancement within the structure of a state which was willing to provide support—at the cost of strict control. For

[76] In 1935, the format was changed again, this time to focus almost exclusively on technical aspects of radio. Once radio was firmly in NS hands, there was no need for open propaganda in *Funk*. An insert entitled "Hitler Youth Radio" appeared between 1937 and 1939.

many in 1933, this was an attractive proposition. Personal gain or perceived benefit to the hobby could easily outweigh conscience, particularly within a context of legalized state terror. Others, of course, acted out of fear, though the promise of benefits to the hobby and the nation could help individuals rationalize their acts. Those without political experience who lacked the backing of an organization were, in any case, helpless in the face of the aggressive and determined minority of National Socialists and sympathizers who intended to force a reorganization of the clubs, and who did not stick to the rules. We can't forget the larger context: outside the doors of these comfortable and civilized clubs there was real terror and violence on the streets. No one in Germany in 1933 would have been ignorant of the fact that open opposition to the new regime might come at a real personal cost to life and limb. Nor did the threat of violence go away. More or less open threats of violence to the uncooperative were made by the leaders of the RDR and the Propaganda Ministry in speeches and published writings in the radio press throughout 1933 and 1934.[77]

The *Gleichschaltung* of the radio hobby was quite similar to that of other civil associations, with two major differences. First, the Propaganda Ministry and its head, Josef Goebbels, were powerful actors in the new regime who placed great importance on dominating all aspects of radio. Second, the RDR already existed, and was an effective tool for the Nazification of the radio hobby. It has no parallel in the *Gleichschaltung* of most other civil associations. The strident RDR began well before 1933 to move the tone and nature of the discourse on radio in a Nazi and nationalist direction, and then provided the perfect vehicle for seizing actual control over and centralizing the radio clubs, ostensibly in the name of "rationalization". The combined result of all these forces destroyed the rich and active ecology of radio hobby organizations in Germany.

Expulsion of Jews

A separate word must be said about the expulsion of Jews from the radio clubs. The Nazis were an explicitly anti-Semitic party, and their German-Nationalist partners were often not far behind. The exclusion of Jews from German society was a fundamental goal of the Nazi "national

[77] For example, see: Alf Krüger, "Den Unbelehrbaren!" (poem), *Funk*, No. 17 (April 20, 1934): 1. The poem openly threatened the "Unbelehrbaren" with a visit to a concentration camp.

revolution". Characteristically, the first phase of the *Gleichschaltung* combined formal legal measures against (some) Jews, particularly in state institutions and important professions[78] with "spontaneous" terror against Jews in general, culminating in the April 1, 1933, boycott of Jewish businesses and professionals. While the April boycott was a failure, the open SA terror against Jews, combined with the (still fairly limited) first legislation specifically aimed at Jews served to send a signal, which sparked further measures elsewhere, such as within the radio clubs. Anti-Jewish measures were enthusiastically taken up at all levels, often without legal justification, and often sometimes even against the wishes of higher levels of state and Nazi Party leadership.[79] Yet as in many other National Socialist measures of the time, the very insidiousness and scale of the measures taken makes them hard to document in detail. Beginning with members of the leadership of the radio clubs, prominent Jewish members were expelled at the same time as new club boards of directors were appointed in April or early May 1933. Often at the same time, and if not, then within the next several months, the so-called *Arierparagraph*,[80] modeled on the "Law for the restitution of the German Civil Service" ("*Gesetz zur Wiederherstellung des Berufsbeamtentums*"), was introduced into the by-laws of all clubs and associations, excluding "non-Aryan" or "non-German" members.[81]

[78] Above all, the "Gesetz zur Wiederherstellung des Berufsbeamtentums", April 7, 1933, RGBl. I, 175.

[79] This went so far that in September 1933, Martin Bormann was forced to send a strongly worded circular to all Nazi Party *Gauleitungen* (regional offices), ordering an end to spontaneous measures against Jews and a return to the status quo. Rundschreiben des StdF/ Stabsleiter, gez. M. Bormann, München, an alle NSDAP-Gauleitungen vom 12.9.1933 (Abschrift) BArch NS 6/215 Bl. 7, reprinted in: Aly et al., eds. (Im Auftrag des Bundesarchivs, des Instutits für Zeitgeschichte und des Lehrstuhls für Neuere und Neueste Geschichte an der Albert-Ludwigs-Universität Freiburg) *Die Verfolgung und Ermordung der europäischen Juden durch das nationalsozialistische Deutschland 1933–1945.* Vol. 1: Wolf Gruner, ed. *Deutsches Reich 1933–1937.* (Munich: R. Oldenburg Verlag, 2008): 242, Document 76.

[80] Literally "Aryan paragraph"; this was the clause in the by-laws of private associations (or other legal document), which specified that Jews could not be members.

[81] Note that in some sports, notably some gymnastics or mountain-climbing clubs, Jews were expelled immediately after the First World War. Lorenz Pfeiffer, "auf zur Gefolgschaft und zur Tat!" Deutsche Turnerschaft und Nationalsozialismus—zwischen Selbstgleichschaltung und Selbstbehauptung?", *IWK Internationale wissenschaftliche Korrespondenz zur Geschichte der Deutschen Arbeiterbewegung* 35, No. 4 (1999): 530–548; Deutscher Alpenverein, Oesterreichischen Alpenverein, Alpenverein Südtirol, *Berg Heil! Alpenverein und Bergsteigen, 1918 bis 1945* (Cologne: Böhlau, 2011).

Many Jewish and other now "undesirable" left-wing and democratic members simply resigned from their club before such measures could come into effect. Since membership lists from the clubs no longer exist, and since members were in any case not formally listed by race or political belief, it is very difficult to determine the degree to which Jews were members of the radio clubs before 1933, and to which they were expelled or compelled to leave in 1933 and 1934. This is compounded by the fact that the clubs themselves tended to use euphemisms to describe what was going on. It was never or rarely said that Jews (pacifists, homosexuals, internationalists, socialists, etc.) were to be expelled; instead, clubs spoke of removing "non-German" or "non-nationalist" elements. In fact, little more may be said with certainty about the details of the exclusion of Jews and other racial minorities (and, for that matter, of politically undesirable members as well) from radio clubs at this time, other than that this was a specific purpose of the *Gleichschaltung* and that it did, indeed, happen everywhere, on a massive scale. The situation in sports clubs is slightly better documented today and can help illustrate what went on in other private clubs, such as those for radio.[82] In general, some sports clubs acted well before they were forced to, others waited until absolutely necessary. Thus, in keeping with the long-standing nationalism within German gymnastics, the large gymnastics association "Deutsche Turnerschaft" asked its Jewish members on April 9, 1933, to resign. This came just a day after the proclamation of the "Law for the restitution of the German Civil Service" was issued. When voluntary resignation did not bring the desired result, the association soon began to expel both individual Jewish members and entire Jewish member clubs.[83] On the other hand, many soccer clubs (which often tended to have a strong working-class element) only removed Jewish or socialist members from their public leadership, and didn't pursue the expulsion of individual Jewish members too seriously until they

[82] Markwart Herzog, *Der "Betze" unterm Hakenkreuz. Der 1.FC Kaiserslautern in der Zeit des Nationalsozialismus* (Göttingen: Verlag die Werkstatt, 2006): 68–95, specifically compares the Gleichschaltung of soccer clubs to that of music associations. In general, see Hajo Bernett, "*Der deutsche Sport im Jahre 1933*", *Stadion* 7 (1981): 225–283 and *Weg des Sports in die nationalsozialistische Diktatur. Die Entstehung des deutschen (nationalsozialistischen) Reichsbundes für Leibesübungen* (Schorndorf: Hofmann, 1983).

[83] "Jüdischer Sport im nationalsozialistischer Deutschland", part 3, "Der "Arierparagraph im deutschen Sport", http://www.s-port.de/david/ns/hist_03.html, accessed March, 2012. Note that the Deutsche Turnerschaft was a particularly nationalistic and even völkisch-oriented organization, and had been since the latter German Empire.

were forced to adopt a model set of by-laws proposed by the German Soccer Association (Deutscher Fussball Bund) and the National Sports Authority (*Reichssportführung*) in 1934.[84] In some cases, in order not to disrupt the coming Olympic Games, some clubs were even ordered to keep Jewish members temporarily.[85] Similar to what happened with the radio clubs, the profusion of confessional, local, and regional sports clubs was ended, and most sport types were eventually concentrated under one large and centralized umbrella organization. These, in turn, came under the control of the National Sports Authority (*Reichssportführung*).[86]

The DASD: German "Hams" as a Special Case

The experience of the *Gleichschaltung* of the small part of the radio hobby interested in transmitting is a special case. Because of its international reach and contacts, and because of the advanced technical skills of its members, the DASD, the organization of those interested in shortwave transmission (the "shortwave amateurs" or "hams") within the DFTV, followed a significantly different trajectory from that of the other parts of the radio hobby. It was certainly subject to the same sort of Nazification as the other clubs. The same combination of internal and external pressure was present. But for the DASD, both inward and outward pressures were stronger, for there was more at stake, even though the DASD was such a comparatively small part of the radio hobby. Two elements contributed to this, and caused the DASD to be able to maintain a form of autonomy through the end of the Third Reich as an ostensibly private organization. The first element grew out of the particular economic and foreign-political context of the early 1930s for Germany, which created a strong need for the new Nazi state to maintain international good will and international peace. The potential dangers of the situation were made clear by the failure of the April 1933 boycott of Jewish businesses instigated by Goebbels and other radical elements within the Nazi hierarchy. The openly anti-

[84] Daniel Koerfer, *Hertha unter dem Hakenkreuz. Ein Berliner Fußballclub im Dritten Reich* (Göttingen: Verlag der Werkstatt, 2009): 34–44. Soccer clubs had to report the final adoption of the new by-laws to the police by June 30, 1934. Only minor local variations from the model by-laws were allowed.

[85] Herzog, *Der "Betze"*, 50–52.

[86] Dieter Steinhöfer, *Hans von Tschammer und Osten. Reichssportführer im Dritten Reich*, Turn- und Sportführer im Dritten Reich 2 (Berlin, Munich, Frankfurt: Bartels & Wernitz, 1973); see also Bernett, *Weg des Sports*, 33–52

Semitic and anti-democratic measures taken by the Nazis in the Spring of 1933 led to a wave of criticism in the foreign press in 1933, and some foreign groups sought to organize a boycott of German goods and of participation in the coming Olympic Games. The Nazi leadership was quite sensitive about its image abroad at this time, and no one more than Goebbels himself (the czar of media policy and the prime instigator of the April boycott).[87] In this context, the Propaganda Ministry assigned the DASD and its members the role of countering anti-German and anti-Nazi sentiment abroad through their direct contacts with foreign amateurs. It was hoped that individual DASD members would act as seemingly independent and neutral witnesses to the normality of life in Nazi Germany.[88] That this ran counter to the culture and explicit law of international amateur radio did not bother the Nazis in the least, and also found favor within at least parts of the existing DASD leadership.

The second element was the strategic nature of radio technology, and the resulting long-standing interest of the German military in the DASD and the radio hobby. We have already seen how the military, by training thousands in telegraphy and radio technology, played an important role in establishing the preconditions for the rise of the radio hobby in the early

[87] German and Nazi leaders feared economic damage from a potential foreign boycott, which might delay economic recovery and cause domestic problems. Moreover, even though Hitler had secretly begun a program of rearmament to overthrow the limits of the Versailles treaty in 1933, the military was still weak, and thus the Nazis were particularly interested in avoiding bad press abroad for fear of provoking opposition. Conversely, they were hoping to gain major propaganda benefits from the coming Olympic Games, which they would now be hosting. All of these things left them particularly sensitive to foreign opinion in the early 1930s. The April 1933 boycott must also be seen in the context of internal rivalries within the Nazi Party and affiliated organizations, which go far beyond the scope of this study. See: Heinz Höhne, *Die Zeit der Illusionen. Hitler und die Anfänge des Dritten Reiches 1933–1936* (Düsseldorf, Wien, New York: ECON Verlag, 1991): 75–95.

[88] Reichsverband Deutscher Rundfunkteilnehmer (R.D.R.) E.V. Berlin (i.A. Dr. Heinz Weiss), *Rundfunk im Aufbruch. Handbuch des deutschen Rundfunks 1934 mit Funkkalender* (Mit einem Gleitwort von Reichsminister Pg. Dr. Goebbels) (Lahr/Baden: Verlag Moritz Schauenburg K.G., 1933). This book was given to all regional directors and governing board members of the DASD in the Summer of 1933 at a compulsory, week-long workshop in Berlin called by the RMVP. It can be considered a written blueprint for the new Nazi conception of radio, in general, and contained explicit reference to the new role of the DASD. In particular, see Pg. Hermann Schäfer, "Der Amateursender: Seine Organisation und Seine Aufgabe", 58–61. For the compulsory workshop for DASD leaders, see "Die Sende-Erlaubnis marschiert!", *CQ* 8, No. 8 (August 1933): 57; "180 Sendelizenzen erteilt!", *CQ* 8, No. 9 (September 1933): 65.

1920s. Certainly, it was also one of the largest employers or "consumers" of trained radio operators and telegraphers (along with the Police and the state-run stations for merchant shipping). The interest of the military in both broadcast and amateur radio went back to the early days of the Weimar Republic.[89]

Moreover, this was not simply a passive interest. German military efforts to subvert the Versailles Treaty restrictions were widespread from the beginning. The German Navy is known to have covertly provided a subvention to the DASD prior to ending the program in 1931, and to have again set up a covert training program in 1932 in cooperation with at least some DASD members.[90] Once the Nazi Party came to power in 1933, they, together with the armed forces, build on and greatly expanded these earlier efforts. Because of the illegality of these and other measures to subvert the Versailles Treaty restrictions, they were surrounded in secrecy; the Reichswehr had a strict policy of producing as few documents on secret armament measures as possible,[91] and in any case, what few documents as may have existed were often destroyed in the later Allied bombing of Germany.[92] Equally secretive was the rearmament carried out after

[89] "Neu A IV 4311 g/36 geh. vom 3.Dez.1936 Betr. Einbau des Freiw. Wehrfunks in die Wehrordnung", BAMA RM20/1983, 154–156. This letter documents that there was interest in the DASD from both the Navy and the Air Force.

[90] See "In7 No 20330/34 g.Kdos. In 7 Ia of 8. November 1934 (Abschrift) an A.H.A. Betr.: Regelung aus dem x = Fonds, I.A. gez. Fellgiebel", BAMA, RM20/1978. This document demonstrates that the German Navy secretly gave money to the DASD down to 1931, and suggests that payments in the old amount resume immediately. The 1932 cooperation is mentioned in: "Abschrift I² 1007/16.11 "Vertrag" (Geheime Kommandosache) Zwischen dem Reichswehrministerium, Marineleitung, vertreten durch den Chef des Allgemeinen Marineamts und dem Deutschen Amateur-Sende-Dienst", in: BArch R55/229 Reichsministerium für Volksaufklärung und Propaganda, DASD, 73–75.

[91] See: Rüdiger Bergien, "Staat im Staate. Zur Kooperation von Reichswehr und Republik in der Frage des Grenz- und Landesschutzes", *Vierteljahreshefte für Zeitgeschichte* 56, No. 5 (2008): 643–678; Jun Nakata, *Der Grenz- und Landesschutz in der Weimarer Republik, 1918 bis 1933. Die Geheime Aufrüstung und die deutsche Gesellschaft* (Freiburg i. Br.: Rombach, 2002) and Rudolf Absolon, *Die Wehrmacht im Dritten Reich*, (4 Vols.), Schriften des Bundesarchivs 16, vol. I, (Boppard a/R: Harald Boldt Verlag, 1969–1979.): 35–39. Much of the secret rearmament and training in 1933 to 1935 was funneled through a special organization set up within the Nazi Party's SA, called the "ChefAW" or "Chef des Ausbildungswesen". See: Thilo Vogelsang, "Der Chef des Ausbildungswesens (Chef AW)" in: Institut für Zeitgeschichte, *Gutachten des Instituts für Zeitgeschichte*, Vol. 2, (Stuttgart: Deutsche Verlagsanstalt, 1966): 146–156.

[92] Particularly important was the bombing and partial destruction of the Army Military History Archive (*Heeresarchiv*) in 1944.

the German government openly announced it was rearming in 1935. Only a handful of documents on the military ties to the DASD from this period survive, but they are enough to demonstrate that the military, and particularly the Navy, remained extremely interested in the DASD during the period of covert rearmament and beyond.[93]

The reasons for this interest lie not simply in the inherently strategic nature of radio as a medium of communications, nor is it solely due to the willing cooperation of the DASD leadership, though the fact that most DASD presidents were former career officers surely helped.[94] Along with these other factors, the high level of competency and training which the DASD had been able to achieve and maintain must not be forgotten. DASD members were respected for their knowledge of radio, their "traffic discipline" (the disciplined way they handled two-way Morse code traffic), and their very high level of Morse code proficiency.[95] In short, the technical skills of the radio amateurs, as guaranteed by the DASD, made it a prize not to be ignored by the German military. While Goebbels could largely ignore or confound the long-standing claim to a monopoly over control of radio which was held by the Postal Ministry, he could not ignore the strong interest of the military, particularly at a time when Hitler was pushing both covert rearmament and seeking the cooperation and support of the military for his political program. The many clubs of radio listeners and home builders were comparatively unimportant, and the military was not interested in them to any great degree. But the DASD was another matter, and from 1933 onward, it was directly engaged in covert rearmament in partnership with the armed forces.[96] This gave the DASD a special status, and allowed the organization to continue an independent, if tightly controlled and highly instrumentalized existence. The close association of the DASD with both the military and the Propaganda Ministry, its perceived importance for both national defense, and foreign

[93] This relationship continued to develop into the 1930s, and will be detailed in the next chapter.

[94] The first President of the DASD was Lt. Col. (ret.) von Stockmayer, the second was Col. (ret.) Fulda (to 1933). Col. Fulda was replaced in the *Gleichschaltung* of the DASD in the Spring of 1933 with Prof. Dr. Gustav Leithäuser, a university professor and scientist (though to be fair, Leithäuser had also served as a reserve officer in the First World War). In 1935, ret. Vice-Admiral Groos became President of the DASD.

[95] The includes the SS. See the documents in BArch NS19/3918, Pers. Stab RFSS, Amateurfunkwesen.

[96] See the next chapter.

propaganda meant that the organization continued to exist and function through the war years, long after the other elements of the radio hobby had disappeared.

For the most part, it appears that the DASD cooperated with the new regime, often quite willingly. Most members were, after all, good, patriotic citizens, even if they were not all Nazis. The "carrot and stick" aspect of the *Gleichschaltung* is quite apparent here. While there was always the threat of losing everything, and even of physical violence, the new regime promised tangible benefits to the shortwave amateurs of the DASD. They were taken seriously and given a flattering role to play in society as partners of the state. They expanded their privileged position with the Armed Forces. Above all, they were promised the right to legally own and use private transmitters. This was the central issue for the DASD since its founding. There was no greater prize to be had for the German amateurs. Illegal transmission had earlier characterized large parts of the amateur radio movement, but had always been dangerous, and the dangers were only increasing as technology (and the growing power of denunciation) made it easier to track illegal transmissions. The Nazi offer of legal private transmitters (as carefully controlled and ideologically bounded as it was) could not have come at a more welcome time.[97]

As with all other civic associations, the DASD was required to institute a new national leadership, and to introduce new by-laws enshrining the "Leadership Principle". But it was already a hierarchical organization, it was already the principle German organization for transmitting amateurs,

[97] There is some lack of clarity about how many new transmitter licenses were actually awarded, and when. In the Spring of 1933, all private transmitting licenses were first rescinded. The subsequent Spring-Summer licenses were technically all new. The (by now semi-official) journal of the DASD announced in August that the first set of new licenses had been granted as early as May 1933, with a second group to follow. "Die Sende-Erlaubnis marschiert!", *CQ*, Nr. 8 (1933): 57. A later article announced that a total of 180 new licenses had been granted in 1933. "180 Sendelizenzen erteilt!", *CQ*, Nr. 9 (1933): 65. On the other hand, one internal document mentions the issue of up to 314 new licenses in 1933. "Reply of Schäfer to accusations, (n.D.)", BArch R55/229 Reichsministerium für Volksaufklärung und Propaganda, DASD, 54–58. By 1936, there were 531 licensed amateurs: "Ansprache des Präsidenten des DASD auf der Kundgebung am 23, Mai- 1936", *DASD Verordnungsblatt* No. 9 (30 April 1936). This figure lends credence to the larger figure for the 1933 wave of new licenses. All of the 1933 licenses were re-issued by the Postal Ministry in 1935, after a new wave of vetting. See Chap. 6 below, and Leo H. Jung, DH4IAB, "QSL's erzählen deutsche Amateurfunkgeschichte", No. 9 "Beim DASD 1933–1945", *Funk-Telegramm*, No. 12 (2003): 9–13.

and its national leadership was already highly nationally-minded. In this sense, little needed to be changed in the fundamental structure of the DASD. Prof. Dr. Gustav Leithäuser was named the provisional "Leader" ("Leiter") of the organization in the Spring of 1933, and he was hardly a stranger to the club; he was an internationally recognized scientific expert on radio, had attended the 1931 general meeting of the DASD as the official representative of the DFTV and Heinrich-Herz-Society, and had long supported the DASD. In a sense, his appointment was an odd step, since his previous two predecessors had been former professional officers, while Leithäuser was a research professor, and a former official of the Postal Ministry.[98] But in the context of 1933, he was a man who combined a commitment to the Nazi "New Order" with significant prestige within scientific and amateur radio circles, and was thus a good choice to initiate the *Gleichschaltung* of the DASD.

The real key to control over the DASD was the issue of transmission and private transmitters; without the ability to transmit, no real shortwave amateur movement was possible, and thus, there was no need for the DASD to exist. The threat of ending transmitting entirely was the Damocles sword over the heads of all shortwave amateur radio enthusiasts in 1933 and 1934. But on the other hand, the promise of allowing a liberalization in private transmission was a prize so big that the DASD would do almost anything to get it.

The threat was emphasized in a general ban on all amateur transmission in early April 1933.[99] It was kept secret from the general public, and saw only indirect mention in DASD publications. The ban had several purposes: it was a general threat of what might happen if the DASD didn't toe the new line. It also helped to expose undisciplined, even resistant elements within the Amateur community, since anyone on the air after the

[98] For biographical details, see: (Paul) Gehne, "Gustav Leithäuser, dem Fünfzigjährigen. Ein offener Brief", *Funkbastler. Fachblatt des Deutschen Funktechnischen Verbandes E.V.*, No. 51 (December 18,1931): 801–803.

[99] The ban apparently went into effect on April 5 for DASD members, and April 1 for other private transmitters licensed by the postal authorities. "Bericht 39 der Überwachungsstelle für den privaten deutschen Kurzwellenverkehr bei der Pol. Hauptfunkstelle Berlin für die Zeit vom 1.-31.3.1933 (Abschrift!)", BArch R/1501/20404 St10 RMdI KPD-Radio-Propaganda Bd. 5 March 1933–March 1934, 63ff. All previous private transmitter licenses were rescinded at this time.

ban was ipso facto illegal.[100] And aside from the more general threat, it was a handy means of pressuring individual DASD chapters to carry out Nazi demands which were part of the Gleichschaltung of German society. Thus, at least some clubs were told that the ban would remain until they expelled all Jewish members. As an example, on April 1, 1933, the *Oberpostdirektion* Augsburg (Augsburg Postal Administration) notified the local police that the national government had just ordered that any radio club which accepted Jews as members would automatically lose the right to maintain a transmitter. As a result, the Post had repealed the transmitter permit of the Augsburg branch of the *Süddeutscher Radio-Club.* The transmitter remained sealed until the club changed its by-laws to exclude Jews. The right to transmit was only restored on May 19, 1933.[101]

Meanwhile, the DASD was gradually purged of what were euphemistically called "undisciplined" elements—meaning anyone who was unwilling to work with the new government, or who persisted in illegal transmission without a license. A so-called *Oberste Aufnahmekommission* ("Supreme Membership Commission", OAK) was established under an associate of Goebbels and the RMVP, SA-Sturmbannführer Hermann Schäfer, to vet all DASD members and their licenses, before they were re-issued.[102] It later had to vet and approve all new applicants for transmitter

[100] Police monitors believed they had identified at least one opposition group, which continued to transmit despite the ban. "Württ. Innenministerium, Württ. Politische Polizei 5/1441/33/gf. An die Nachrichtensammelstelle im Reichsministerium des Innern, Berlin Betr. Oppositionsgruppe 'Eiserner Ring' oder 'Stählerne Reihe' (Eilt sehr!) vom 8. Juni 1933", BArch R/1501/20404 St10 RMdI KPD-Radio-Propaganda Bd. 5 March 1933–March 1934, 76. In fact, the "opposition group" turned out to be made up of NSDAP members and their allies, who were seeking to Nazify the DASD. "Nachrichtensammelstelle im Reichsministerium des Innern IAN 2162 g/8.6.33 an das Württembergische Innenministerium, Württ. Politische Polizei Betrifft: Oppositionsgruppe 'Eiserner Ring' oder 'Stählerne Reihe'(eilt sehr) of 15.6.1933", BArch R/1501/20404 St10 RMdI KPD-Radio-Propaganda Bd. 5 March 1933–March 1934, 77 and "Auszug aus einem Schreiben von Luwo Schäfer, München 13, Elizabethstrasse 31 IV, an die Oberpostdirektion München, Oppositionsgruppe (Vertraulich! Quelle darf nicht angegeben werden!)", BArch R/1501/20404 St10 RMdI KPD-Radio-Propaganda Bd. 5 March 1933–March 1934, 78. The *Gleichschaltung* was a complicated time.

[101] DF1MD [Wolfgang Schmidt], "DARC Ortsverband Augsburg", http://www.darc.de/distrikte/t/01/geschichte/1927-1933/, accessed May 29, 2012. The page cites documents from the Stadtarchiv Augsburg.

[102] The OAK contained representatives of the ministries and organizations, which were most interested in amateur radio, such as the military, Interior Ministry and SA. Schäfer, as head of the commission and Goebbels' personal representative, seems to have set the agenda,

licenses. It was the most important tool of the RMVP to control and purge the DASD.[103] To make the job of vetting all members easier, a moratorium was declared on new membership applications, similar to what was done slightly later for the other radio clubs.

The complete ban on transmission in the Spring of 1933 may be a sign that the RMVP first did not know quite what to do with the DASD, or it may be a sign that it had not yet gained enough power in competition with other government and Party institutions to make its own decisions. It certainly was also a tool to bring the DASD into line with the new regime. But by the late summer of 1933 at least, the Propaganda Ministry had set on a role for the DASD and began to wrap it in a deadly embrace. The DASD leadership received its marching orders in August 1933. The national officers and all of the regional "Traffic Officers" ("*Gruppenverkehrsleiter*") were ordered to attend a week-long indoctrination session in Berlin during the Great German Radio Exposition. Meetings and special receptions with leading officials of the Propaganda Ministry and National Broadcast Association flattered the men, while the speeches and writings of these same officials made it clear that the DASD would be expected to support the regime completely. They took a series of classes in technical aspects of radio, but more importantly, also in Nazi ideology, and were given a crash-course in how to be good propaganda representatives of the new order. After this intense week of training they were then expected to go back to their regions and spread the word to the ordinary members.[104] At this point, the DASD had lost any independence

and immediately installed himself at the DASD main office. See RMVP documents in: BArch R55/229 Reichsministerium für Volksaufklärung und Propaganda, DASD. See also the next chapter, and Leo H. Jung (DH4IAB), "QSL's Erzählen Deutsche Amateurfunkgeschichte", Part 9: "Beim DASD 1933–1939", *Funk-Telegramm* No. 12 (2003):9–13, here pp. 9–10.

[103] Ostensibly part of the DASD national leadership, the *Oberste Aufnahmekommission* actually was subordinate to the RMVP, despite the participation of a number of government ministries. Schäfer was clearly Goebbels' man, and tasked with the *Gleichschaltung* of the organization in a National Socialist sense. Happily, for the DASD, he did not last long. His fate will be described in the next chapter.

[104] "Die Sende-Erlaubnis marschiert!", *CQ* 8, No. 8 (August 1933): 57; "180 Sendelizenzen erteilt!", *CQ* 8, No. 9 (September 1933): 65. The role intended for the DASD in particular, and radio in general, was set out in a book given to all of the attendees. See: Reichsverband Deutscher Rundfunkteilnehmer (R.D.R.) e.V. Berlin (i.A. Dr. Heinz Weiss), *Rundfunk im Aufbruch. Handbuch des deutschen Rundfunks 1934 mit Funkkalender* (Mit einem Gleitwort von Reichsminister Pg. Dr. Goebbels) (Lahr/Baden: Verlag Moritz Schauenburg K.G., 1933).

it had once had and became a tool for propaganda and rearmament. On the other hand, it had gained much in return.

The big prize for the DASD, besides the simple continuation of the organization, was the long-awaited increase in the number of legal amateur licenses. This was publicly announced with great fanfare by the Propaganda Ministry and DASD just in time for the Great German Radio Exposition of 1933, but had in fact been authorized by postal regulations in May.[105] It is even likely that the Postal Ministry was moving toward expanding the number of amateur transmitter licenses well before the Nazis took power, and that the Nazis simply took credit for a development, which was already underway.[106] In any case, the promise of a liberalization of the license policy, which was rumored from the Spring,[107] was crucial in obtaining the support of the DASD—along, to be sure, with the threat of violence or at least an end to the organization. The lucky few who obtained new licenses—or the reissue of their former ones—in May–November 1933[108] had all been vetted by the OAK. They tended to be experienced DASD members, for the simple reason that the OAK, and particularly its Chairman *SA-Sturmbannführer*[109] Schäfer, had no real understanding of what it took to be a successful amateur radio operator. The OAK could vet names for political reliability but had no way of knowing who to consider. The names of those who got licenses in this first wave

[105] Leo H. Jung (DH4IAB), "QSL's Erzählen Deutsche Amateurfunkgeschichte", Part 9: "Beim DASD 1933–1939", *Funk-Telegramm* No. 12 (2003): 9–13. "Die Sende-Erlaubnis marschiert!", *CQ* 8, No. 8 (August 1933): 57; "180 Sendelizenzen erteilt!", *CQ* 8, No. 9 (September 1933): 65.

[106] The Police were already working under the assumption that the Postal Ministry was preparing to issue more private transmitter licenses to amateurs by at least December 1932. "Bericht 25 der Überwachungsstelle des privaten deutschen Kurzwellenverkehrs bei der Pol. Hauptfunkstelle Berlin für die Zeit vom 1–3. XI 1932" of 17. December 1932 (Geheim!)", BArch R/1501/20062 Reichsministerium des Innern KPD-Radio-Propaganda, 1930–1934, 248–249.

[107] "Bericht 28 der Überwachungsstelle des privaten deutschen Kurzwellenverkehrs bei der Pol. Hauptfunkstelle Berlin für die Zeit vom 1.2.-28.2 1933" of 17. March 1933 (Geheim!)", BArch R/1501/20062 Reichsministerium des Innern KPD-Radio-Propaganda, 1930–1934, 256–257; "Die Sendeversuchserlaubnis Rückblick und Ausblick", *CQ*, No. 5 (1933): 33.

[108] A handful of DASD members seem to have received provisional transmission permits even earlier, in the late Spring of 1933. This was likely done in order to test the discipline of the newly coordinated DASD, and also to maintain the illusion abroad that Nazi Germany was a normal state in which hams continued to practice their hobby.

[109] This was one of many peculiar ranks found in Nazi paramilitary organizations. It was the equivalent of Major.

were thus all recommended to the OAK by the DASD leadership, who naturally picked core DASD members with long experience, even though most of this experience was via illegal transmitting during the Weimar Republic.[110] Exceptionally, there was no exam requirement for these licenses, another reason why only experienced DASD members were involved.[111]

In fact, the licensing policy for amateurs in the Third Reich was not at all generous. The Postal Ministry remained in control even in 1933, and as later enshrined in the "Bekanntmachung über Versuchsfunksender" of February 10, 1935.[112] Issue of new licenses was temporarily suspended already in November 1933.[113] The authorities remained unwilling to issue too many amateur radio licenses, and the issue of more licenses after the first wave was much slower, and was made contingent upon membership in the DASD, a practical and theoretical exam, and most importantly, a thorough vetting by the Police and the Nazi Party. After [1937] the licensing process essentially stopped. German amateurs continued to be dissatis-

[110] Körner, *Amateurfunk*, 104. The Navy also insisted that the old DASD members who had secretly cooperated with the Navy in 1932 be included. See: "Reply of Schäfer to accusations, (n.D.)" in: BArch R55/229 Reichsministerium für Volksaufklärung und Propaganda, DASD, 54–58.

[111] While the new license holders had to be DASD members, a handful came from the former ARB, though not many. Leo H. Jung (DH4IAB), "QSL's Erzählen Deutsche Amateurfunkgeschichte", Part 8: "Der DASD bis 1933", *Funk-Telegramm*, No. 11 (2003): 28–29. Note that the military, in particular, insisted on the issue of transmitting licenses to suspected political unreliables in some cases, in order to lead them to commit crimes for which they could be punished. "Reichswehrminister ??33 g. Wehramt Jn/7.IV Betrifft. I[2] 1007/16.11 "Versuchsfunksendeanlagen für Funkfreunde" (Geheim) to Reich Minister of Interior, Reich Minister of Propaganda and Public Enlightenment, Reich Minister for Post, Prussian Minister of the Interior, Reich Aviation Minister, Supreme SA Leadership, Inspection for Training, of 16 November 1933" In: BArch R55/229 Reichsministerium für Volksaufklärung und Propaganda, DASD, 68–72.

[112] "Bekanntmachung über Versuchsfunksender", §2 FAG of 10. February 1935 and "Bedingungen für die Errichtung und den Betrieb einer Versuchsfunkanlage für Funkfreunde" (Anlage 1 zum Amtsblatt Vf. Nr. 53/1935), as cited in: Leo H. Jung (DH4IAB), "QSL's Erzählen Deutsche Amateurfunkgeschichte", Part 9: "Beim DASD 1933–1939", *Funk-Telegramm*, No. 12 (2003): 9–13.

[113] Reichswehrminister??33 g. Wehramt Jn/7.IV Betrifft. I[2] 1007/16.11 "Versuchsfunksendeanlagen für Funkfreunde" (Geheim) to Reich Minister of Interior, Reich Minister of Propaganda and Public Enlightenment, Reich Minister for Post, Prussian Minister of the Interior, Reich Aviation Minister, Supreme SA Leadership, Inspection for Training, of 16 November 1933 In: BArch R55/229 Reichsministerium für Volksaufklärung und Propaganda, DASD, 68–72.

fied and frustrated at the inability of many to obtain a license. Meanwhile, the accusation of illegal transmission and accompanying threat of persecution was continually used keep the DASD in line.

For the DASD, The benefits of cooperation with the new state were so great, that it was easy to overlook the price: loss of true independence—something which was willingly given up in exchange for becoming a state partner—the expulsion of a handful of left-wing, Jewish or simply ill-disciplined members, the new duties of self-policing in a National Socialist sense, and the need for an ideological security check. None of these things raised much opposition in the face of the great attraction of finally being able to be a full member of the international amateur radio community, with one's own "voice" on the air. And if any of these costs threated to become too painful, they could always be soothed with the balm of the rationalization that one was serving one's country. A truly Faustian bargain. We will see in the next chapter how this played out.

Conclusions About the *Gleichschaltung* of Radio Clubs

The *Gleichschaltung* put an end to the independent existence of the radio hobby, and soon to the very existence of most clubs. The first to go were the working-class clubs, but the middle-class ones followed only a few weeks later. The process took place through a combination of internal and external pressures. Internally, it came from German-National, conservative, and National Socialist club members, coupled with "anticipatory obedience" ("*vorauseilener Gehorsam*") on the part of many club leadership councils, which sought to avoid the worst through the appearance of eager compliance. Externally, the Propaganda Ministry sought to seize control over the radio clubs as a relatively minor part of the larger project of gaining control over broadcast radio, which value as a political instrument was quite clear to National Socialists around Goebbels. One of their most powerful tools in the Nazification of the radio hobby was the long-term subversion of the hobby by the RDR. It helped move the tone and nature of the debate in a Nazi and nationalist direction, and then provided the perfect vehicle for seizing control over and centralizing the radio clubs, ostensibly in the name of "rationalization". Though the process of Nazifying the middle-class radio clubs was longer and more convoluted than the way the working-class clubs were dealt with, the end result was the same: an end to the independent and rich ecosystem of the radio hobby. Throughout this process, it cannot be forgotten that while many greeted the changes warmly, the threat of great physical violence was also always present. While this violence was used mainly against the Left, the fact that it was there cannot be ignored, and its threat certainly played a central role in the Nazification of radio at every level.

The "shortwave amateurs" (ham radio enthusiasts) under the DASD were a partial exception to this rule. Due to their high level of technical skill and their great importance for rearmament, and due to their ties with international organizations, they were able to escape dissolution, albeit at the price of the loss of their independence. But they could not escape their own *Gleichschaltung*, and in the end, made a truly Faustian bargain with a state now firmly controlled by the Devil of National Socialism. We will follow this process in the next chapter.

CHAPTER 6

The Radio Hobby in the Service of National Socialism, 1935–1945

> We National Socialists want to be clear that the coming community of centrally organized broadcast radio listeners, [together with] the working groups of listeners, technicians, and amateurs, etc. will be a National Socialist one, meaning it will be structured through command and obedience![1]
> Public appeal to all German radio clubs, 1933.

By 1935, the Nazification of German society was essentially a fait accompli. It continued, of course, right down to the end of the Second World War and the total collapse of the Nazi system; in many ways, it was only fully actualized during wartime. Yet by late 1934 or early 1935, there was no going back, and no serious focus for active domestic opposition left. The upheaval and open brutality of the "Seizure of Power" and Nazification (*Gleichschaltung*) gave way to a new normality. Granted, this normality was built upon a progressive regimentation of society, the exclusion of racial and political "enemies", and the routine terror of a well-functioning police state. Yet there was also growing employment, rising wages, and so

[1] "*Wir Nationalsozialisten lassen keinen Zweifel darüber, daß die kommende Gemeinschaftsarbeit der zentralorgnisierten Rundfunkteilnehmerschaft, der Fachgruppen der Hörer, der Techniker und Amateure usw. eine nationalsozialistische, d.h. nach Befehl und Gehorsam gegliederte Ordnung sein wird!*" (emphasis in the original). "Aufruf an alle deutschen Funkverbände und Vereinigungen!", *Funk und Bewegung* 1, No. 5 (June 1933): 2.

B. B. Campbell, *The Radio Hobby, Private Associations, and the Challenge of Modernity in Germany*, Palgrave Studies in the History of Science and Technology, https://doi.org/10.1007/978-3-030-26534-2_6

a growing expansion of at least passive support for the Nazi regime.[2] In the Third Reich, the broad and vibrant radio hobby culture of the 1920s was significantly changed and, for all intents and purposes, it was destroyed by the *Gleichschaltung*, as we saw in the previous chapter. The Nazification of society was deadly poison for radio clubs and private associations.

But as we have also seen in the previous chapter, in the long run, the decline of the radio hobby clubs also had less immediate reasons. Technological and economic changes also played a role in the decline of the radio hobby in the 1930s, even though this was accelerated greatly by the Nazis. Simply put, it was increasingly unnecessary to build one's own radio receiver or participate in a radio club by the mid-1930s, even if there had still been radio clubs available. Radio ownership in Germany by 1935 was widespread and growing,[3] and thus economies of scale were able to reduce the cost of a radio receiver, a process visible since the early 1930s. With rearmament and the progressive drop in unemployment in the mid- and late-1930s, more and more people could afford a manufactured radio. Prices for (at least some) manufactured radio receivers were now within reach of ordinary people, particularly with a little push from the new government in the form of the "People's Radio". Do-it-yourself (DIY) radio-building was still encouraged under the Nazis, but mainly within the Hitler Youth. It was less and less an occupation for adults.

DIY radio-building and other hobby activities were still possible in the Third Reich, but they were simply robbed of their institutional base and the associated social and physical space.[4] In the Third Reich, those interested in radio and sound technology for their own sake still had access to technological information through the journal *Funk* (chief among several), which continued to publish until it fell victim to the growing short-

[2] Tim Mason, *Social Policy in the Third Reich: The Working Class and the 'National Community'* (Providence, Oxford: Berg, 1993); Karsten Steiger, *Kooperation, Konfrontation, Untergang. Das Weimarer Tarif- und Schlichtungswesen während der Weltwirtschaftskrise und seine Vorbedingungen.* Beiträge zur Unternehmensgeschichte 5. (Stuttgart: Franz Steiner Verlag, 1998): Section 7, 278–287.

[3] Hans-Jörg Koch, *Das Wunschkonzert im NS-Rundfunk*, Medien in Geschichte und Gegenwart 20 (Cologne, Weimar, Vienna: Böhlau, 2003): 52–53.

[4] I distinguish here between "social space" and "physical space". Radio hobby clubs provided both. Physical space means labs and workshops ("maker spaces") but also space where hobbyists could meet and exchange ideas (chiefly clubhouses and pubs with a relationship with a given club, but also exhibitions and fairs). Social space as I define it means clubs and other institutions which largely provided the physical spaces, as well as journals and a social identity as radio enthusiasts.

age of paper in wartime.[5] Radio parts were still freely available and remained so into the early war years,[6] and the "do-it-yourself" community could still make radios at home, even if they were robbed of their former institutional framework and collaborative workspaces. Small numbers of Germans surely did continue to build and modify their own radios in private. There may even have been a slight upsurge in such DIY activity during wartime, to compensate for the disappearance of commercially produced radios due to the shift to wartime production.[7] Yet the vibrant club culture was destroyed, and people found their time increasingly filled with longer work hours and the myriad NS-oriented activities and busy work required to show allegiance to the regime. Among other activities, Germans were encouraged to engage in "service" in National Socialist—or government-sponsored organizations like the Brownshirts (SA), Imperial Civil Defense Association (*Reichsluftschutzbund*), Hitler Youth (HJ), or Association of German Girls (*Bund Deutscher Mädel*, BDM), collecting money for the "National Socialist Winter Relief" (*NS Winterhilfswerk*), or taking part in regimented "Strength through Joy" (*Kraft durch Freude*) activities for their free time.[8] The free-for-all private civic associations gave way to the regimented and constrained organizations, which bound individuals evermore tightly to the regime, and kept them busy with approved activities. In the Third Reich, there was little time for hobbies.

Of course, the Nazis aggressively encouraged radio ownership as a tool for propaganda, and the Nazification of the media meant that radio own-

[5] *Funk* last appeared in the Spring of 1944 (21, No. 7/12 April–June 1944). It was followed by a short-lived omnibus journal *Funktechnik*, which sought to combine *Funk* and three other radio journals (*Funk*, *Funkschau*, *Funktechnischer Vorwärts*, and *Bastelbriefe der Drahtlosen*) in a special wartime publication. It managed to appear as late as January 1945.

[6] In 1943, the DASD mail order department was forced to close. Members were admonished to think about the needs of industry and the war effort when making their last orders. *Nachrichtenblatt Landesverband Reichshauptstadt im Deutschen Amateur-Sende- und Empfangsdienst e.V.*, 9, No. 1 (January–February 1943).

[7] This is a supposition. Information on such activities by individuals, especially during wartime and under the suspicion that any radio manipulation might be subversive, is naturally hard to obtain.

[8] For longer work hours, Karsten Steiger, *Kooperation*, 278–287, and especially Table 15, 283. The "*Winterhilfswerk*" (WHW) was a Nazi Party-sponsored welfare and relief organization originally established during the Depression to help the poor and unemployed make it through the cold winter. Collecting money and goods for the WHW became an important test of conformity and obedience for Germans. See: Herwart Vorländer, "NS-Volkswohlfahrt und Winterhilfswerk des Deutschen Volkes", *Vierteljahrshefte für Zeitgeschichte* 34, No. 3 (1986): 341–380.

ers were subtly pushed more and more toward government-controlled media and government-sponsored ideas. It was still possible to listen to broadcasts from abroad, and this was perfectly legal until wartime. Published radio program listings in journals continued to give program schedules for foreign stations until the outbreak of war, and the various models of the *Volksempfänger* were perfectly capable of receiving foreign stations with the right antenna. Kits to allow the addition of long-distance shortwave reception to *Volksempfänger* were even produced. But increasingly, the Nazification of both the radio clubs and the media meant that it was easier just to listen to local stations.[9] There was enough residual paranoia about uncontrolled use of radio technology in government and Party circles in the 1930s that the decline in do-it-yourself radio construction was not mourned. The large numbers of those in the radio hobby who were simply interested in listening to broadcast radio and who consumed radio media for entertainment (and sometimes education) were very easy to accommodate within the Nazi culture of radio. Goebbels was quite willing to expand the amount of pure entertainment broadcasts offered by German broadcast radio, a long-standing public demand. The star system and glamour of broadcast radio and the illustrated media went on as before. The only difference was that the stars were just all "Aryan" now, and tended to hobnob publicly with Party leaders in the illustrated magazines. That didn't fundamentally weaken the star system, it reinforced it.

In short, simple broadcast radio listeners didn't need a hobby culture anymore when they had their own radios and a wealth of Nazi-dominated illustrated journals and newspapers to feed their interests, and when good manufactured radios were now affordable for most Germans.[10] Technical maturation, greater social regimentation, and (paradoxically) a trivialization of content caused the number of hobbyists interested in the technological aspects of radio, and perhaps in radio technology as a proxy for

[9] The technical simplicity of the *Volksempfänger* meant that the stronger local German stations could inevitably be received more clearly than foreign stations, but this was a technological limitation rather than an ideological one.

[10] This is, of course, a very brief summary of a much more complicated phenomenon of NS media politics. In general, see Corey Ross, *Media and the making of Modern Germany: Mass communications, Society and Politics from the Empire to the Third Reich* (Oxford: Oxford University Press, 2008). Note that the radio journals which did not disappear with the end of the radio hobby organizations were not free of Nazi propaganda, it was simply packaged in a glossy and subtle way. See also Ansgar Diller, *Rundfunkpolitik im Dritten Reich* (Munich: DTV, 1980).

modernity, to decline. Yet there was one small group of radio hobbyists still left. Their organization was not disbanded by the Nazis, but rather encouraged. These were the hobbyists concentrated in the part of the radio hobby we call "ham" radio.[11] This chapter and the following one will focus on them.

The DASD and Amateur Radio Under the Nazis

Because of its international reach and contacts, and because of the advanced technical skills of its members, the "*Deutsche Amateur Sende- und Empfangs- Dienst*" (German Amateur Transmission—and Reception Service, DASD), the organization of those who were interested in transmitting their own messages and who participated in the international amateur radio movement, followed a significantly different trajectory from that of the other parts of the radio hobby. It was certainly subject to the same sort of Nazification as the other clubs, as we saw in the previous chapter. The same combination of internal and external pressure to conform to the new order of the Third Reich was present. But for the DASD, both inward and outward pressures were stronger, for there was more at stake, even though the DASD was a comparatively small part of the radio hobby.

Already in the previous chapter, we have seen that the small portion of the radio hobby which was interested in transmitting (as opposed to just receiving) was treated in a significantly different way by the Nazis. Amateur radio in the Third Reich was thoroughly Nazified, but it was able to continue to exist and to have a limited degree of autonomy after 1933. Indeed, as long as it showed itself to be disciplined and useful, it was even encouraged and protected. The strategic value of radio technology was the main reason, although the regime also remained interested in the propaganda value of amateur radio well into the war years. If radio technology by the 1930s had grown beyond the capabilities of most do-it-yourselfers, this was not the case for the amateur radio community, which at any rate had a large number of engineers and scientists among its members. They not only understood the technology and were capable of contributing to its technical development, they were also the masters of the actual practice of radio transmission and reception, making them a highly important strate-

[11] Remember, "ham" radio and "amateur radio" are terms used in English to refer to those hobbyists who not only listen to radio receivers, but also build and use radio *transmitters*.

gic resource—as long as they behaved. They were thus the objects of great interest by the state, great scrutiny by the police and Nazi Party, and were granted both support and a limited autonomy, though both waxed and waned as a function of external conditions and the infighting of various state and Party authorities.

In many ways, the DASD in the Third Reich acted like just any other club (*Verein*).[12] It worried about interference from outside authorities and worked to build its membership. Its members sometimes forgot to pay their dues, and the DASD often worried about its finances, seeking to recruit "Supporting Members" ("*Fördende Mitglieder*") who would donate money and help with networking.[13] Its officers were often overworked with club business. Local chapters were often late in sending in reports and paperwork. Headquarters had to remind the members again and again to follow the rules.[14] And the club worried about how it was perceived by the public. Anyone who has been active in a club will recognize many or all of these themes. And yet, the DASD was anything but an ordinary club during the Third Reich.

The pattern set immediately during the *Gleichschaltung* for the way state and Party authorities treated the DASD continued in the following years. Several government and Nazi Party agencies continued to have an interest in amateur radio, and they continued to compete with each other for influence over it. There were two constants in the way state and party agencies viewed amateur radio. There was a varying degree of admiration for, and recognition of, the technical knowledge and ability with which the DASD was able to equip its members. This was no small advantage, since radio remained a vitally important strategic tool. On the other hand, there was also nearly universal suspicion of amateur radio operators and fear of the kinds of mischief they could work if left to their own devices. If amateur radio was too important to ignore, it was also far too dangerous to escape tight control and scrutiny. Beyond these two givens, the interest of the various agencies with a stake in amateur radio varied.

The Nazi Party, or more specifically, the Ministry of Popular Enlightenment and Propaganda (*Ministerium für Volksaufklärung und*

[12] Except, of course, for the fact that normal radio clubs no longer existed.

[13] *DASD Verordnungsblatt* No. 2, (February 5, 1936) describes the different types of membership.

[14] For example, all of these themes appear in *DASD Verordnungsblatt* No. 9 (April 30, 1936).

Propaganda; RMVP), was the strongest stakeholder in the radio hobby. It continued to see in the DASD and its amateur radio operators a means of influencing foreign opinion about Germany. The need for this was perhaps most acute during the anti-Jewish boycott and other measures during the 1933 *Gleichschaltung*, but continued even into wartime. Goebbels was also interested in amateur radio and the DASD for the simple fact that he was interested in all facets of radio. If he clearly preferred the top-down structure of broadcast radio as a tool for propagandizing the (passive) masses of radio listeners, he was smart enough, and suspicious enough not to cede any influence over any form of radio technology, even a fairly small and relatively unimportant group of hobbyists.

In one sense, the international contacts of the DASD, though they made it suspect, also tended to protect it. The DASD was tied into an international web of amateur radio organizations, which existed within a framework of international law and treaties, which Germany did not wish to be seen violating, at least too openly. At least until the advent of war, and in many ways even thereafter, the international presence of amateur radio set certain limits to how amateur radio could be treated. In particular, the DASD was the German representative to the IARU, the International Amateur Radio Union.[15] All major world states were participants in the IARU. The German participation in this framework was thus just one of the many ways the German state was tied into the international order, and was thus a symbol of national sovereignty. Fearing isolation in the 1930s, the Nazis and the German state authorities were careful to continue their participation in the IARU via the DASD, which thus had to at least appear to remain independent. They didn't like international organizations, and were suspicious of the international contacts of the DASD and its members, but were eager to maintain appearances abroad.[16]

The military remained interested in amateur radio as a means of training radio operators, and it too was a very powerful force within the state. None of the other competitors for control over radio technology could get beyond the trump card of national defense and none, not even the Propaganda Ministry, dared to compromise the interests of the military

[15] Originally, the DFTV was the formal IARU member organization, but this was transferred to the DASD sometime in 1933.

[16] I 2/100 7 "Aufzeichnungen" of June 15, 1934, BArch R55/229 Reichsministerium für Volksaufklärung und Propaganda, DASD, 110–112. Given the general xenophobia of the Third Reich, the DASD was at pains to downplay its international ties when it came to its self-representation to the public.

early in the regime. The degree of interest in radio varied among the different branches of military service. The Army recognized the value of DASD training in principle, but does not seem to have done much more than show interest. It did accept DASD certification of radio skills and sometimes (but not always) made an effort to assign DASD members to communications units.[17] The German Air Force took a middling position, and mainly tried to recruit DASD members for the ostensibly civilian *Flugwacht* (national system of aircraft spotters) before the coming of war. The Navy, on the other hand, had provided covert support to the DASD before 1931, was involved with the DASD in covert training again in 1932, and then immediately expanded its covert ties in 1933 after the Nazis came to power. It took a very active interest in the DASD, and made it an accomplice in covert training. The interest the Navy had in the DASD is demonstrated by the fact that it exercised a strong influence over the choice of the DASD President, who was named by the Propaganda Ministry, but only after consulting the armed forces.[18]

Navy interest in wireless (radio) went back to the very origins of radio technology, and not just in Germany. Of all the services, the navy had the greatest need for highly skilled radio operators, if only because radio was much more important to it than other means of communication. Radio

[17] A pending agreement with the Army was announced in 1936, at the DASD annual membership meeting. "Ansprache des Präsidenten des DASD auf der Kundgebung am 23, Mai- 1936", *DASD Verordnungsblatt* No. 9 (April 30, 1936). On the other hand, the Army seems to have often ignored the qualifications of DASD members during the early war years.

Note that there were long-standing practical reasons for Army ambivalence toward wireless: through the Second World War, although it did of course make use of radio technology, the German Army remained tied to wired (not wireless) communications whenever possible, and over land, it almost always was possible. Most Army radios were short-range, tactical radios working in the ultra high frequency (UHF) and very high frequency (VHF) portions of the radio spectrum. The Army thus had comparatively little need for shortwave specialists, unlike the Navy. Werner Niehaus, *Die Nachrichtentruppe, 1914 bis heute. Entstehung und Einsatz.* (Stuttgart: Motorbuch Verlag, 1980):39–285.

[18] In the case of both Otto Groos and Heinrich Gebhardt, the Navy took the initiative to propose their appointment to the Propaganda Ministry. "Der Reichswehrminister a IV n 1794/34. Geh. (Geheim!) to Herrn Reichsminister für Volksaufklärung und Propaganda, Betrifft: Deutscher Amateur- Sende-Dienst, Besetzung des Präsidenten, of 23. September 1934, signed Blomberg, Generaloberst", BArch R55/229 Reichsministerium für Volksaufklärung und Propaganda, DASD, 143; "R.M.f.V.u.P. 2 1007/22.6. ?ef. Min. Rat. Rüdiger, Reg. Rat. Hushahn, an den Herrn Reichskriegsminister und Oberbefehlshaber der Wehrmacht (Geheim!) of 9. September 1935" BArch R55/229 Reichsministerium für Volksaufklärung und Propaganda, DASD, 177.

was simply the only reliable way to communicate with and between ships at sea over long distances, and the navy continued to use Morse code instead of telephony much more than the other services. Because of the distances involved in communicating with ships at sea, the Navy was particularly interested in the shortwave part of the radio spectrum, just where radio hams had the most practical experience.

Specific to the German situation between 1933 and 1935 was the need to keep ties between (civilian) amateur radio and the armed forces secret. Hitler and the military services all immediately embarked on a highly secret and broad-gaged plan to rearm when the Nazis took power in 1933. Immediate rearmament was an area where the Nazi Party and the armed forces were in complete agreement, but the Versailles Treaty meant that German rearmament on this scale was illegal. Given earlier Allied intransigence about allowing any sort of rearmament, the relative German military weakness in 1933, and the resulting great fear that German rearmament might bring a crushing Allied response, strict secrecy was an absolute necessity.[19] Only in 1935 did the Nazis feel strong enough to announce rearmament publicly, and even then, the true scope of it was still concealed.

The consequence was that the German Navy developed very close ties to the DASD as a partner in covert rearmament. Very quickly after the Nazis took power and began rearmament, the Navy and the DASD negotiated a secret agreement creating the so-called Special-Group M (*Sondergruppe M*) within the DASD.[20] The members of these units were ostensibly ordinary DASD members, but in fact, they were naval personnel (either active duty or reserve) who were trained in radio technology and procedures by the DASD. They were issued special transmit permits and call signs which bypassed the regular testing and permit procedure (which was controlled by the Reich Postal Ministry), but which were indistinguishable from normal civilian call signs. Thus, if heard on the air, they seemed to be only innocent amateur radio hobbyists, while, in fact, they were a part of the covert rearmament and expansion of the Navy. Once German rearmament was made public in 1935, the *Sondergruppe M* was no longer necessary, and it was disbanded in June 1937, and its mem-

[19] Edward W. Bennett, *German Rearmament and the West 1932–1933* (Princeton N.J.: Princeton University Press, 1979).

[20] "M" stood for "*Marine*", "Navy" in German. *DASD Verordnungsblatt* No. 11 (May 23, 1935).

bers, some 500 in total, mostly left the DASD.[21] The Navy and the DASD continued to work hand in hand, however. Beginning in 1932 (thus, before the Nazi takeover), special units or groups of young civilians were secretly formed by the Navy to be trained as future naval radio operators. They were later called the Voluntary Naval Military Communications Group (*Freiwillige Wehrfunkgruppe Marine*; FWGM). Though it is not completely clear, the DASD seems to have been involved from the very beginning in 1932.[22] From 1933 to 1937, the FWGM program was formalized, and FWGM groups were concealed as a part of the DASD. At this time, the mission was to provide volunteers trained in radio skills for the Navy and to prepare men to take the Navy "*Seesport-Funkzeugnis*" (Maritime Sports Radio Certificate) exam.[23] From 1933 on, the FWGM was first recruited out of members of the SA and HJ, ideally those belonging to naval communications formations of either organization, as well as regular DASD members.[24] Later, the participation of the SA was more or

[21] See *DASD Verordnungsblatt* Nr. 32 (II/10) (May 22, 1937). At the time of the dissolution, the DASD stated that this led to a loss of approximately 400 members. *DASD Verordnungsblatt* Nr. 40 (II/18) (November 5, 1937). Ironically, the experience of working with the DASD led some of these men—who all of course had an interest in radio—to later reenter amateur radio as a hobby after the war. An example is Willi Fock, DL3NX. See "Vorläufige Mitgliederliste FWGM/Sondergruppe M—Stand: 21.08.2005", Dokufunk Archive, copy available at: http://dokufunk.org/upload/m_liste(1).pdf. True, this did not benefit the DASD directly, since it was disbanded after the end of the war, but it did benefit the hobby and its new postwar organization, the DARC (Deutscher Amateur Radio Club/German Amateur Radio Club).

[22] Heinz Mehler *F.W.G.M./Marine-Wehrfunk. Ein Beitrag zur Geschichte der Organisation.* M.S., Mainz, 1993, Dokufunk Archive, 5–7.

[23] *Mitteilungsblatt der DASD Landesgruppe A* (n.D., 1936). In general, see: Mehler *FWGM/Marine-Wehrfunk*. For a memoir of a young worker who took the *Seesport-Funkzeugnis* exam and became a navy radio operator, see the memoir by Bernhard Schröder, (born 1923 Osnabrück-Eversburg), http://schroeder-eversburg.de/memoiren.html. He was seconded during the war from the Navy to the German Foreign Ministry (possibly a cover for German Military Intelligence) and worked in both Spain and France.

[24] In 1935, the FWGM saw an influx of men from naval radio units of the SA after the winding down of the Chef.-AW organization within the SA. Schnellbrief an Marineverteiler, vom Chef des Führungsamtes (gez. Jüttner) "Der Oberster SA-Führer, Führungsamt, F 3 a Nr. 66 073, Betr. Freiwillige Wehrfunkgruppen (F.W.G.M.) of 16. December 1935", BA-MA RM20/1975. Both the HJ and the SA had special naval units, particularly in coastal regions, but also in areas with major river traffic. The idea was to attract those with a maritime background or interests, and also to maintain maritime skills for use by the Navy. Both

less dropped, but members were still drawn from the HJ.[25] Until June 1937, FWGM groups existed within the DASD and while there, they were considered regular members of the DASD. During this time, FWGM members were issued special transmitting licenses and call signs, such that if they were heard on the air, they would seem to be ordinary DASD members, much like the similar *Sondergruppe M.* At first, the instructors were mainly DASD members, who also built most of the radio equipment, according to DASD standard plans. Even after FWGM groups formally left the DASD, loose ties continued.[26] The program continued after ties with the DASD were loosened, and its instructors were mostly taken from naval personnel until wartime. Once war broke out, it became harder and harder to find qualified teachers who were not already on active duty, and the DASD was again called on to help, albeit informally.

The Hitler Youth (HJ) also had a strong interest in the DASD as a tool for training radio operators, even beyond the FWGM groups.[27] At first, in 1933 and 1934, it seemed as if the HJ was trying to take over the DASD altogether, as just one small instance of the much broader 'imperialism' of nearly all National Socialist organizations trying to usurp and take over state responsibilities and private interests in the period after the "seizure of

the SA and HJ also had special communications units. For the SA, see "OSAF ChM. 181/33 Betr: Marine-Stürme der S.A. of 12.7.33; "Erfahrungsbericht über die Ausbildung in Nachrichtensport" (Geheim)" (n.d.)[1933 or 34]; "SA der NSDAP, Marinestandarte 7, Gr. Kurpfalz Br.B. Nr. 64/35a geh betr: Ausbildung in Nachrichtensport (Geheim) of 12.4.35", both in BArch, Sammlung Schumacher 405; and "FOa Das Nachrichtenwesen der S.A. of 4.4.38", BArch NS23/vorl.194.

[25] In 1933, the SA had shown great interest in the DASD, and even attempted to take over control of it from the RMVP. The armed forces worked against this behind the scenes, and the decline in status of the SA after the Röhm Purge of June 1934 put an end to it. See: "I 2/100 7 "Aufzeichnungen" of June 15, 1934", in: BArch R55/229 Reichsministerium für Volksaufklärung und Propaganda, DASD, 110–112. The internal competition between different Party and government interests was a characteristic element of the Nazi system. In this regard, the SA and HJ interests in the DASD were quite similar and parallel in their ambitions, the difference being that the SA was brought to heel in 1934, whereas the HJ retained more internal power within the Nazi system.

[26] The difference between the FWGM and the Sondergruppe "M" was that the latter was made up of naval personnel, while the FWGM was mainly made up of young people who had yet to perform military service. It was thus part of a more long-term strategy for forming future radio operators well in advance of their military service.

[27] The Hitler Youth was both a Nazi Party and German state organization after 1933. Like most of the Nazi Party paramilitary affiliates after the Nazi takeover, it existed in a liminal state as both an organ of the government and of the Nazi Party.

power".[28] The combination of the armed forces and the Propaganda Ministry quickly made this impossible. Instead, the HJ soon signed an agreement (called a "contract") with the DASD in November 1935,[29] whereby it was allowed to nominate selected HJ members of age 14 or older with an aptitude for radio for special DASD membership. (They needed a special agreement, since 18 was the usual minimum age for the DASD.) These boys participated in DASD training and activities, paid dues, and were considered DASD members. Where possible, they were grouped into special Youth Groups (*Jugendgruppen*) in each local DASD chapter. Each DASD regional section (*Landesgruppe*) had a Youth Leader (*Jugendgruppenleiter*) who was appointed by the DASD president at the suggestion of the HJ national leadership, and who was ideally simultaneously the Radio Advisor in the appropriate HJ regional organization.[30] This agreement with the HJ brought thousands of new members to the DASD. The HJ (and therefore, the Nazi state) hoped in this way to prepare the military communications experts of the future, while the DASD hoped to escape too much control from the outside, and simultaneously, to gain new members and funding.

This formal association with the HJ is characteristic of the kind of compromises the DASD had to enter into in the Third Reich. On the one hand, there was close cooperation with the HJ, and hundreds of HJ members were brought into the DASD for training in radio, a move which helped to keep the DASD numerically strong and relevant into the future. On the other hand, the DASD had to bow to the wishes of the HJ leadership, and tolerate special HJ subgroups within the organization. In hindsight, there are two ways to look at this arrangement: on the one hand, it shows just how deeply entwined the DASD was with the Nazi system. On the other hand, many HJ members found a passion in their radio activities, and, sometimes, too, a refuge from too much militarization and Nazi ideology.

[28] The SA pursued a similar policy until the Röhm Purge temporarily sidelined that organization. Robert L. Koehl, "Feudal Aspects of National Socialism", *American Political Science Review* LIV, No. 4 (December 1960): 921–933.

[29] "Abkommen zwischen dem Deutschen Amateur Sende- und Empfangsdienst e.V. (…) und der Hitlerjugendbewegung e.V., Reichsjugendführung(…) March 21,1935", *CQ-MB* No. 12 (December 1935): 148. This was also reprinted in: DASD *Verordnungsblatt* No. 6 (April 3, 1936).

[30] *DASD Verordnungsblatt* No. 6 (April 3, 1936). Many members of this young HJ cohort (along with some former members of the Sondergruppe "M" and FWGM) would later build the backbone of the postwar amateur radio organizations in both East and West Germany.

While they were supposed to remain active in the HJ and participate in its activities during their radio training, the DASD classes allowed some HJ activities to be skipped, and a young boy who was so inclined could get out of many obligatory activities.

Two other key players around radio as a hobby were the Ministry of the Interior/Police and the Reich Postal Ministry. Unfortunately, the broad destruction of their files during the Second World War makes it very difficult to tease out the details of their involvement in amateur radio during the Nazi period. It is likely to have been mainly negative, due to the fear of amateurs' ability to contact the outside world or wreak havoc on Police communications networks. We have already seen that the Postal Ministry was originally the dominant authority in radio in Germany. Though forced to cede control over broadcasting content to the Propaganda Ministry and confine its authority over radio ostensibly to technical matters in 1933, the Postal Ministry was adamant in asserting its legal authority within these limits. Its trump card in the battle to control radio was the fact that according to law, only the postal authorities could provide the license needed to own and use a radio transmitter. Though the Propaganda Ministry had forced the Postal Ministry to expand the issue of private transmitter licenses in 1933, it did so very unwillingly, at least in this form, and in fact the licenses announced with great fanfare in 1933 were reissued in 1935 with closer Postal Ministry control.[31] The Reichspostministerium (RPM) remained extremely suspicious of amateur transmitters, and sought to restrict them and choke them with red tape whenever it could. The police also had a powerful word to say, for the Nazis made the issue of both a DASD radio listener's license (the so-called DE-Number, a first step to a full transmitting license) and the license to own and use a private transmitter contingent on possession of a police certificate, certifying that the police found the applicant politically trustworthy and of good moral character.[32] The police could thus exercise some indirect control over the DASD through these certificates. The police remained quite suspicious of amateur transmitters because they feared both subversion and unauthor-

[31] Leo H. Jung, DH4IAB, "QSL's erzählen deutsche Amateurfunkgeschichte", No. 9 "Beim DASD 1933–1945", *Funk-Telegramm* No. 12 (2003): 9–13.

[32] Later, this political vetting was expanded. From April 1937 on, all members of the DASD had to be vetted for political trustworthiness by the Nazi Party in their region (*Gau*) of residence, as well as by the police. This was done through the regional Party apparatus and the *Reichsamtsleitung Rundfunk*. See *DASD Verordnungsblatt* No. 28 (II/6) (April 1, 1937).

ized interference with the police radio networks. They worked closely with the Postal Ministry to track down and arrest illegal transmitters.[33] There was some reason to fear the illegal use of radio, as we shall see below, but the fear of uncontrolled technology in the hands of ordinary citizens was deep and long-standing.

The DASD alone had no chance against any of these powerful organizations with an interest in radio. If it was able to maintain its very existence (when so many other parts of the radio hobby were essentially eliminated) and even a very small degree of autonomy, it was because no single organization was able to take it over before the outbreak of war, due to the fact that its many powerful suitors were also its principle benefactors and protectors—against the others. It is probably too much to claim that the DASD was able to play these organizations against one another, for it was too weak to really have the power and influence to have much agency in this sense. But certainly, the fact that none of these powerful state and Party agencies were willing to cede their interests in amateur radio to the others meant that for all its weakness, the DASD continued to exist as a nominally independent private organization through the end of the war, one of very few such private organizations to do so. There was a real price to pay, however. Between 1935 and 1939, the structure and activities of the DASD were progressively regimented and structured in ways which suited the state and Party.

Post-*Gleichschaltung* Consolidation

The aftereffects of the confusion and turmoil of the *Gleichschaltung* in 1933 and 1934 were felt for a long time. A more hierarchical structure had to be built up within the DASD.[34] There was a rapid change in leadership at the top. Prof. Dr. Leithäuser, who took over as President of the DASD in 1933 during the *Gleichschaltung*, was able to mitigate some of the damage inflicted on the DASD by the Nazis, but rapidly used up his leadership capitol. SA-Sturmbannführer Schäfer, the head of the "*Oberste Aufnahme Kommission*" (Supreme Membership Commission, OAK)

[33] See, for example, "Bericht 33 des Überwachungsstelle für den privaten deutschen Kurzwellenverkehr bei der Pol. Hauptfunkstelle Berlin für die Zeit vom 1.-31.7.1933", BArch R/1501/20400 St10 RMdI KPD-Radio-Propaganda Bd. 2 Mai1931–Juli1933, 2–6.

[34] Based, of course, on the Nazi Leadership Principle. See *DASD Verordnungsblatt* No. 10 (May 29, 1936).

turned out to be an absolute disaster, and Leithäuser expended much of his energy dealing with his excesses. Schäfer was a typical product of the *Gleichschaltung* era. His appointment within the DASD was designed to both cement Goebbels's influence and to guarantee a fast and ruthless Nazification of the organization. He was brutal and venal, even likely a thief.[35]He was a rabid National Socialist and SA member, though one with deeper ties to Goebbels than to the Berlin SA.[36] Like many Nazis, he was a profiteer, and sought to use the instability of the *Gleichschaltung* for his personal gain. He imagined himself the coming president of the DASD, with a major role in national defense. Worst of all, he was a megalomaniac with a big mouth, which he opened in public, to his ultimate downfall. By

[35] Characterization of Schäfer is based on the following documents: "Reply of Schäfer to accusations", (n.D.), BArch R55/229 Reichsministerium für Volksaufklärung und Propaganda, DASD, 54–58; "A.E. II S "Abschrift" "Bericht" of November 3, 1933, by Krim.-Ass. Kretschner [?] and Krim.-A.-Anw. Transfeld on a speech given by Schäfer at a Werbeabend of DASD on November 2, 1933", BArch R55/229 Reichsministerium für Volksaufklärung und Propaganda, DASD, pp. 64–65; "DRVP I[2] 1007/113.12, ?? Min.Rat. Rüdiger, ?? Min. Amtm. Protz, "Vermerk" of December 15, 1933", BArch R55/229 Reichsministerium für Volksaufklärung und Propaganda, DASD, 80–81; "S.A. Der N.S.D.A.P. Gr. Niedersachsen Abtl. II. A., Schr. N., Betr. Sturmbaf. Schäfer, Bezug. O.S.A.F. V. 20.12.33 II Nr. 1742/33, Anlagen B.B. Nr. 1802/34 to O.S.A.F. of February 5, 1934, signed der Führer der Gr. Niedersachen m.d.F.b. Brif. [Illegible]", BArch R55/229 Reichsministerium für Volksaufklärung und Propaganda, DASD, 84; "Der Reichswehrminister, Chiffrierstelle (Geheim)(Abschrift) Niederschrift der wesentlichsten Gedanken der Aussprache zwischen Herrn Direktor Hadamowsky, Oberstleutnant Fellgiebel und Hauptmann Oschmann am 6. November 1933 of 7. November 1933, signed, Hauptm. Oschmann" BArch R55/229 Reichsministerium für Volksaufklärung und Propaganda, DASD, 85; "III. An den Reichsminister Dr. Goebbels über Herrn Min.Rat Rüdiger und den Herrn Staatssekretär, in Hause of 20 February 1934" BArch R55/229 Reichsministerium für Volksaufklärung und Propaganda, DASD, 91–92; "I 2/100 7 "Aufzeichnungen" of June 15, 1934", BArch R55/229 Reichsministerium für Volksaufklärung und Propaganda, DASD, 110–112; "R.M.f.V.u.P., Beauftragter,(Geheim), signed B?? of June, 22, 1934", BArch R55/229 Reichsministerium für Volksaufklärung und Propaganda, DASD, 134; "Letter from Prof. G. Leithäuser, President, DASD, to Herrn Dr. Collatz, Ministerium für Volksaufklärung und Propaganda (Abschrift) of June 25, 1934": BArch R55/229 Reichsministerium für Volksaufklärung und Propaganda, DASD, 135 and "Dem Herrn Minister habe ich folgendes zu Berichten", of July 12, 1934, BArch R55/229 Reichsministerium für Volksaufklärung und Propaganda, DASD,140. Interestingly, there are no remaining SA records concerning Schäfer.

[36] His SA rank of "*Sturmbannführer*" corresponds to a Major in the Army. On the SA in Berlin and its relationship to Goebbels in 1933, see: Stefan Hördler, ed., *Der SA-Terror als Herrschaftssicherung. "Köpinecker Blutwoche" und Öffentliche Gewalt im frühen Nationalsozialismus* (Berlin: Metropol Verlag, 2013).

the Fall of 1933, Schäfer had alienated the DASD leadership with whom he had to work, and worse, had drawn the concerted opposition of the armed forces, SA, and Interior Ministry. He was accused of improper bookkeeping and even theft, and worse, he publicly revealed much of the secret cooperation between the DASD, the military, and the SA, bringing an internal investigation for high treason. Despite the support of high officials in the Propaganda Ministry, he was forced out of the DASD and expelled from the SA in 1934.[37] This was an annoying defeat for Goebbels, who reluctantly had to concede the shared influence of the other stakeholders in the DASD and, in particular, that of the armed forces.

A leadership vacuum existed within the DASD in the Fall of 1933 and into the Spring of 1934, until the Schäfer affair was settled. As part of the reconsolidation, the DASD was legally re-founded as a civil private association (*eingeschriebener Verein*), naturally, with a new Nazified set of bylaws. It remained under the direct control of the RMVP, with a strong influence from the other stakeholders, notably the Military.[38] Leithäuser resigned as DASD President in late summer or early fall 1934. His academic and research duties became more pressing, and he had used up all of his leadership capital in navigating the Schäfer affair. Under pressure from the Navy, he was replaced in October 1934 by the just retired Vice Admiral Dr. h.c. Otto Groos.[39] When Groos was again called up for active military service in 1935, he was in turn replaced as DASD President by Rear Admiral (ret.) Heinrich Gebhardt.[40] Groos was not particularly dynamic or interested in the DASD, and in any case, did not stay long as President. It was really only under Gebhardt that the DASD found any

[37] See note 35 above. Körner speaks of this episode in strangely elliptical fashion. *Amateurfunk*, 113–114.

[38] "Deutsche Amateur-Sende-Dienst, Prof. Dr. Leithäuser, Präsident des DASD, to the Ministerium für Volksaufklärung und Propaganda, z. Hd. des Herrn Collatz of 15 June 1934": BArch BAK R55/229 Reichsministerium für Volksaufklärung und Propaganda, DASD, 124–125.

[39] Groos was a noted naval historian, as well as a professional officer. He retired from the Navy in 1934, and was DASD President from October 12, 1934, to September 30, 1935, when he was recalled to active service. He was later promoted during the war to full Admiral, and held important positions in the Navy High Command.

[40] Heinrich Gebhardt was born in 1885. A professional naval officer, he retired in 1934 as a Rear Admiral (*Konteradmiral*). He was preparing for reactivation by the Navy in the Spring of 1939, but died suddenly of blood poisoning in July 1939, without returning to active service.

sort of stable leadership, and even then, it can't be said that Gebhardt developed any particular loyalty to the DASD.

The severe purges of DASD members, the increased scrutiny by the police and Nazi Party, and the growing regimentation of club activities all took a toll on the interest and commitment of the members in the transition period from 1933 to 1935. Regional and local DASD leaders were troubled by the number of members who had become inactive or whose participation in club activities had decreased significantly.[41] In some cases, club activities had to be suspended for lack of interest or simple lack of numbers.[42] The DASD had to build up both members and organization from the mid-1930s onward, in order to remain dynamic.

The much-touted issue of new amateur transmitting licenses in 1933 was quickly followed by a moratorium on new licenses by the end of the year. It was clear that the process had been a little too hurried, and order had to be restored here as well. The Postal Ministry finally issued a new set of regulations for amateur licensing in February 1935,[43] and on this basis, all the amateur licenses were reissued by the Postal Ministry after another round of political vetting. At the very least, the DASD was now specifically named in the new regulations as the sole non-governmental organization within which amateur licenses might be obtained, giving it a certain legal status. Moreover, despite the strict vetting of license candidates, when the dust settled roughly three times as many amateur licenses had been issued in the first two years of the Third Reich, as compared to the Weimar Republic, and there was the promise of further licenses to come. German amateurs were never satisfied with the number of licenses actually granted, but the simple fact that the number had been dramatically increased went a long way toward keeping many active in the hobby.

The problems with the top leadership of the DASD notwithstanding, the mid-1930s saw a consolidation of the inner organization of the DASD. Like all other organizations, the DASD in the Third Reich was forced to have a strictly hierarchical organization, and to conform to the Nazi's "*Führerprinzip*" (Leadership Principle). Below the National President and DASD headquarters staff, there were 19 regional divisions

[41] See, for example, *Mitteilungsblatt des Landesverbandes A des DASD e.V* (August–September 1935), or *Landesgruppenleiter* Walter Trott, *Rundschreiben 2/35 der Landesgruppe A* (March 5, 1935).

[42] *Landesgruppenleiter* Walter Trott, *Rundschreiben 2/35 der Landesgruppe A* of 5.3.1935.

[43] "Bedingungen für die Errichtung und den Betrieb einer Versuchsfunkanlage für Funkfreunde" Anlage 1 zum Amtsblatt des Reichspostministeriums Vf. Nr. 53/1935.

called "*Landesgruppen*".[44] Heads of the *Landesgruppen* were appointed by the President, and they, in turn, appointed the Treasurer and the heads of the regional units and the local chapters, in keeping with the Nazi Leadership Principle. Each *Landesgruppe* was then subdivided into "*Bezirksgruppen*" (with an appointed Leader) and local chapters or "*Ortsgruppen*" (with an appointed leader). Each *Landesgruppe* had a Technical Coordinator (*technischer Referent*) and a "Traffic Coordinator" ("*Betriebsreferent*").[45]

The DASD headquarters was fairly small (15 paid employees by 1936), but had important duties. The key members of the headquarters were the President (Gebhardt), Herrn von Bülow (Gebhardt's deputy, and liaison to the Nazi Party) Herrn Graff (Technical Office), Herrn Garnatz (QSL Department but also Wehrmacht relations), and Herrn Queck (Mail Order Department). Rolf Wiegand was publisher of *CQ*, and Dipl. Ing. Slawyk was in charge of "foreign relations".[46] The few paid employees were supplemented by a larger number of members who did unpaid work for the headquarters, often of a technical nature. As a group, the headquarters staff were a remarkable concentration of skilled radio specialists. They not only provided leadership and important organizational services for the club and its members, but also sometimes served as technical advisors to the government and industry.

Within the headquarters there was:

- a laboratory for technical research,
- a Technical Department (with Technical Advisors in all 20 Regional Divisions),
- a "Control Station" ("*Leitfunkstelle*") with a "Control Transmitter" (*Leitfunkstation*) (with redundancy of equipment and electrical supply in case of emergency),
- a QSL card distribution office,

[44] *CQ* No. 10 (October 1933). The number varied, and later grew to 21, with the inclusion of Danzig and the annexation of Austria. "DASD Organisationsplan", *CQ* No. 10 (October 1938).

[45] *DASD Verordnungsblatt* No. 7 (April 7, 1936). Later still, new *Landesgruppen* were added as the Nazi empire expanded into France (the annexation of Alsace and Lorraine) and eastern Europe (the annexation of western Poland). DASD organization in these conquered territories remained relatively undeveloped.

[46] Along with Gebhardt's 1936 address, on the organization of DASD headquarters, see: "Amtliche Mitteilungen der DASD-Leitung", *CQ*, No. 1 (January 1937):14–15.

- an office for the scientific analysis of transmission and reception logs kept by DASD members,[47]
- a "Foreign Division" (*Auslandsabteilung*, under OM Slawyk), and
- a Mail Order Division (*Warenabteilung*).

The headquarters supervised the national Traffic Handling Service (*Betriebsdienst*), and carried out nationwide surveillance of shortwave transmissions (*Betriebsdienstüberwachung*) as a form of self-policing. It sent out a periodic newsletter of orders and regulations (*Verordnungsblätter*) and published the monthly journal *CQ*. Regional chapters also published a regular newsletter of their own. The Technical Department set standards, checked parts, and worked to standardize the equipment of DASD members. It also helped to keep all members on legal frequencies, by carrying out national tests of frequency accuracy. It had a slide library for training, and subscribed to 75 foreign radio journals. In the name of standardization and instruction, since January 1, 1936, each issue of *CQ* contained an enclosed file card with standard plans and/or tech information.

Much of this organization—the National President and headquarters; the regional, intermediate, and local chapters; and traffic nets—had existed before 1933, but after the Nazi takeover, they were tightened, subject to more strict discipline, and made compulsory. DASD members were now required to attend meetings of their local chapter, and to participate regularly in the disciplined passing of messages from one station to another, called by the DASD "*Betriebsdienst*" or "Traffic Nets".[48] Those who did not participate in meetings and other activities like the traffic nets, and did not make rapid progress toward obtaining the necessary radio skills and licenses, were dropped from the organization for lack of interest.

In order to fit into the Third Reich, the DASD national leadership sought to eliminate or at least greatly reduce the individualistic, even anarchic, freewheeling, and often-rebellious spirit of the pre-1933 DASD. The practice of illegal transmission was a particular target. The Nazified DASD leadership also sought to de-emphasize the hunt for individual radio contacts with far-flung domestic and foreign amateurs in order to obtain transmission confirmation reports (QSL cards) and, on that basis, win

[47] It was first called the "*Logauswertungstelle*"—"Radio Log Evaluation Point", and later, in mid-1936, the name was changed to "*Naturwissenschaftliche Forschungsstelle*" or "Physical Science Research Office".

[48] See below, pp. 217–218.

national and international awards and certificates of achievement for numbers and distance of successful radio contacts. This was subsumed as "radio sport", and was (at least publicly) labeled too individualistic and detrimental to a more regimented practice of radio skill as a form of national service. Given the strong interest which persisted among DASD members, and the centrality of "radio sport" to the larger international culture of ham radio, it could not be completely eliminated, however. The search for QSL cards and thus contact with foreign amateurs continued down to the outbreak of war (and in a much more limited form, even afterwards). As much as they hated the individualism and international scope of the hunt for QSL cards, the Propaganda Ministry sought to regiment it and use it for propaganda purposes. The old pre-1933 spirit of individualism and refusal of external restraint, which characterized the DASD (and in fact, international ham radio in general), did not completely die, but was thoroughly suppressed. What the Nazis wanted was not individualism and anarchy, but collectivism and control.

What Did German Radio Amateurs Do in the 1930s?

What did German shortwave enthusiasts do in the 1930s? In some ways, they did exactly what they had done in the 1920s, but in the Third Reich, there was a distinct shift toward regimentation and the needs of the state and Nazi Party. Any "ham" has two main goals: to learn about radio technology and perfect his or her own abilities and equipment, and to make on-air radio contacts with other radio enthusiasts. German hams were no different. In the 1920s and 1930s, most German hams sought to make as many contacts over as great a distance as possible, just as all hams did worldwide. This was both a test of ability to get the most out of what was often rather rudimentary equipment, and it was simply exciting: the idea of sitting in one's own home and making contact with a like-minded person a half a world away was (and can still be) thrilling.[49]

Concretely, German amateurs had to first learn the basics of radio electronics and become proficient in Morse code. They also had to build their own transmitters and receivers and other ancillary equipment; commercial ones were not produced in Germany in the 1920s and 1930s, at least not

[49] Even a strict National Socialist like Admiral Gebhardt had to acknowledge this fact. See "Ansprache des Präsidenten des DASD auf der Kundgebung am 23, Mai- 1936", *DASD Verordnungsblatt* No. 9 (April 30, 1936).

to any degree, and in any case, like foreign imports, were far too expensive for all but a handful of amateurs. They also had to learn the written and unwritten rules and procedures governing amateur radio contacts. Finally, they had to obtain a license and a call sign (a set of unique letters and numbers, organized according to an international plan, which identified the individual amateur and his or her region or nationality). To do this, they had to pass two exams, and be vetted by both the Police and the Nazi Party multiple times.

In Germany, there were two basic types of license. Both required the passing of a technical and practical exam. The first one most German amateurs earned, and numerically the most common type, was the so-called DE-License.[50] It got its name from the first two letters of the call sign or identifier used by holders of this license class, "DE".[51] Because of the very restrictive German laws on owning an amateur transmitter, this license or permit was essentially no more than a license to listen to shortwave transmissions. The DE-number was issued by the DASD according to its own criteria, and had very little legal standing in Germany. It simply documented that a budding amateur had learned the fundamentals of electronics and radio theory, Morse code, and basic on-air procedure, and possessed a receiver capable of receiving the shortwave bands, and, from 1935 onward, a functioning frequency tester (*Wellenmesser*). Both of these, first, had to be built by the candidate and licensed through the Post Office.[52] Prospective DASD members were required to take and pass the DE-exam within a set time in order to remain members. Requirements for the DE-test became increasingly elaborate and bureaucratic form the mid-1930s onward, but the fundamental fact remained: it did not authorize radio transmission.[53]

The second level of exam prepared the applicant for one of the rare transmitter licenses and was a prerequisite for it. It was administered by

[50] For the origins of the DE-License, see pp. 128–129 above.

[51] So, for example, the "DE-Number" 2653/F was issued to Günter Balz in Berlin-Wilmersdorf in 1934. *Nachrichtenblatt der Landesgruppe Berlin des D.A.S.D.* (December 1934).

[52] Rolf Wiegand (DE0065 D4cxf) *DASD-Fibel. Wie werde ich Kurzwellen-Amateur?*, Herausgegeben in Einvernehmen mit dem Rundfunkamt der Reichsjugendführung vom Deutschen Amateur-Sende- und Empfangsdienst eV. (Berlin: Weidmannsche Buchhandlung, 1936).

[53] Ibid.

the DASD under authority granted by the Postal Ministry.[54] Admission to the exam once again required political vetting by the police. Once the exam was passed, and once the applicant was finally awarded (by the Postal Ministry) one of the limited numbers of hotly sought-after transmitter licenses, he or she was also granted a proper international radio call sign. In Germany, at this time, they all began with "D", and were sometimes called a "D-Number".[55] Only those who held this type of call sign were authorized to transmit legally, and thus to participate in two-way radio contacts. Transmission was only allowed using Morse code (telegraphy); as earlier, voice communications (telephony) were forbidden, and legal transmitters had to be built in such a way that telegraphy was the only mode of transmission technically possible. As might be imagined, the lucky few who held a D call sign represented an elite within the DASD. In the Third Reich, it also meant that these men and women had been thoroughly vetted for their political reliability several times.[56]

On a day-to-day basis, German amateurs would first study for the DE-test, learning the basics of electronics and practical radio theory. They would also have to build a radio receiver capable of receiving shortwave transmissions. By 1935, most would build a receiver according to a standard DASD plan, and could count on help from fellow members. Moreover, DE-candidates would have to learn Morse code and basic on-air procedures, either through self-study or through DASD courses. By the mid-1930s, they would also have to document their ability to receive a certain number of transmissions on various frequencies, and participate in listening to the *Betriebsdienst* traffic nets.[57] In the Third Reich, DASD members were expected to pass this DE-exam within six months of joining, and if they did not pass it, they could be dropped from the organization unless they could show good reason why. Before the Third Reich,

[54] See the description "Für den Anfänger—Der Weg zur Sendeerlaubnis", *CQ-MB. Mitteilungen des Deutschen Amateur-Sende- und Empfangs-Dienstes e.V.* No. 7 (July 1935): 103–104. The legal basis was: "Bekanntmachung über Liebhaberfunksender vom 10. February 1935" (Amtsblatt des Reichspostministeriums vom February 13, 1935) and later "Verordnung über Sender für Funkfreunde vom 9. Januar 1939, mit Ausührungsbestimmungen". (Amtsblatt des Reichspostministeriums von 12. Januar 1939, Nr. 5).

[55] For example, the prominent German amateur Alfred Noack held the call sign D 4 BOF in 1934. *Nachrichtenblatt der Landesgruppe Berlin des D.A.S.D.* of December 1934.

[56] The 1939 regulations only required vetting by the police. Prior to this, between November 1937 and the outbreak of war, a second level of vetting was done by the Nazi Party. See below.

[57] For more on the *Betriebsdienst*, see below.

some members were slower to take the DE-exam or avoided it altogether. Those who already possessed a DE-number would then need to study more complicated radio theory and the practical construction of a (short-wave) transmitter to prepare for the licensing exam. They would only be authorized to actually build a transmitter once they were granted a transmit permit and call sign by the postal authorities. Radio amateurs also participated in their club activities, attending (or not) regular meetings and technical presentations on various topics of radio theory and practice. Those studying for an exam would also attend classes organized by the club if they were offered locally. We have seen how, in the 1920s, radio club members participated in the wide range of social activities offered by the clubs, which were a major attraction for new members.[58] This social side was severely curtailed from the mid-1930s onward. The DASD in the Third Reich did not (openly) waste time on social activities; its role in national defense was far too important.[59] Radio in the Third Reich became a very serious business, even as a hobby.

Finally, all German hams were expected to participate in the so-called *Betriebsdienst* or Traffic Nets. This was an organized and strictly disciplined system to pass messages from station to station in a set order and on a fixed schedule, using Morse code. This simulated the transmission of messages in an emergency, and was seen as one of the most important means of training DASD members to a high standard, not least with the goal of aiding national defense. It thus existed on the intersection of "radio sport" and rearmament. Though this was a kind of practical training mainly intended for the D-license holders, DEs were also expected to participate by accurately writing down transmissions they received. By mid-1936 the DASD had a network of 110 (mostly individual) amateur transmitting stations in Germany, which engaged in weekly traffic passing.[60] By 1936 there was an official in the DASD headquarters and one in each regional division of the DASD charged with organizing and supervising the *Betriebsdienst*. The headquarters also had a "Control Station" (*Leitstation*) which was active in the *Betriebsdienst*, among other activities.

[58] See above, Chap. 1.

[59] Naturally, there was still a social side to club activities, but it was not at all advertised openly.

[60] Schedules were printed in the newsletters of the various DASD Regional Sections. For example, "Betriebsdienst", *MBT: Mitteilungsblatt der Landesgruppe Mittelrhein* 1, No. 1 (January 1935): 6, or "Funkplan", *Nachrichtenblatt Landesverband Reichshauptstadt im Deutschen Amateur Sende- und Empfangsdienst e.V.* 9, No. 3 (April 1943): 8.

Points were awarded to the operators for discipline, accuracy, and speed. In the Third Reich, the DASD emphasized this training more and more as time went on, continuing it even during wartime. Participation became mandatory, and the penalty for failing to participate was expulsion for lack of interest.[61] It was a real thorn in the side of the amateurs who were more interested in making individual contacts and one of the factors driving the progressive emphasis on ever-stricter discipline. Amateurs in other countries engaged in similar traffic nets, but their participation, as in the case of the US, was optional, and it was not used as a disciplinary tool as it was in Germany. In this way, a practice accepted within the international amateur community became in the Third Reich a tool for discipline and preparation for national defense on a large scale. Many former DASD members later complained bitterly about being forced to participate in this training, even though all agreed that it was very effective.[62]

1936 Presidential Address to Members: A Snapshot of the Reconsolidated DASD

An address given by retired Rear Admiral Gebhardt, the then President of the DSAD to the yearly meeting of the club on May 23, 1936 gives a good snapshot of the state of the organization under the Third Reich and after its Nazification was complete.[63] It shows an organization trying desperately to maintain its existence, at the price of a thorough purge, increasing imbrication with Nazi militarization and propaganda, and a growing emphasis on strict discipline and conformity. Later, postwar histories of amateur radio in Germany like to portray this as mere outward conformity by a still largely independent and purely hobby-oriented organization. Certainly, that is one part of the truth. But how far can a club go in this outward conformity before it effects the inner substance? How long before conformity becomes commitment? That is the real question here.

[61] A high degree of participation was made possible not least by progress in radio design, which allowed the "rejection" of nearby signals and the reception of a single signal on a fairly narrow slice of the radio spectrum. This allowed a greater number of simultaneous transmissions on narrower pieces of the spectrum and thus greater participation.

[62] Körner, *Amateurfunk*, 135–136, praises the efficacy of the *Betriebsdienst* as training and as a political argument for the continued existence of DASD.

[63] "Ansprache des Präsidenten des DASD auf der Kundgebung am 23, Mai 1936", *DASD Verordnungsblatt* No. 9 (April 30, 1936).

To understand the new place of the amateur radio hobby in the Third Reich, it is instructive to look first at the non-members who attended the meeting. Most of them represented organizations which, one way or another, were interested in the DASD and sought to either use it or control it. Their presence at the meeting signaled their claim to an interest in the DASD. They were there to demonstrate this, to keep an eye on the proceedings, and also to show rival state and Party organizations that the interests of their own organization had to be respected. Their presence demonstrates the role and importance of the DASD within the totalitarian state, but also indicated the inner-system rivalries which pulled at the DASD from different directions.

Significantly, the first guest recognized by the president of the DASD in his speech was a Herr Droysen from the *Reichsministerium für Volksaufklärung und Propaganda* (RMVP, the Propaganda Ministry). The RMVP was the master of the DASD and exercised the most far-reaching control over it. He was seconded by two men, Herrn Dudda u. Bruchmann, both from the *Reichsrundfunkkammer* (The Reich Radio Chamber, RRK). The DASD was a member of the RRK, which was, in turn, subordinate to the RMVP.

Next in the list of honored guests were three officers from the Armed Forces, Lieutenant Commander (*Korvettenkapitän*) Schmolinske from the Navy, and Army Captains Kopp and Runge, all representing the Ministry of War. They were present because of the close cooperation of the DASD with the military. They were there to underline the armed force's interest in the DASD and to jealously watch over the other organizations seeking to influence it. They were joined by retired Navy Lieutenant Schrimpff, ostensibly a civilian who was a leading official in the newly formed "*Freiwillige Wehrfunkgruppe Marine*" (Maritime Voluntary Military Radio Group, FWGM), which was formed within the DASD to conduct training of navel communications personnel.

Following the military representatives were guests from the Postal Ministry, *Amtsrat* Schulle, and the Ministry of the Interior, (Police) Lieutenant Tramps. The Postal Ministry was also represented by *Oberpostrat* Reipert and by *Obertelegraphen Inspektor* Wittke from the *Reichspost-Zentralamt* (the Central Office of the Postal Service, charged with technical oversight). The Postal Ministry was at the meeting as the legal authority in charge of all civil radio and telegraph communications, and not least, as the legal authority in charge of granting individual amateur transmitting licenses. As always, the Postal Ministry was a jealous guardian of its prerogatives, and remained highly suspicious of radio amateurs and the DASD. The Ministry of the Interior, and therefore the police, was also very suspicious of amateur radio. It is significant that the Interior Ministry sent a police officer to represent it and not a civilian official. It should also be mentioned that there was a separate guest from the Gestapo, (SS) *Sturmbannführer* (Major) Riechers. The Gestapo, as the political police, was particularly interested in ferreting out regime opponents and suspected them in the (once very cosmopolitan) DASD. All of these "honored guests" were there to demonstrate their power, and to keep an eye on a group of potential troublemakers.

Next came a number of guests from other organizations which cooperated with the DASD in the training of their members in radio. Three men came from the Hitler Youth (Herrn Panek, Saunat, and Schreiber) and one each came from the Reich Labor Service (*Oberstfeldmeister* or Labor Service Colonel Dr. Seipp), and two from the Reich Civil Defense Association (*Reichsluftschutzbund*, RLB), *Luftschutz Oberführer* (a rank between Colonel and Major General) Jährig from the RLB headquarters, and *Bezirksgruppenleiter*[64] Grunell from the RLB Regional Office in Berlin. The Hitler Youth, in particular, had big plans for cooperation with the DASD, and their presence in such numbers was a sign of their interest in the club.

The Nazi Party also sent a representative, though only relatively low-ranking "*Gaufunkstellenleiter*" Böker from the Berlin regional party organization.[65] It is clear that the Nazi Party Organization itself was not a player in radio questions—that role was taken care of from a Nazi Party perspective by the Propaganda Ministry. But the *Funkstellenleiter* did have an interest in Radio, and wanted to be in on the proceedings.

Naturally, two of the three most recent former DASD presidents (and honorary members) retired Colonel Fulda and Prof. Dr. Leithäuser were in attendance, as would have been common in nearly any club. The immediately previous DASD President, Vice Admiral Dr. H.c. Groos (also still an Honorary Member) was not present, however. He had only fairly recently been called back to active duty out of retirement, and was ostensibly not able to attend because of his military duties. On the other hand, he was seen by Goebbels as uncooperative and slow to act, and had been

[64] Not a rank, but the head of a regional office, in this case, Berlin.

[65] *Gaufunkstellenleiter* were officers within the Gau (region) leadership, whose job it was to deal with radio issues. It was not a particularly high-ranking position.

more or less forced out of the DASD. It is no surprise that he was "unable" to come.

Of course, the press was also in attendance, for everyone involved—and not least, the DASD itself—had an interest in using the meeting to make a good impression on the public. Herr Bachmann, the Head of the Press Office of the National Broadcast Radio Authority (*Reichssendeleitung*) was present, as were *Oberregierungsrat* Dr. Gehne, from the radio journal *Funk*, Dr. Mickeley from the *Völkischer Beobachter* (the official organ of the Nazi Party), and Herr Sinn from the local *Charlottenburger Zeitung*. Finally, there was also a Herr Drossel from the Film and Picture Office of the Berlin City government (*Film und Bildamt der Stadt Berlin*). They were working with the DASD to prepare a film on radio technology to train high school physics teachers to teach it in their classes.

From the list of outside attendees, it is clear that the DASD was no longer an independent organization. It was closely watched, and tightly controlled by some of the most powerful institutions in German society. If it had any independence at all, it was because all of these powerful institutions were as jealous of each other as they were interested in the DASD.

The 1936 presidential address contains a great deal of information about the history and future of the DASD—from a post-*Gleichschaltung* perspective. It is clear that Gebhardt felt he needed to offer a look back at the history of the DASD from the perspective of its position in the National Socialist state. He was at pains to cast the DASD as having been patriotic ("*vaterländisch*") from the very beginning. Yet he was also very open about the special fascination of radio as a hobby, which stood somehow outside of a political framework:

> It is the ability of often derisively small and primitive equipment to have a mysterious long-range action, around the entire globe, which has such a terrifically strong effect on people. Amateur radio operators are able to exchange ideas with people whom they never see face-to-face, and whose lands and habits are completely strange, whose lives are lived in ways completely differently from their own. (…) The magic of the vastness, the feeling of having radio-friends all over the world, has crept into the quiet room of the amateur.[66]

[66] "Ansprache des Präsidenten des DASD auf der Kundgebung am 23, Mai 1936", *DASD Verordnungsblatt* No. 9 (April 30, 1936).

He was also quite open about the often-illegal activities and "anarchic" spirit of the DASD and the radio hobby in general before 1933. Nevertheless, the address stated forcefully several times that those days were over, and that the DASD's commitment to Germany required the strictest of discipline. He had a special caution for any DASD members who still engaged in illegal transmission or "*Schwarzsenden*". This was actually an admission that such activities still went on, despite the best efforts of the DASD itself, the Postal Ministry, and the Police to stop them.[67] Gebhardt was very candid about the reasons why the DASD and most of its members supported the National Socialist regime: the allowance of private licenses for amateur transmission. The address is also surprisingly open about the price which was paid for this privilege, the ideological and racial purge of the club:

> We can identify the main benefit of the first period of the "National Upheaval" for the DASD as the ideological and moral conversion of both its members and its duties and goals to the conceptual world of National Socialism and the Germany of Adolf Hitler. Inevitably, this resulted in the fact that a number of members who did not belong in our ranks racially and ideologically had to be expelled, while, on the other hand, a flood of new and enthusiastic amateurs came to us and brought fresh new blood to our organization.[68]

This is a plain admission that the DASD was purged of racially and ideologically undesirable members. From the tone of the address, it is clear that not only was this price recognized, it was paid willingly in exchange for the right to privately transmit.

In his discussion of the current status of the club, Gerhardt reminded his audience that the DASD was formally recognized as an officially approved club, and the legal right of amateurs to transmit with a private license was guaranteed by the Postal Ministry in its "Declaration on Experimental Transmitters of February 13, 1935". He further stated that the DASD monopoly as the only non-governmental organization able to grant this license in the name of the Postal Ministry had also been guaranteed, as a sign of its importance in the Third Reich. Yet in the same paragraph, he also stated that the DASD had been built into the National

[67] See below.

[68] "Ansprache des Präsidenten des DASD auf der Kundgebung am 23, Mai 1936", *DASD Verordnungsblatt* No. 9 (April 30, 1936).

Socialist system through its *Gleichschaltung* and its subordination under the Ministry for Popular Enlightenment and Propaganda and its membership in the Reich Radio Chamber. Henceforth, the DASD would consult the Propaganda Ministry on all fundamental questions, and the President of the DASD and his Deputy would be named by the Ministry. Again, the message was crystal clear: the DASD had become a subordinate part of the Third Reich in exchange for its continued existence. To survive, the DASD as an organization and all of its members would have to change their goals to correspond to those of the state. The days of individualism, of hobbyists focusing on purely amateur radio concerns were over, and national goals and tasks now had to take the fore. The individualistic, even cosmopolitan or internationalist radio amateur of the past had now been replaced by a new spirit of national community, which ostensibly gave DASD members a feeling of responsibility to the state to perform at the highest level. The consequences also included, according to Gebhardt, the installation of the Nazi Leadership Principle (*Führerprinzip*) and the imposition of strict discipline. As he puts it, this was a logical consequence of the important work the DASD was doing for rearmament. Again and again, Gebhardt stressed that discipline was important in the current order. This was done as a warning to DASD members, but also as a statement to the watching representatives from the Propaganda Ministry, Postal Ministry, Police, Military, and Nazi Party that the DASD would do what was expected.

According to Gebhardt, by mid-1936, the DASD had roughly 4000 members. Of these, 531 held active private transmitter licenses. This was a substantial increase over the roughly 120 private licenses which existed prior to the license expansion promised by the Third Reich. More were to come, for 116 license applications were still pending, held up by the lack of the required political Certificate of Suitability (*Unbedenklichkeitszeugnis*) granted by the police.[69] A full 40 license applicants had been refused their license for political reasons by 1936. Nearly that many (39) had withdrawn their applications, mostly because the applicant had entered the Army or the Labor Service,[70] and a further 55 had their licenses revoked (most either temporarily for military service or permanently for lack of

[69] This is separate from a similar certificate provided by the Party, which was required from 1937 onward.

[70] All men serving in the armed forces or Labor Service for less than two months were forced to formally leave the DASD. Leaving the DASD also meant that they had to give up their permit to transmit and surrender their transmitter, if they had one. (This did not apply to those sent into the DASD by the Navy).

interest). Though the Third Reich was not the el dorado of open licensing which DASD members hoped for, it had significantly raised the number of legal private transmitters. For the hobbyists of the DASD, this was a powerful argument for cooperation with the regime. Clearly, though, licensing standards (both political and technical) remained quite strict, and those who did not conform either were prevented from obtaining a license or saw their license quickly revoked. This was a powerful argument in favor of conformity. According to the presidential address, membership in the DASD was growing in 1936. This stands in contrast to internal communications, which demonstrate uneven growth across Germany, and some fear of declining membership if active recruitment did not take place.[71]

One difficulty was that, while many new members joined, there were also a large number of members who left, either voluntarily or involuntarily. Some who left the DASD were simply too busy with work to continue; this is partially a reflection of the (rearmament-related) upswing in the German economy after the Depression. Others left voluntarily because they began their military or labor service—both compulsory by 1935. DASD membership (and all other private memberships) ended during service in either organization. Then there were the continued number of involuntary expulsions, most for "lack of interest", which meant either failure to pay dues or failure to participate in the ever-growing number of compulsory activities, or for "lack of discipline", often a euphemism for illegal transmission. Furthermore, DASD members were also still being expelled for racial or political reasons in the wake of the *Gleichschaltung* well into the mid-1930s. In business year 1935–1936, 796 DASD members left, the vast majority expulsions. What the numbers in the address and other items in the various circulars sent from headquarters indicate is that in the mid-1930s, a sea change was happening in the DASD. Jews and politically questionable elements were purged in the *Gleichschaltung* and continued to be hunted out and eliminated. By the mid-1930s there was also increasingly less place in the DASD for those who couldn't follow the rules or who were not willing to place club business and national rearmament before their love of the hobby and their pursuit of as large a number of individual radio contacts as possible. Non-conformity or individualism of any kind had no place in the DASD under the Third Reich and was increasingly being weeded out.

[71] See the various numbers of the *Verordnungsblatt des DASD* for 1936.

Moreover, though the number of transmitting licenses were increased in the Third Reich, the new licenses did not come close to meeting demand. A tightening of their award and the slow workings of the Postal and Party bureaucracies led many to resign from the DASD in frustration in 1937.[72] Counterbalancing these losses were significant numbers of new members who were willing to observe the discipline and constraints of the radio hobby under National Socialism, or who were brought to the DASD for training as a part of the formal cooperation with the Navy and Hitler Youth. These men and boys were full members of the DASD for the duration of their training (and paid dues!), but most did not come to the DASD as "pure" hobbyists, their purpose was more utilitarian and narrow. But radio was magic, and clearly, the DASD hoped that many of these fellow travelers would catch the fever and stay members, or come back after the end of their military service. It is hard to say how many of the temporary members became real hobbyists, but the post-Second World War experience suggests that, in fact, many did become enamored with the magic of radio, a small but perhaps not insignificant victory for hobby over totalitarianism.

Yet these numbers are difficult to interpret, since many surely left because they had taken on new jobs and simply had too little time for a hobby, or else because they had joined the military and were not allowed to remain DASD members. Rearmament brought a lot of new opportunities, which took men away from their hobby. DASD members would continue to be pulled away from the hobby by military service and professional responsibilities as Germany was progressively mobilized for war. Some DASD members certainly left voluntarily, either because they saw the writing on the wall or because they were put off by the increasing emphasis on discipline.

Four Major Tasks

In the mid-1930s, the DASD set itself four major tasks:

1. Continuance and improvement of the purely "sportive" or hobby aspects of radio
2. Help with rearmament
3. Technical development of radio
4. Advancement of science through collaboration with universities and industry.

[72] Körner, *Amateurfunk*, 143–145.

Radio Sport Even at the formal yearly full-membership meeting of the DASD, with so many observers from the military, police, and Nazi Party, the DASD president (appointed by Goebbels, remember!) could not resist or avoid speaking of the magic of radio and the thrill of the hobby.[73] Certainly, he was playing to the crowd, but the remarks still are a major key to understanding the hobby culture of radio. Radio is magic. Voices from around the world come wirelessly in people's homes, and in the case of amateur radio, one might converse with complete strangers from the other side of the globe. The kind of excitement radio generated is a fundamental element of explaining why so many young men (and women) would join the DASD and remain with it despite the growing discipline and repression. Nevertheless, Gebhardt seems to have defined "radio sport" not in terms of individual ability to make long-range contacts, but rather more as collective performance of DASD members in national and international competitions.[74]

Technical Efforts of the DASD and the Advancement of Science Amateur radio depended on equipment which was either built completely from scratch or heavily modified from commercial or military-surplus radios. The market for amateur equipment in Germany in the 1930s was far too small for there to be any commercial production, and in any case, before 1935, legal private transmission was nearly non-existent. Radio amateurs were thus all "makers", in the modern sense. This also meant that the amateur radio community also had to be largely responsible for advancing the technical state of their equipment, which meant that amateur radio always had a strong element of "citizen science". The DASD, both before and after 1933, sought to encourage the activity of citizen scientists within the radio hobby, not only because it was fun, but also because it gave amateurs a powerful argument for their own usefulness and importance. Characteristically, after 1933 the emphasis on citizen science was intensified, but went hand in hand with a greater emphasis on discipline, and a stronger focus on certain key questions. Not coincidentally, the scientific problems the DASD sought to address were also of interest to the military.

[73] "Ansprache des Präsidenten des DASD auf der Kundgebung am 23, Mai 1936", *DASD Verordnungsblatt* No. 9 (April 30, 1936).

[74] For example, see: DASD President Gebhardt, "Zur Jahreswende", *CQ* No. 1 (January 1938): 1–5. See also: W. Schierenbeck, "Deutsche Gründlichkeit—Auch im DASD!", *CQ* No. 3 (March 1938): 37.

Once again, under National Socialism, the fascination of the hobby was restrained and directed toward state ends. As one of the four main foci of its attention, the DASD worked in various ways to contribute to the technological development of radio. This was one of the main responsibilities of the Technical Office within DASD headquarters. There were also Technical Advisors (*Technische Referenten*) in every Regional Division (*Landesgruppe*). Moreover, various DASD members volunteered to form ad hoc working groups to study particular topics, for example, the workings and propagation of the new 10-meter band or even the 5-meter band, which was even newer.[75]

One of the principle tools for scientific cooperation and research maintained by the DASD was the special office for the study of reception and transmission logs kept by all DASD members. It was located in Breslau, and was run by DASD member Dr. Kunze. Just before the 1936 annual meeting, the name was changed to "Physical Science Research Office" ("*Naturwissenschaftliche Forschungsstelle*"), and it was expanded to 12 full-time employees. Its task was to study the links between propagation of radio waves and meteorological conditions, with a goal of producing a guide to the best transmission times for each amateur radio-frequency band. It was also interested in investigating the phenomenon of fading (*Schwunderscheinungen*). All DASD members were required to keep logs of radio stations they had heard or with whom they had made a two-way contact. These logs had to be forwarded to this office monthly (even though many members were slow to turn them in). According to Gebhardt's 1936 address, it received on average 50,000 reported reception reports per month by the beginning of 1936. The DASD understood that these questions were too big for citizen scientists alone, and that they could only be addressed in cooperation with professional scientists. The Physical Science Research Office was intended to concentrate this coop-

[75] From the mid-1930s, the DASD worked on researching propagation on the then still poorly understood very high frequency (VHF) portion of the radio spectrum, particularly on the 10-meter and 5-meter bands. (German: UKW; roughly 30 MHz–300 MHz.). See, for example, "Ansprache des Präsidenten des DASD auf der Kundgebung am 23, Mai 1936", *DASD Verordnungsblatt* No. 9 (April 30, 1936). Portions of these bands had been opened for amateurs at the 1927 International Radiotelegraph Conference, and soon gained the interest of industry and the military. This research continued into the war. See, for example, *Nachrichtenblatt Landesverband Reichshauptstadt im Deutschen Amateur-Sende- und Empfangsdienst e.V.* 9, No. 1 (January–February 1943).

eration in a central point. According to Gebhardt, DASD cooperation with Technical Schools and Universities was growing, and should increase the scientific precision and value of ham observations. But this also means that, in the future, there would have to be more discipline and concentration in the observations: the old amateur practice of letting everyone observe everything had to stop. In other words, science in the Third Reich had to be disciplined and directed by the appropriate institutions. Characteristically, the language used to describe and encourage DASD participation in these scientific observations was a mix of military and romantic metaphors:

> The DASD needs men who know how to employ, at the decisive point, fantasy mixed with logic, and thus to make a daring individual advance into the unknown no-man's-land of science, and thus give new goals to the collective work which marches behind them according to plan in a broad front. (...) In this belief the DASD seeks to fight for a place in science.

This is a manifesto in favor of citizen science, albeit one wrapped in fascist language of struggle and bravery. It also stakes a claim to being more than just a leisure activity: amateur radio was also part of science, and therefore something which ought to be kept around, and whose value ought to be recognized. Later, in 1937, the DASD signed a cooperation agreement with the National Research Council (*Reichsforschungsrat,* a national research consortium), further linking amateur radio to the scientific establishment.[76]

In the 1930s, the DASD also made a great effort to raise the technical standards of its members, and to persuade them to standardize their home equipment. It did so by developing reference plans for key pieces of equipment and publishing them in its club magazine, *CQ*. It first published a standard frequency meter (necessary for on-air discipline), which was followed by other plans/standards, complete with parts lists. In the name of both standardization and instruction, from January 1, 1936, onward, each issue of *CQ* contained an enclosed file card with standard plans and/or technical information.[77] Though typically very short, *CQ* contained technical articles, and after 1935, there were special issues devoted to various

[76] DASD President Gebhardt, "Zur Jahreswende", *CQ* No. 1 (January 1938): 1–5.

[77] Nearly a full run of *CQ* is available in several university libraries, and online at: http://www.afu-df3iq.de/index-Dateien/page0024.htm. It was not only sent directly to DASD members, but was included as an insert in the journal *Funk*.

topics, as well as the inserts containing the DASD's standardized plans for radio equipment. According to Gebhardt in the 1936 Presidential Address, roughly 4000 individual copies were printed each month. Copies were also included as an insert in the magazine *Funk*, which itself had a print run of 9300 per month.[78] Also as a teaching tool, the DASD published, in cooperation with the HJ, a beginners' guide to radio called the *DASD-Fibel*. The cost was RM 1.30 for DASD members. It was designed to prepare DASD members for the "DE-Prüfung", the first mandatory license test for DASD members. Other published materials (*Kurzwellentechnik*, now in a 2nd. edition) were designed to take the member further, all the way to the "D-Prüfung", the prerequisite for a license to transmit.[79]

The Technical Division built a lending library of instructional slides for teaching purposes. DASD headquarters subscribed to around 75 foreign radio journals/magazines, to keep on top of technical developments worldwide. In this same effort, the lab at DASD headquarters was expanded in the mid-1930s, again largely in the name of standardizing and testing radios and parts. Similarly, the Control Transmitter (*Leitstation*) at headquarters was also expanded, not least because the National Traffic Service was run from there, but also so that it could provide standardized test transmissions to help members measure the accuracy of their equipment. The Postal Ministry and the DASD worked together to provide standard frequency measurement tools and special standardized frequency transmissions, so that all DASD members' equipment can be certified to be working on frequency.

One can look at these efforts to standardize and modernize both individual "stations" and the DASD headquarters station in two ways. It is certainly true that making sure members used up-to-date equipment was in everyone's interest, not least because it helped amateurs to transmit clean signals on an accurate frequency, and thus helped to prevent interference and out-of-band transmission (transmission on the wrong frequency outside of the range reserved for amateur use). On the other hand, it is hard not to see the connection to the growing political regimentation of the club. Standardizing equipment was also a way of standardizing

[78] "Ansprache des Präsidenten des DASD auf der Kundgebung am 23, Mai 1936", *DASD Verordnungsblatt* No. 9 (April 30, 1936).

[79] Deutscher Amateur-Sende- und Empfangsdienst, Hauptverkehrsleitung, Technische Abteilung, *Kurzwellentechnik, Ein Leitfaden für den Amateur* (Berlin: Rothgiesser & Diesing, 1931, 2nd. edited edition 1935). Wolf Wiegand, DE0065, D 4 cxf, *DASD-Fibel. Wie Werde ich Kurzwellen-Amateur?* (Berlin: Weidmannsche Buchhandlung, 1936).

behavior. Helping members obtain and use quality, modern equipment was also preparing them to use modern military equipment. And increasing the performance of the headquarters station provided the members not only with a reference standard in terms of frequency and operation, but also ensured that it would be able to run the disciplined mandatory traffic nets of the *Betriebsdienst.* These were all either modern, commonsense efforts, or more sinister, totalitarian ones. Both interpretations are true.

Cooperation with Wehrmacht Among the four tasks the DASD set itself in the mid-1930s, cooperation with the military was the most important. After all, why wouldn't the DASD want to collaborate with the military? Most German amateurs were also German patriots, and in any case, the more left-wing or cosmopolitan elements had all been purged. Moreover, collaboration with the Military was a welcome counterweight to the Ministry of Propaganda and the state agencies which were suspicious of amateur radio (the Postal Ministry, the Interior Ministry). Military cooperation was a central part of the balancing act which the DASD had to perform in order to guarantee its sheer survival. Besides, the sincere recognition of the skills of DASD members (particularly by the Navy) opened the adventurous prospect of service in military communications. It was flattering, and might just lead to a job, which, in the aftermath of the Great Depression, was not negligible. The problem was, by the mid-1930s, the military's interest in this cooperation began to wane.

Cooperation with the Navy continued in the latter 1930s via the "*Freiwillige Wehrfunkgruppe Marine*" (Voluntary Naval Military Radio Group), which focused on pre-military radio training. As Gebhardt said in his address, the DASD would now encourage young men to train for and obtain the newly instituted "Nautical Sport Radio Certificate". At first, the DASD helped with the training and supplied training equipment, including training kits to teach Morse code and standardized radios. Gradually, the role of the DASD declined, as instruction was taken over by active personnel, and by the latter 1930s, had essentially ended.[80]

While the long-standing relationship with the Navy meant that DASD men had a clear path to communications positions with the Navy, the

[80] This pre-military training for the Navy ended in 1939, when the SA was given responsibility for all forms of pre- and post-military training by an order of Adolf Hitler.

cooperation with the Army and Air Force was less clear. The DASD certainly wanted to achieve a similar guarantee that DASD membership and training might lead automatically to service in communications with the Army. While Gerhardt promised in 1936 that such an agreement with the Army was in the works,[81] this seems to have been largely wishful thinking; at least in the early years of the war, the DASD complained of the great waste of talent caused by drafting DASD members into non-communications units. At least talks had taken place between the Army and the DASD; Gebhardt also expressed the wish that a similar system would be agreed upon with the German Air Force, but had to admit that nothing was yet in the works. The only firm cooperation with the Air Force Gerhardt could cite was participation of DASD members in ostensibly civilian flying training sponsored by the Air Force as a form of covert rearmament.

Nevertheless, in 1936, Gebhardt expressed the strong wish that the DASD would strengthen its cooperation with the armed forces and would continue to contribute even more to German rearmament in the future. Interestingly, in this context, he claimed that similar participation in military training was already being done by the amateur radio organizations in the other major powers and specifically mentioned the US, England, France, Poland, and the Soviet Union. Once again, in the German mind, the Great War continued to cast a shadow over the present.

In retrospect, by the mid-1930s, relations between the DASD and the armed forces were at a turning point. In the early few years of the Nazi regime, rearmament had already begun, but only in the strictest secrecy due to the restrictions of the Versailles Treaty. During this time, the DASD was not only a necessary technical partner, but also a very convenient screen behind which rearmament could be hid. That changed in 1935, when German rearmament was made public. From this point onward, there was less and less need for a civilian fig leaf like the DASD. The armed forces were gradually building up a cadre of their own radio specialists who could take over training, and as rearmament became more and more open, there was less reason to hide behind the DASD. Hence, we see, by

[81] Somewhat after this address, the DASD began to issue a new "Certificate of Radio Proficiency" (*Funkzeugnis*) to its members. A loose agreement with the Army called for it to be recognized when making assignment to specialty formations. This was superseded for the military in 1939 when the SA was given all forms of pre-military training, and replaced the DASD *Funkzeugnis* with its own certificate. The DASD certificate was still issued and used to prove radio skills in private industry.

mid-1936, a weakening of direct cooperation with the armed forces, and an added emphasis on cooperation with other organizations such as the Hitler Youth, Labor Service, and Civil Defense Association. This did not prevent the DASD leadership from continuing cooperation of various kinds with the armed forces, it merely reduced the leverage which the DASD enjoyed.

If the DASD was still working on developing close formal relationships with all three arms of service in the mid-1930s, it was expanding its work with the Hitler Youth and other official and semi-official state organizations. On November 7, 1935, the DASD signed a formal cooperation agreement with the Hitler Youth.[82] According to this agreement, the DASD would allow selected members of the Hitler Youth to join the DASD, and would thus make a special exception to its membership rules, which generally limited DASD membership to those 18 and over; HJ members would be allowed to join at age 14. The DASD agreed, where numbers permitted it, to form separate youth groups within the DASD for these new members. By mid-1936, such cooperation existed in all 20 regional DASD divisions. Even though the DASD complained that this was often a significant sacrifice for the organization, since many HJ members did not have the means to pay their dues, it was nevertheless worth the effort, since "herewith the wellsprings are tapped, which will continually bring it [the DASD] the best young German blood and fresh energy". HJ members were named by the HJ headquarters in each *Landesgruppe* to head the Youth Group.

Similarly, the DASD was beginning to work with the "National Civil Defense Association" ("*Reichsluftschutzbund*", RLB), a semi-private association sponsored since 1933 by the Air Force, whose task was to improve civil defense awareness and training within the general population. Gebhardt

[82] The Hitler Youth (including the girl's wing, the *Bund Deutscher Mädel* or BDM) was originally a private organization of the Nazi Party. Like the SA, it then became a state-sponsored organization in 1933, when the Nazis took power. It was first voluntary, but membership was later made compulsory in 1936. See: Michael H. *Kater, Hitler Youth* (Cambridge: Harvard University Press, 2004); H.W. Koch, *The Hitler Youth: Origins and Development, 1922–1945* (New York: Barnes and Noble, 1996); Gerhard Rempel, *Hitler's Children: The Hitler Youth and the SS* (Chapel Hill, NC: University of North Carolina Press, 1989); Arno Klönne, *Jugend im Dritten Reich* (Cologne: Diederichs, 1982); Dagmar Reese, *Straff, aber nicht stramm—herb, aber nicht derb: Zur Vergesellschaftung von Mädchen durch den Bund Deutscher Mädel im sozialkulturellen Vergleich zweier Milieus*, Ergebnisse der Frauenforschung (Weinheim: Beltz, 1989).

gave few specifics in his address, since the collaboration was new. The DASD also hoped to work more closely with the "National Labor Service" (*Reichsarbeitsdienst*, RAD), which became compulsory for all German men (and voluntary for all German women) at age 18.[83] Little eventually came of this wish, since by law, members of the RAD had to give up any outside memberships, including in the DASD, though exceptions were made for the small cadre of career RAD leaders.[84]

Another new project involved collaboration with the City of Berlin on a project to teach more radio science in schools, and thus get more school pupils interested in radio. The DASD was collaborating with the Berlin "Film and Picture Office" to make a film to teach the science of radio to physics teachers, as a way of helping them to reach their pupils. The DASD was of course thrilled and cooperating as much as it could. (Remember, a representative from the Film and Picture Office was present at the 1936 general meeting.) The first teacher training course had just finished and gone well by the time of the 1936 general meeting, and a second was about to begin. The DASD hoped through this cooperation to be able to recruit pupils of college preparatory high schools in Berlin, and hoped that this might lead to a similar teaching effort across the entire country.

Another mid-1930s DASD project was to try to get German emigrants interested in amateur radio. It had begun working with the Nazi Party's "Foreign Organization" (*Auslandsorganisation*) on this topic, with the hope that radio would become a way of tying these Germans closer to their country and culture. Again, very little seems to have come of the plan.[85] In short, the DASD was keenly aware that it needed to make itself "useful" in the Third Reich in order to survive. In the 1930s, it continually sought tasks which would allow it to thrive and allies who would protect it from prosecution and disbandment.

[83] On the Reich Labor Service in international perspective, see Kiran Klaus Patel, *Soldiers of Labor: Labor Service in Nazi Germany, and New Deal America, 1933–1945*, Publications of the German Historical Institute (Cambridge: Cambridge University Press, 2005).

[84] "Sendelizenzen für Angehörige des Reichsarbeitsdienstes", *DASD Verordnungsblatt* (April 3, 1935): 3–4.

[85] "Ansprache des Präsidenten des DASD auf der Kundgebung am 23, Mai 1936", *DASD Verordnungsblatt* No. 9 (April 30, 1936).

Schwarzsenden[86]: The Threat of Uncontrolled Transmission

For as long as the radio hobby had existed, the authorities had feared illegal radio transmission by ordinary citizens. That this remained a fear in the Third Reich should not come as a surprise, nor should we find it unusual that the accusation of illegal transmission was one of the most powerful weapons the authorities had to keep the DASD in line. What may be surprising is the fact that, despite all the threats, despite the existence of a finely tuned police state, some individuals actually did continue to transmit illegally.

In his 1936 address at the DASD general meeting, President Gebhardt felt obligated to speak about the problem at length:

> At this point I wish to speak again to the problem of illegal transmission. Unfortunately, there are still isolated cases where DASD members simply can't restrain their impatience at not receiving an official permit to transmit, and therefore begin transmitting illegally.
>
> I can't help but get the impression that it is often a superficial arrogance which apparently confuses the current moment with the "system" of the Weimar Republic, in which German radio amateurs, with the toleration of well meaning officials, made a sport out of illegal transmission. My dear illegal transmitters, these days are over. Whoever among our DASD members transmits illegally, even worse, whoever misuses the call-sign of a comrade to do so, will be expelled forever from the DASD and brought before a regular court of law.
>
> I also want to emphasize, that our members and particularly our own Monitoring Central participate energetically in the observation and identification of illegal transmitters, of which there are unfortunately still quite a few in Germany. Our members have made quite a few nice catches in this

[86] Literally, "black transmitting". It came from an earlier expression "*schwarzhören*", to listen illegally. One who made illegal transmissions (or an illegal radio transmitter) was a *Schwarzsender*. There are other words in German using this same construction, notably "*Schwarzfahrer*", someone who rides public or commercial transport without a ticket, or "*Schwarzmarkt*", black market. The practice of illegal transmission was closely related to the larger problem of illegal listening, be it listening to German broadcasts without paying the Radio Broadcast Fee, or, in wartime, the practice of illegally listening to foreign broadcasts. On the latter, see Janina Fuge, *An den Funkpranger gestellt und mit dem Wellendetektiv gejagt. Die Schwarzhörer als Konstante der Rundfunkgeschichte*, Nordwestdeutsche Hefte zur Rundfunkgeschichte 7, edited by Hans-Ulrich Wagner (Hamburg: Forschungsstelle Geschichte des Rundfunks in Norddeutschland, 2009): 9–10.

regard, and moreover, through their monitoring of the airwaves have provided the state monitoring authorities (with whom were are in constant and close cooperation) with valuable material to complete their files in a very useful way.[87]

It is quite clear that illegal transmission was a real worry for the DASD leadership. Illegal transmission was one of the major issues surrounding the DASD—perhaps the major one. The fear that radio hobbyists might make illegal transmissions, and either communicate with Germany's enemies or disrupt important government communications at home, is a clear thread running through the relationship between hobbyists and the German state from the very beginnings of radio. On the other hand, the audience for these repeated condemnations and calls to help root out illegal transmitters is not always clear. While the members had to be warned, the DASD leadership was also at pains to show the Police, Party, and Post that it was not only behaving, but was also a necessary partner in the fight against illegal transmission.

Whatever the real scale of illegal transmission in the Third Reich, making a fuss about illegal transmission gave the Postal Ministry, the Police, and the Nazi Party an issue they could use effectively at any time against the DASD. It was probably the single most heinous act of which the DASD could be accused, and its mention could both justify any measure

[87] "*An dieser Stelle mochte ich auch noch einmal auf die Schwarzsenderfrage eingehen. Leider ereignen sich immer noch vereinzelte Fälle wo auch Mitglieder des DASD ihre Ungeduld, endlich offiziell die Sendegenehmigung zu erhalten nicht bezähmen können und sich als Schwarzsender betätigen.*

Ich kann mich nicht ganz des Eindrucks erwehren, als ob es sich hierbei manchmal um eine vermeintliche Forschheit handelt, die offenbar die heutige Zeit mit jenen Systemzeiten verwechselt, in denen es für die deutschen Amateure ein Sport war unter stillschweigender Duldung wohlmeinender Stellen schwarzusenden. Diese Zeiten, meine Herren Schwarzsender, sind heute vorbei. Wer von unseren DASD-Mitgliedern schwarz-sendet, noch dazu unter missbräuchlicher Benutzung des Rufzeichens eines Kameraden wird für alle Zeiten aus dem DASD ausgeschlossen und ausserdem dem ordentlichen Gericht zügeführt.

Ich möchte hierbei auch hervorheben, dass unsere Mitglieder und besonders unsere eigenen Überwachungszentralen sich eifrig an der Beobachtung und Ermittlung von wilden Schwarzsendern beteiligen, deren es leider immer noch eine ganze Anzahl in Deutschland gibt. Unsere Mitglieder haben auf diesem Gebiet schon manchen schönen Erfolg aufzuweisen gehabt und im übrigen durch ihre Beobachtungen das entsprechende Material der staatlicher Überwachungsstellen, mit denen wir in dauernden und enger Zusammenarbeit stehen, in wirksamer Weise vervollständigt". "Ansprache des Präsidenten des DASD auf der Kundgebung am 23, Mai- 1936", in: DASD Verordnungsblatt Nr. 9 of April 30, 1936.

against amateur radio and provoke the DASD leadership to immediately fall into line. It was thus a convenient line of argumentation.

On the other hand, the temptation for amateurs to transmit, even without a license, was great. While it was certainly thrilling to receive faraway transmissions and fulfilling to participate in large-scale experiments in reception and radio-wave propagation, most DASD members, who only held DE-Numbers, could not legally transmit or participate in two-way exchanges. But everything in club life, and in the wider international practice of radio amateurs, was geared toward being able to transmit as well as receive. It was the difference between passive listening and active participation. The temptation was great to break the law, and many did, particularly because of the double standard which had developed in the 1920s and early 1930s. The older hams all boasted about their illegal transmission and other hijinks during the Weimar Republic, and this culture was hard to fully stamp out until wartime, despite the best efforts of the DASD leadership.[88]

As far as can be determined, the practice of illegal transmission was diminished, yet remained widespread within the amateur radio community into the Third Reich. Amateurs not only did it, many were proud of their disobedience. German hams felt they were part of a very modern international fellowship of radio hobbyists, and were justified in flouting national laws which were rooted in the past. Meanwhile, in both the Weimar Republic and Nazi Germany, the official representatives of the radio hobby were put in the uncomfortable position of having to publicly condemn some of the most skilled and active elements in their own hobby, and thus their condemnation was often not wholehearted.[89] Despite the expansion of the number of legal transmitter licenses, there were never enough to go around. Some amateurs continued to flout the rules and transmit without a permit, even during the Third Reich.

Sometimes, illegal transmission was a deliberate act of political resistance, as when socialist and communist radio hobbyists disrupted speeches

[88] The German system of tightly controlled radio use created the conditions for its own contravention. By supporting the acquisition of technical proficiency, the state also gave German hams the illegal knowledge to contravene restrictions on transmission without a license. It is hard not to think of de Certeau here: within the strategic system of tightly controlled public access to the airwaves, German amateurs developed a tactical knowledge, which allowed them to transmit and contact others in contravention of the law. See Michel de Certeau, *The Practice of Everyday Life* (Berkeley: University of California Press, 1984): 34.

[89] See, for example, the repeated condemnation of illegal transmission in *CQ*.

by Hitler or other leading Nazis.[90] In other cases, the motivations were more complex, though whatever else might have been part of the equation (impatience, the thrill of risk-taking, technical arrogance, boyish enthusiasm, etc.), illegal transmission placed commitment to the thrill of radio above respect for the law. And in the Third Reich, the law had real consequences. Not all illegal transmitters were members of the DASD, but as a public organization dedicated not only to radio as a hobby but quite particularly to transmitting, they were the natural suspects. As a result of both the practice and the accusation of rampant law-breaking, the DASD was forced to continually pressure its members to behave. The issue comes up again and again in internal DASD correspondence. For example, in January 1936, the DASD Regulations Gazette (*Verordnungsblatt*) published a reminder from the National Postal Minister, reminding the organization that any member caught transmitting illegally may never receive a legal transmission license in the future.[91]

Events in 1937 provoked a major crisis for the DASD and a crackdown by both the RPM and the Party. In February 1937, the DASD announced to its members that "again", several DASD members had been caught and punished in the courts for illegal transmission. The DASD president sent a long, angry, and practically tearful report to the membership, reminding them what was at stake both personally and for the organization. All were asked to help catch illegal transmitters and told in no uncertain terms that such people, and even those who knew about it and chose to look the other way, had no place in the organization.[92] To drive home the danger that suspicion of engaging in illegal transmissions posed for the organization, members were also cautioned not to mention any previous illegal transmission they may have engaged in before the Third Reich.[93] A subsequent Regulations Gazette provided detailed instructions on exactly how

[90] For example, see the story of DASD member Rolf Formis, who secretly ran a transmitter for the "Black Front" ("Schwarze Front") and who was killed in Czech exile by Gestapo agents. Körner, Amateurfunk, 127–132. Katrin Paehler, *The Third Reich's Intelligence Services: The Career of Walter Schellenberg* (Cambridge: Cambridge University Press, 2017), 112.

[91] "Liebhaberfunksender: Der Reichspostminister teilt unterm 19- Dezember 1935 mit": in: *DASD Verordnungsblatt* No. 1 (January 10, 1936). In a similar fashion, roughly a year later, DASD members were reminded that anyone convicted of illegally transmitting could never be readmitted to the club. *DASD Verordnungsblatt* No. 32 (II/10) (May 22, 1937).

[92] *DASD Verordnungsblatt* No. 25 (II/3) (February 19, 1937).

[93] Ibid.

to report any suspected case of illegal transmission, "however improbable it might seem". DASD members were instructed to make a report first to DASD headquarters, unless there was a risk that the criminal might escape. In that case, members were told to contact the postal authorities directly, not the police, since the Postal Service had the authority in these cases, though even then, a detailed report also had to be sent to DASD headquarters.[94] The arrest of the illegal transmitters was met with the issue of a new law specifically banning illegal radio transmission and imposing strict punishment, including either penitentiary or prison sentences. This was the "Law Against Illegal Transmitters" ("*Gesetz gegen die Schwarzsender*") of November 1937.[95] The DASD was ordered to reprint the law in its entirety in *CQ*, and a full copy was also sent to each head of a DASD regional branch.[96]

It is hard to say at this point whether or not the degree of illegal transmission actually warranted this crackdown or whether it was simply another move in the intricate competition between the ministries interested in radio. Anecdotal evidence suggests that arrest and prosecution for illegal transmission before 1939 might not always have resulted in a draconian punishment (this was also the case in the Weimar Republic), though a lot would have depended on the motivation for the crime: oppositional transmissions were punished harshly.[97] But declaration of war in 1939 changed this scenario, and illegal transmission was punished much more harshly in the context of war.[98] Nevertheless, given the amount of concern,

[94] *DASD Verordnungsblatt* No. 26 (II/4) (February 26, 1937).

[95] Gesetz gegen die Schwarzsender vom 24. November 1937 (RGBl. Nr. 130 vom November 29, 1937). Note that transmitting without a license was already illegal. The new law greatly increased the potential punishment, however.

[96] "Das neue Schwarzsendergesetz", in the rubric: "Amtliche Mitteilungen der DASD-Leitung", in: *CQ* No. 1 (January 1938):13–14. To reinforce the message, the announcement of the new law was immediately followed by a second article announcing the sentencing and expulsion of two former DASD members (both still in school) for illegal transmission. "Verurteilung von Schwarzsendern", in: ibid., 14. Their sentencing occurred before the new law, so that they were only sentenced to fines. See also "Gesetz gegen die Schwarzsender", *Verordnungsblatt des DASD* No. 42 (December 10, 1937): 2.

[97] Hans Bauer, Nürnberg, ed. *Nachrichtenblatt-Nr. 11 der Landesgruppe Bayern-Nord des DASD e.V.* (November, 1935). The newsletter discusses the case of Norbert Braun, as announced in *CQ* No. 10 (October 1935). It goes on to say that since his sentence was so minimal, the "authorities" have taken steps to increase penalties for illegal transmission in the future, which will certainly include prison sentences.

[98] "Verordnung über Sender für Funkfreunde" of January 9, 1939 (Amtsblatt des Reichspostministeriums Nr. 5 of January 12, 1929).

there were relatively few cases of DASD members being caught and punished. Make no mistake, there were those foolish enough to transmit illegally even in the Third Reich. But due to the nature of the "crime", it is also hard to determine the real amplitude. One surviving document lists 104 separate cases of known transmission by illegal transmitters in 1937 alone. Only one of these instances give a name and address of the guilty party caught for making it.[99] Names of the guilty were sometimes published in CQ or in the DASD's internal newsletter. One example is Wolfram Körner (DE3559/N) Stuttgart, who was one of those arrested and expelled from the DASD in 1937.[100] After the war, many histories or memoirs written by German hams proudly mention the practice, though often the details are often imprecise. Given the fairly widespread illegal transmission made in the Weimar Republic, and the equally widespread nature of the practice under Allied occupation after the Second World War,[101] and given the several cases of men either caught or admitting to the practice afterwards,[102] it is clear that some, perhaps many had the courage or the stubbornness to transmit illegally in the Third Reich, despite the danger. But given the very thorough monitoring conducted by the Police, armed forces, and postal authorities even before the Third Reich (and by the DASD itself after 1933!), illegal transmissions never escaped attention of the authorities.

Fear of illegal and uncontrolled transmission certainly increased as the international situation worsened and as war became evermore likely. Already in December 1937, all DASD members were sent a copy of the new Law Against Illegal Transmission.[103] In September 1939, when war

[99] "Zusammenstellung der bis zum 1. November 1937 beobachteten Schwarzsender" in Dokufunk Archive.

[100] *DASD Verordnungsblatt* No. 33 (II/11)(June 1, 1937). Körner was a leading West German ham, and is best known for writing a history of amateur radio in Germany. See: Körner, *Amateurfunk*. For another example, see the case of Norbert Braun, in *CQ* No. 10 (October 1935).

[101] See the next chapter (Chap. 7).

[102] As one example, see the memoir by Rudolf Binz, DL3SO, "Mein Funker-Leben. (Privates und Unprivates)", Part 1, https://z37.vfdb.org/wp-content/uploads/DL0SO_Teil1.pdf and Part 2, https://z37.vfdb.org/wp-content/uploads/DL0SO_Teil2.pdf. Binz, later a high official in the West German Postal Ministry, admits to building and using an illegal transmitter in the early and mid-1930s, even once he was a DASD member and Hitler Youth radio operator (*HJ-Bannfunkwart*). (Part 1, 1–7; Part 2, 7).

[103] "Schwarzsendergesetz and Ausführungsbestimmungen. In: Reichsgesetzblatt, Teil I, Gesetz, Seite 1298 und 1299; Verordnung zur Durchführung desselben, Seite 1303 und 1304". Announced in *DASD Verordnungsblatt* No. 42 (II/20) (December 10, 1937).

was declared, an immediate ban on most amateur transmissions was immediately put into effect.[104] Also in 1939, the Nazis passed new legislation, which made listening to enemy broadcasts illegal; this could and often did carry the death penalty, though actual punishment varied widely.[105]

Fear of illegal transmission continued through the war. In December 1940, DASD members were again given a special warning about illegal transmission.[106] And in the summer of 1941, a dire warning from the High Command of the military was passed on the DASD members by the president: "Illegal transmission is treason. Every illegal transmitter risks the death penalty".[107] The message went on: "Whoever transmits illegally in wartime puts himself outside of the national community (Volksgemeinschaft), and must assume that they will be punished as a traitor with time in a penitentiary (*Zuchthaus*) or the death penalty". Not only was this quite a stark warning, its use of specific National Socialist language made it even more threatening. Putting oneself "outside of the national community" meant, in effect, losing one's humanity. No one could doubt the seriousness of the message.

At the end of the day, what is perhaps most important here is not whether or not German radio amateurs transmitted illegally during the Third Reich or any other time, but rather why the problem seemed to provoke such hysteria. Part of the worry certainly comes from a deep fear in any dictatorship of undisciplined acts carried out by the population. This was true for both the German Empire (whose spirit remained alive in the RPM) and the Third Reich, and later for the German Democratic Republic (GDR). It was no accident that this fear reached such a high pitch in the case of radio, for technology gave power to citizens which, at

[104] As we shall see below, and unlike most other countries, Germany did not ban all amateur transmission for propaganda and scientific reasons. But only those issued special wartime transmission permits were allowed to be on the air.

[105] On September 1, 1939, the "Verordnung über außerordentliche Rundfunkmaßnahmen" was issued. Fuge, *Schwarzhörer*, 18–21.

[106] *Verordnungsblatt des DASD* (December 1940). The DASD Leitung sent special warnings against Schwarzsenden directly to members. Leitung des DASD, "An alle Mitglieder und ehemalgen Mitglieder des DASD, of 6. December 1940", Dokufunk archive schn_020a. The warning mentioned six illegal transmitters who were recently prosecuted and sentenced.

[107] "Schwarzsenden ist Landesverrat. Jedem Schwarzsender droht die Todesstrafe", *CQ* No. 7/8 (July–August 1941). By this time, there was a new DASD President, SS-Gruppenführer and retired Lieutenant General Sachs.

least in the 1920s and early 1930s, could not really be matched by the authorities. On the other hand, there was nothing unique about the German fear of uncontrolled use of radio technology. Wartime brought a similar hysteria about illegal transmission to the US, and all amateur transmission was banned between late 1941 and 1945.[108] Though amateur transmissions were curtailed during wartime in the Third Reich, they continued, perhaps because, in the final analysis, Nazi Germany was better able to control its population. Two things came together in the mid-1930s, which gave government more of a fighting chance to catch illegal transmitters. One was the simple fact that techniques of radio direction finding became easier and more precise in the mid-1930s. To put it another way, government mastery of technology evened the playing field. The second element was the simple fact that as time went on, the Third Reich (as later the GDR) was able to develop a fine-grained police state. Most illegal transmitters were caught due to denunciations by neighbors.

Into the Latter 1930s

In the latter 1930s, the DASD worked hard to build and strengthen as many bridges as possible to the NSDAP and the state. The strategy was clearly to make itself as useful as possible, in the hope of avoiding further restrictions or even complete dissolution. Cooperation with the Hitler Youth continued, a new agreement was signed with the National Research Council (*Reichsforschungsrat* a national research consortium), and new projects were in negotiation with the NSDAP and SA.[109]

As we have seen, the threat of sanction was always there, particularly concerning the question of illegal transmission, but also because of growing suspicion in Germany of foreign contacts, as the Nazi dictatorship deepened, and Germany became evermore inward looking. The price the DASD continued to pay was the necessity to expand its supervision of its own members. They were increasingly expected to train actively, show conformity with the National Socialist regime, and exercise ever-greater discipline.

In retrospect, 1937 marked a turning point. Perhaps it was the fact that illegal transmission continued, perhaps it was a sign of the waning interest

[108] There was an exception for amateurs who worked in support of the military. See Haring, *Ham Radio*, Chapter 5.

[109] DASD President Gebhardt, “Zur Jahreswende”, *CQ* No. 1 (January 1938):1–5.

of the military in the DASD, perhaps it was a shift in the constellation of interests behind and opposed to amateur radio, perhaps it was all three. Sometime in late 1936, perhaps because of renewed incidents of illegal transmission, the authorities, and particularly the party, seem to have cracked down hard on the DASD. The granting of license to transmit was slowed nearly to a stop, the conditions were tightened,[110] and the Party ordered that all DASD members go through a political vetting carried out by the respective Nazi Party regional administration (Gauleitung).[111] The DASD Regions each had to send a contact man to the local *Gau* Administration, preferably a party member, who would be responsible for settling any problems which might arise between the Party and the DASD.[112] In effect, from this point onward, DASD members (who had already been severely purged during the *Gleichschaltung*) had to be vetted twice for political reliability: once by the police (when they applied for membership, and again if they applied for a transmission license) and once by the Nazi Party. This was a close and uncomfortable degree of scrutiny,[113] about which the DASD could do nothing.

The moratorium on the transmit license process cut to the heart of the DASD's reason for being. Obtaining such a license was certainly the main reason why many had joined the DASD in the first place. The fact that the authorities not only kept adding more and more conditions and restrictions, but also more or less stopped the award of new licenses for a time altogether without saying why, caused great anger and even despair in the ranks of the DASD. In frustration, a number of long-time members

[110] DASD President Gebhardt, "Zur Jahreswende", *CQ* No. 1 (January 1938): 1–5; *DASD Verordnungsblatt* No. 2 (III/2) (22.1. 1938). For all intents and purposes, there was a moratorium on the granting of new licenses in 1937, until the new Law on Illegal Transmitters was issued.

[111] Remember that the Nazi Party was organized into regional divisions, called *Gaue* (1 *Gau*, 2 *Gaue*). The head of the Gau, the *Gauleiter*, was a particularly powerful member of the party who enjoyed direct access to Hitler, and a great deal of autonomy in his *Gau*.

[112] *DASD Verordnungsblatt* No. 22 (December 18, 1936); *DASD Verordnungsblatt* No. 28 (II/6) (April 1, 1937).

[113] It is interesting that this new round of vetting was to be done at the regional (*Gau*) level. This may indicate the growing influence of the *Gaue* vis-à-vis the Propaganda Ministry, though it was their representative in the *Gau* who whose office actually was supposed to conduct the checks. On the other hand, the Nazis always placed great value on personal contacts and reputation, and it may simply indicate that the vetting was to be done by people who could be expected to know something about the DASD members firsthand.

resigned in 1937, convinced that they would never be issued licenses.[114] Others, of course, were increasingly being pulled away from their hobby activities by longer hours at work, or now obligatory service in the RAD and military, or were simply put off by the increasing discipline. New licenses and a "final regulation" of the license question were promised in January 1938,[115] but DASD members had long grown suspicious of such promises.

Since the DASD was largely financed out of member dues, this was a serious problem. The Navy's cancelation of the "Sondergruppe M" also led to the departure of some 400 members.[116] This, plus a temporary slowdown in cooperation with the Hitler Youth, and the increasing use of Navy, as opposed to DASD instructors for the FWGM groups, also raised the fear of irrelevance.[117] To make matters worse, DASD President Heinrich Gebhardt died unexpectedly in July 1939. Thus, on the eve of the outbreak of war (though this was not clear at the time), the DASD was left leaderless. It remained so until May of 1940. This meant that the DASD was left without direction, and without a high-ranking champion and defender within the Nazi system, during the crucial first year of the Second World War.

Amateurs in Wartime

It is hard to speak of hobby activities during wartime: who has time? Wouldn't a hobby detract from the war effort? Yet the DASD continued to function, right down to the bitter end. Some people did have time for a hobby, even in wartime. But on the whole, we need to see the DASD during the war less as a hobby organization, and more as a semi-official auxiliary of the state. Despite the fact that many, probably most of its members were called up for military service and thus had to leave the

[114] DASD President Gebhardt, "Zur Jahreswende", *CQ* No. 1 (January 1938):1–5; Körner, *Amateurfunk*, 138.

[115] DASD President Gebhardt, "Zur Jahreswende", and subsequent editorial note, both in *CQ* No. 1 (January 1938): 1–5.

[116] *DASD Verordnungsblatt* No. 32 (II/10) (May 22, 1937). The DASD members who belonged to the Sondergruppe M all paid regular dues, though, in fact, the money came directly from the Navy, and not out of their individual pockets. Since the DASD had roughly 4200 members in 1937, this was a major blow. See also DASD President Gebhardt, "Zur Jahreswende", *CQ* No. 1 (January 1938): 1–5.

[117] *DASD Verordnungsblatt* No. 37 (II/15) (August 27, 1937).

DASD, it remained active. The outbreak of the war and the immediate call-up of so many men disrupted training efforts at first. As the war progressed, there was a gradual slowing down and thinning out of the whole organization. More and more members went on active duty in the armed services and thus left the organization. For those who stayed at home (either because of age, infirmity, or because they held positions in industry or the Party vital to the war effort), free time shrank. Work hours increased during wartime. Party activities continued to take up time, even for non-Party members. Everyday life also became more and more difficult, as people had to spend more and more time acquiring food, heating material, and other necessities.[118] From 1942 onward Allied bombing took an increasing toll, robbing people of sleep and increasingly destroying basic infrastructure.[119] Certainly, in the big cities, civilians had to spend increasing amounts of time clearing rubble and repairing buildings—if the buildings remained standing at all.[120] With a new president in 1940, the DASD reorganized, resumed its training activities for the HJ, and even took on new, war-related tasks, but the war made maintaining normal club activities increasingly difficult.

In this context, it is amazing just how active many German "hams" remained. At the outbreak of war, all amateur transmission was banned, and transmitters were confiscated. Quickly, though, a limited number of transmission licenses were reissued, as so-called K (*Kriegs-*) call signs. Even before the war, the DASD as an organization was active in practical tests of propagation, not least on the ultra-shortwave bands. These continued into wartime, often disguised as routine traffic-handling practice nets by men and women using K call signs, so that Allied radio intelligence would not know the depth of German research into use of these frequencies.[121] On-air practice was still highly important for training new

[118] Gesine Gerhard, *Nazi Hunger Politics: A History of Food in the Third Reich* (London: Rowman & Littlefield, 2015) Chapter 2.

[119] On the Allied bombing of Germany, see Jörg Friedrich, *Der Brand. Deutschland im Bombenkrieg -1945* (Spiegel-Verlag, 2007). Note that many civilians also participated in the system of civilian civil defense alongside their jobs, meaning that time outside of work hours was increasingly occupied.

[120] See the report from *Bezirksverband* 6 (Berlin Südosten), *Nachrichtenblatt Landesverband Reichshauptstadt im Deutschen Amateur-Sende- und Empfangsdienst e.V.* 9 No. 3. (April 1942).

[121] As mentioned above, from the mid-1930s onward, the DASD worked on researching propagation in the VHF portion of the radio spectrum, particularly on the 10-meter and 5-meter bands. (German: *UKW*; roughly 30 MHz–300 MHz) This research continued into

radio operators, and the K-License holders played a major role in propagation studies. For this reason, an entire section of these licenses were issued for use exclusively on the 10-meter band.[122] The propagation of radio waves at these low, ultra-shortwave frequencies was not well understood at the time, and this was of particular interest to military.[123] Issue of the K call signs depended on the military authorities, not the postal authorities, and were merely rubber-stamped by the Propaganda Ministry. The DASD recommended particularly capable members to the military for licensing. Eventually, just under two hundred of these permits were issued. Some German "hams" even continued to make foreign contacts, though the transmission of QSL cards began to be severely restricted and eventually to break down.[124] Nevertheless, the Nazi authorities felt it was important to preserve a show of normalcy as a propaganda tool. At the end of the war, the British decided to issue a very limited number of special wartime transmission permits of their own, and there were apparently cases where German hams and British hams exchanged polite contacts, and possibly even QSL cards in the middle of wartime.[125] There is some evidence that German Military Intelligence (the *Abwehr*) was involved, and may have used the K-licenses holders to monitor the airwaves for illegal transmissions.[126]

the war. See, for example, *Nachrichtenblatt Landesverband Reichshauptstadt im Deutschen Amateur-Sende- und Empfangsdienst e.V.* 9, No. 1(January–February 1943).

[122] Körner, *Amateurfunk*, 170–172. A list of all still licensed German stations as of August 1944 is reproduced at: http://www.viehl-radio.de/homeda/chronik/kriegslizenzen.pdf. The provenance is sadly not listed. Note the division into shortwave and 10-meter permits.

[123] See, for example, "Auswertungsbericht zum QM-Relais-Test vom 20.12.42-3.1.1943", *Nachrichtenblatt Landesverband Reichshauptstadt im Deutschen Amateur-Sende- und Empfangsdienst e.V.* 9, No. 3 (April 1943).

[124] The DASD staff devoted to "nonessential" tasks such as the transmission of QSL cards was severely cut in wartime.

[125] See Körner, *Amateurfunk*, 153, 163, 170–172, here: 172. On the British "Plan Flypaper" to counter the presence of German amateurs on the air during the war, see Southgate Radio Amateur News, "Plan Flypaper", at: http://www.southgatearc.org/news/july2006/plan_flypaper.htm (last accessed June 15, 2018).

[126] See the report on SS-*Obergruppenführer* and DASD President Ernst Sachs' postwar interrogation by the Office of Strategic Services (OSS): "Report from Headquarters, United States Armed Forces European Theater, Military Intelligence Service Center, APO 757 to Counter Intelligence War Room, Subject: Re CI War Room brief FF 603.185/W.R.C.Sa. of 19.November 45", United States National Archives (NARA) RG226, Box 33 Office of Strategic Services Entry 119A; London X-2 PTS Files; London X-2 PTS-3 Files Thran, Friedrich THRU Ulrich, Wolfgang, Folder 866 PF 603.125 OGRUF. SACHS.

Club activities for those not at the front were still mandatory, and remained rather varied, at least until 1943. There were local chapter meetings at least monthly, and regional meetings quarterly.[127] Local and regional club administrations held regular talks on technical subjects, and continued to offer classes in Morse code, basic electronics, and radio theory, both for their own members and for the Hitler Youth. More than ever, efforts by the DASD to train and maintain radio skills were appreciated in wartime, and the pressures from the Postal Ministry and Nazi Party seem to have backed off during the war: everyone had other things to do.

Individual club regions (*Landesverbände*) continued to publish newsletters until 1943 and, in isolated cases, even later.[128] The DASD continued to publish *CQ* down to the end of the war, though it radically changed and reduced its format in 1943.[129] The ability to continue to publish newsletters and, indeed, the right to continue club activities was a sign that the DASD and its activities were considered important to the war effort: paper was in increasingly short supply, and was allocated on the basis of the importance of its use to the war effort.

One key to the continued ability of the DASD to obtain scarce paper allocations (at least until 1943) was the fact that all clubs were expected to maintain contact with their members at the front for the sake of maintaining morale. Clubs were expected to send copies of their publications to frontline soldiers, even though they were technically no longer members.[130] Given wartime censorship, and the draconian Nazi measures against "defeatism", no one had the ability to speak freely, and certainly not in a semi-official publication. But sometimes, using very coded language, DASD members could comment on what was going on.[131]

[127] For example, *Nachrichtenblatt Landesverband Reichshauptstadt im Deutschen Amateur-Sende- und Empfangsdienst e.V.* 9, No. 4. (October 1943).

[128] For example: Stellv. LVF/K Hans Goldmann, *Mitteilungen des Landesverbandes Niedersachsen, LV/K im DASD*, Bremen,(Christmas, 1943) or *Nachrichtenblatt Landesverband Reichshauptstadt im Deutschen Amateur-Sende- und Empfangsdienst e.V.* 9, No. 3 (April 1943).

[129] The last known *CQ* was Vol. 18, No. 11 (November 1944).

[130] See *DASD Verordnungsblatt* No. 5 (March 1941): 1.

[131] For example, *Nachrichtenblatt Landesverband Reichshauptstadt im Deutschen Amateur-Sende- und Empfangsdienst e.V.* 9, No. 3 (April 1943). Note that in some cases, the telegraphic abbreviations typical of amateur radio traffic were used to further hide the meaning conveyed.

In 1943, worsening supply conditions, and the shift to "total war", led to the ending of all local and regional DASD newsletters, except for the main club magazine *CQ*.[132] Of course, the DASD had one means of maintaining contact with its members at the front, which no other German club could match: at least some frontline former DASD members serving in radio units continued to monitor radio traffic on the home front, and in at least one case, even sent in signal reports.[133] Still, the call-up of more and more men for the armed forces, and the pressures of other wartime activities, led to a slow decline in DASD work, even for training and other war-related purposes, between 940 and 1942. By then the DASD still maintained the *Betriebsdienst* during wartime for training, and in case alternate communications networks were ever needed by the state, but it was increasingly difficult. Failure to participate in the *Betriebsdienst* for those still left at home could still lead to the loss of one's transmitting license.[134]

The DASD found new life and renewed purpose after the appointment of Ernst Sachs as the President in May 1940. Sachs was an interesting, if sometimes controversial, figure, and he played an important role in maintaining and even reinvigorating the DASD through the end of the war.[135] On the one hand, he was a former Army Lieutenant General, and had been one of the leading figures in developing the communications branch of the German Army since before the First World War. He was a real telecommunications specialist, and was head of the German Army's communications school when he retired in 1936. He was not only acceptable to the armed forces, the Nazi Party, and Propaganda Ministry, he was

[132] It seems as if some Regional Leaders continued to send a rudimentary newsletter even afterwards. See, for example, *Nachrichtenblatt des Landesverbandes Reichshauptstadt* 9, No. 4 (October 1943).

[133] For a soldier actually sending in reception reports for Betriebsdienst traffic, see *Nachrichtenblatt Landesverband Reichshauptstadt im Deutschen Amateur-Sende- und Empfangsdienst e.V.* 9, No. 3 (April 1943). The soldier's name was Gerhard Polke. For a soldier mentioning that he listened to Betriebsdienst traffic from the front, see Stellv. LVF/K Hans Goldmann, *Mitteilungen des Landesverbandes Niedersachsen, LV/K im DASD*, Bremen (Christmas 1943).

[134] "*Hallo ob's! hr N.B.B." Nachrichtenblatt des Landesverbandes B-Pommern des DASD e.V.* No. 6 (June 1938) "Nur für Mitglieder des DASD bestimmt!" [Only for Members of the DASD!].

[135] See his "SS Personalakte", United States National Archives (NARA) Microfilmed Records Received from the Berlin Document Center, SS Officer Personnel Files. Microfilm Publication A3343, Series SSO. In particular, see his *SS-Führer Stammkarte* and *Lebenslauf*.

also a *Gruppenführer* (Major General) and later *Obergruppenführer* (Lieutenant General) in the SS. This meant that Sachs had much more clout and influence than his predecessor, Gebhardt. This was particularly true, given the growth of the SS in power and importance throughout the war: by 1943, even Goebbels had to take SS wishes into account. Moreover, Sachs, who was also by this point Waffen-SS Chief of Communications Troops (*Chef für Nachrichten Verbände im SS-Hauptamt*) and (surprisingly) head of the German National Carrier-Pigeon Society (*Reichsverband Deutscher Brieftaubenzüchter*), was also personally quite close to Himmler.[136] In short, Sachs had the clout and backing to get things done, and he reinvigorated the DASD by plugging it directly into the war effort.

One of his most important acts was to send a questionnaire in 1944 to all DASD members serving in the armed forces, to determine whether they were being used in a way commensurate with their knowledge of radio.[137] Many were not: the Army (unlike the Navy) had not paid attention to the skills of DASD members, and many had found themselves as infantrymen, truck drivers, or filling other non-technical positions. Germany could not afford this waste of talent, and by 1944, Sachs had the influence to obtain reassignments for many of these men. If the Army would not listen, he and the SS had enough influence by this point to have the men transferred to the Waffen-SS.[138] Of course, for the men involved, this was a double-edged sword: as grateful as many of them undoubtedly were to be transferred out of frontline units into (slightly) safer rear-echelon communications positions, the transfer to the Waffen-SS did leave them with some explaining to do after the war.[139] The SS was quite happy

[136] "SS Personalakte" Ernst Sachs (DOB 24 December 1880) in: United States National Archives (NARA) Microfilmed Records Received from the Berlin Document Center, SS Officer Personnel Files. Microfilm Publication A3343, Series SSO. In particular, see Himmler's letter to Sachs of August 1944.

[137] A copy of this questionnaire may be found in: NARA Record Group 226, Box 33 Office of Strategic Services Entry 119A; London X-2 PTS Files; London X-2 PTS-3 Files Thran, Friedrich THRU Ulrich, Wolfgang, Folder 866 PF 603.125 OGRUF. SACHS.

[138] After the failure of the July 20 attempt on Hitler's life, Himmler was named Head of the Reserve Army (Ersatzheer), placing him in charge of all troops not in combat areas, and effectively giving him control over the armed forces. At the same time, the SS also took over the Abwehr.

[139] On the SS in general, see Robert L. Koehl, *The Black Corps. The Structure and Power Struggles of the Nazi SS* (Madison: University of Wisconsin Press, 1983) and Heinz Höhne, *The Order Under the Death's Head. The Story of Hitler's SS* (New York: Ballantine Books, 1977 (1966)). On the Waffen-SS and the difference between the General-SS and the

to be able to obtain significant numbers of trained communications specialists. Himmler had continually sought to expand the armed SS troops at his disposal, often in the face of Army opposition.[140] Until mid-1944, the Army always had the upper hand, since it controlled the manpower allocations to the different services; in order to continue to expand, the SS had to find sources of manpower that the Army either didn't control or didn't want, though it was able to successfully push the Wehrmacht for ever larger allocations of men once war began.[141] By 1943, SS influence had grown to the point that men within Germany began to be drafted into the Waffen-SS, and by mid-1944, after the failed July 20, 1944, attempt on Hitler's life, the SS took over control of manpower from the armed forces. This is just when DASD men began to be transferred out of the Army and into the Waffen-SS as a result of Sachs' pressure.

According to former DASD members, despite his SS loyalties, Sachs quickly became emotionally attached to the organization, and apparently successfully fought for mild penalties for those men accused of "radio-crimes" during the war, such as the (inevitable) illegal transmission. Körner, in his history of amateur radio in Germany, speaks extremely positively, even warmly of Sachs, whereas he has nothing good to say about his predecessor, Gebhardt.[142]

With Sach's clout and energy, and the gradual expansion of SS intelligence activities, the DASD gained renewed importance in the second half of the war. Training of new radio operators in the Hitler Youth ceased its decline, and expanded. The technical contribution of the DASD to the war effort also expanded, often in surprising ways. Remember that more or less all DASD members had to build their own equipment, and some had become quite skilled in not only the assembly, but also the design of radio equipment, particularly that designed for use on the shortwave

Waffen-SS, see Koehl, *Black Corps*, Chapter 6. In general, see George H. Stein *The Waffen-SS: Hitler's Elite Guard at War, 1939–1945* (Ithaca and London: Cornell University Press, 1966) and Bernd Wegner *Hitlers Politische Soldaten: Die Waffen-SS 1933–1945. Studien zu Leitbild, Struktur und Funktion einer Nationalsozialistischen Elite* (Paderborn: Ferdinand Schöningh, 1982).

[140] In general, see Wegner, chapters 6–8.

[141] For example, the Army controlled the military service of all German nationals, but did not have the same control over ethnic Germans in areas occupied by Germany or on territory belonging to German allies. To fuel Waffen-SS expansion, Himmler instituted a military service obligation for foreign ethnic Germans in July 1942.

[142] Körner, *Amateurfunk* 138; 143–146; 165–169.

bands, where there was less practical experience in industry. Remember, too, that a number of DASD members were either academic researchers or active participants in the radio and electrical industries. DASD headquarters had consciously built up a network of technical specialists and advisors among club members and also possessed a state-of-the-art lab and workshop. Particularly once Sachs took over as President, the DASD was called on more and more to provide special technical expertise. Again, this was not unrelated to Sachs' position in the SS.

The DASD as Reservoir of Technical Expertise

Despite the prominence of the German electrical industry and the high level of training in the German military, the DASD continued to be an important source of technical expertise during the war, particularly when it came to the shortwave and ultra-shortwave bands and to efficient design and construction of transmitters and receivers. Neither industry nor the military had a comparable base of practical experience. Of course, many of the technical experts within the DASD were already employed full time in industry or education or had been before they were called up for military service, and were not exactly amateurs when it came to radio-frequency engineering. Nevertheless, what DASD membership brought to the table was practical experience, both in DIY construction and in practical on-air technique. DASD members, and sometimes, the DASD as an organization, were called upon during wartime to provide services and expertise unique to the radio hobby. The electronics industry was the first to demand the return of experts in radio-frequency work and ask for the allocation of men with these skills, including DASD members. Individual DASD members served in several capacities where their special expertise was sought. Many continued their work in the electrical and radio industry.[143] Many became radio operators not only for the German armed forces (though many were not used in this capacity until late in the war), but also for the diplomatic corps.

Yet perhaps the most surprising contribution of the DASD to the war effort was in aiding the various German intelligence organizations. First, the DASD headquarters and lab were involved in secret work for the

[143] Ernst Fendler, DL1JK (ex D4IDF, etc.) "DL1jk Chronik" in: Dokufunk Archive. Fendler first served in the Army, and then worked for Telefunken during the Second World War. He was known as a VHF specialist.

armed forces and intelligence services. When the German foreign intelligence agencies wanted compact transceivers for their agents abroad, they came to the DASD to construct them.[144] This was not as big a step as it might seem, since the DASD was asked to construct material for the German government as early as the mid-1930s. For example, the DASD was asked to supply simple and robust receivers, transmitters, and other equipment to the Navy for use in training, not least of SA and HJ men in the FWGM.[145] The DASD was not the only source of this equipment, and contracting was eventually shifted to private industry, yet it was still quite significant that when it came to solving practical problems in radio design which required innovation or unusual applications, the shortwave amateurs of the DASD were often the best-suited inventors. Nor was this all: one account of the DASD during the war even credits its wartime technical director and another member of the headquarters staff with the invention of a successful device to distinguish between "chaff" and the radar signature of Allied warplanes.[146]

Moreover, as the war progressed, the DASD members were increasingly recruited for intelligence work as individuals. When war began, Germany had two major foreign intelligence agencies. The first, oldest, and most powerful was the *Abwehr*, which was a part of the military.[147] There had long been ties between the DASD and the Abwehr, and many DASD men worked for the Abwehr during the war.[148] The second foreign

[144] For Army intelligence, "letter from Dipl. Ing. E. Graff to DASD Headquarters of 28.11.1939", BArch, NS19/3918, Pers. Stab RFSS, Amateurfunkwesen, in which he asks DASD for money to pay overtime for workers occurred while building two "Geräte" "für OKW Abt. Ausl. IV f". For DASD production of radios for the SS/SD, see BArch record group NS19/3917, Persönlicher Stab, Reichsführer-SS, Havelinstitut.

[145] "Der Funkbeauftragte, Rundschreiben Nr. 1/35 (Geheim) of February 4, 1935, gez Schmolinske", BA-MA RM20/2975 OKM/Marinekommandoamt der Reichsmarine und Kriegsmarine.

[146] Körner, *Amateurfunk*, 169. The two amateurs were Hans Plisch and Hans Ehrlich. To be fair, some DASD members, including technical staff in the headquarters, were well-trained engineers or scientists with "day jobs" in industry. What set them apart from others is the practical do-it-yourself experience they had as amateurs.

[147] On the Abwehr, see David Kahn, *Hitler's Spies: German Military Intelligence in World War II* (New York: Macmillan, 1978).

[148] For one example, see the correspondence between W.F. Körner and Herbert Wulfhoop (ex D4jth) about the latter's service in the Abwehr, in Dokufunk Archive, DASD15 (DASD Allgemeines Schriftgut). For another example, see NARA Record Group 226, Box 4, Office of Strategic Services Entry 119A; London X-2 PTS Files; London X-2 PTS-3 Files Bovensiepen PF 602.627, folder 105: Bodigheimer, Ferdinand DF 602.900; Bödigheimer,

intelligence agency was under the wing of Himmler, Heydrich, and the SS. This was within the "Reich Security Main Office" (RSHA—*Reichssich erheitshauptamt*). It was founded in 1936 by amalgamating the fire departments, uniformed police, criminal police, Gestapo, and *Sicherheitsdienst* (Security Service, SD).[149] Within this office, both the Gestapo and SD carried out intelligence functions, with the Gestapo mainly limited to domestic and counterintelligence, though, of course, there was considerable overlap. Office VI of the RSHA developed into a Party-controlled foreign intelligence agency competing with the Abwehr. An agreement between the RSHA and the Abwehr known as the "12 Commandments" regulated the responsibilities of the different agencies. The SS continued to test and try to push beyond these limits, until after the July 20 attempt on Hitler's life, the once proud Abwehr was absorbed by its once smaller rival, Office VI (Amt VI) of the RSHA.[150] The head of Amt VI, Walter Schellenberg, was hugely ambitious, despite the fact that Amt VI was both incompetent and poorly organized.[151] Be that as it may, during the war, both the Abwehr and Amt VI had huge unmet needs for both trained radiomen and compact, robust, and simple shortwave transceivers which could be easily hidden. The DASD was asked on several separate occasions to construct spy transceivers and other equipment for both Military Intelligence (the *Abwehr*) and Party intelligence agencies connected to the SS (RSHA Amt VI).[152] The DASD also supplied growing numbers of trained radio operators with crucial practical experience on the short waves. While the Abwehr (with its links to the German Navy) first drew on the resources of the DASD for personnel, the RSHA increasingly turned to the DASD late in the war, thanks not least to Sachs' control over the organization. Sachs ties to the SS were crucial in this regard. By some accounts, the main purpose of his wartime questionnaire to all DASD members was designed to funnel radio specialists into the SS and RSHA.[153]

Ferdinand, Hauptmann, Nachr. Abt. III. Bödigheimer had been an early ham, and was a DASD Regional Leader until he was called up for military service, whereupon his wife took over as Regional Leader.

[149] Michael Wildt, *Generation des Unbedingten Das Führungskorps des Reichssicherheitshauptamtes* (Hamburg: Hamburger Edition, 2003).

[150] In general, see Paehler, *Schellenberg*.

[151] Paehler, *Schellenberg*, 354–359.

[152] See note 145 above.

[153] "Report from Headquarters, United States Armed Forces European Theater, Military Intelligence Service Center, APO 757 to Counter Intelligence War Room, Subject: Re CI

It helped that the RSHA maintained a center for clandestine communications in southwest Berlin, not too far from the DASD headquarters. This was the "Havelinstitut", which was used for both monitoring and communications with agents. Many former DASD members worked there, and its director was eager to obtain more.[154]

Conclusion: Amateur Radio Under National Socialism

Few hobby clubs worked as closely with the Nazi regime as the DASD, but on the other hand, few were allowed to survive down to the end of the war, certainly with as little direct Party or State control. The DASD, because of its technical expertise and practical experience, was better placed than nearly any other group to argue that it was important for national defense and national security. But the survival of the hobby organization and the continued ability of its members to practice their hobby came at a high cost.

The organization had to submit to a severe purge of members, reorganization in ways dictated by the Party, and continued scrutiny by police, Party, and security services. Moreover, it was asked to collaborate with the regime in several ways. Some were perhaps still "honorable" or "normal", in the case of the contributions to national defense, but close work with the Police to seek out illegal transmitters, collaboration in the persecution of its own members who held subversive ideas, and particularly the work with the SS and intelligence services were potentially rather different.

Can the DASD really be faulted for working so closely with the regime? Did it even have a choice? Certainly, choices for everyone were limited once the Nazis came to power. The DASD did manage to survive as a

War Room brief FF 603.185/W.R.C.Sa. of November 19, 45", in: NARA RG226, Box 33 Office of Strategic Services Entry 119A; London X-2 PTS Files; London X-2 PTS-3 Files Thran, Friedrich THRU Ulrich, Wolfgang, Folder 866 PF 603.125 OGRUF. SACHS.

[154] "RFSS/Persönlicher Stab: Havelinstitut, 05. Dezember 1943"; (EAP 161-b-12/384), frames 2655410–2655435, NARA, Captured German Records Microfilmed at Alexandria, Virginia, Records of the Reich Leader of the SS, and Chief of the German Police [RF-SS]. Microfilm Publication T175. See particularly the " wish-list" attached to: SS-Sturmbannführer Siepen to SS-Obersturmbannführer Brandt of December 5, 1943 "Erfahrungsbericht über Aufbau und bisherige Leistung des Havelinstituts; Drahtlose Verbindung in rückwärtige russische Gebiete, Invasionsfunknetze in besetzten Gebieten, Unternehmen Zeppelin und anderes; Bitte um Freistellung von Fachkräften", frames 5413–5434. See also BArch record group NS19/3917 (Persönlicher Stab, RFSS-Havelinstitut), which is only partially identical to the previous folder.

private organization, albeit now state-sponsored and controlled. In retrospect, its relationship with the NS regime was certainly too close, but that could not necessarily have been foreseen from the beginning, nor could it have been avoided without civil courage bordering on the suicidal. For the individual member, things were a bit different; an individual always had the option to drop out, though that would mean abandoning the hobby, and potentially risking questions about why one no longer wished to be a part of an organization working so closely with the regime. The technical knowledge and skills of its members gave the DASD a certain leverage or value within the Third Reich, but it was not enough to counter the weight of the entire system. Compromise and complicity were the best possible outcome, more was not achievable; the question was always: how much compromise? One major variable lay in the attitudes of the membership; hobbies and hobbyists, even technologically or scientifically based ones, are not isolated from their larger culture. There had always been a very strong element of German patriotism in the DASD, despite the willingness of its members to break the law in order to fully practice their common hobby. A little illegal transmission did not at all mean that the DASD members were disposed to put the entire system in question. The DASD leadership was full of former military officers, and most members came from precisely those middle-class professions which had always been paragons of patriotism in Germany. Remember, the organization had collaborated with the German armed forces in covert rearmament long before the Nazis ever took power, though this was certainly not known to most members. That it later collaborated to such an extent with the Nazis, who after all, represented the German state, should really not be a surprise.

What is important here is that, without the technical expertise and experience of the individual DASD members, which resulted from their hobby activities, they would never have had the choice to collaborate or not. Without this potential and hard to replace value to the regime, the DASD would never have survived the *Gleichschaltung*, and the persecution of its members might have been even more severe. Technical expertise gave the DASD a small but crucial bit of power and leverage vis-à-vis the regime, but also made it the target of the regime's notice in the first place. In a modern dictatorship, some degree of complicity can only be avoided through deliberate resistance or flight. Given that, the big question is this: is there any such thing as a little complicity?

The story of the radio hobby under the Nazis is not completely negative. Many Germans were able to pursue their hobby until the outbreak of war, a few even long after. Many Germans were introduced to radio for the first time in their military, HJ, or SA training, and some of these men either found their way to the DASD before 1945 or returned to the hobby in the postwar era. While the increasing discipline prevalent in the DASD in the Third Reich was a negative experience for many amateurs, those who survived the war were a tight group. Just as after the First World War, this large reservoir of men (and a few women) trained in radio, and the availability of cheap surplus equipment after the war, would be very important in re-creating the hobby after the German loss. It is to this postwar story and the rebirth of the radio hobby that we will now turn to in the following chapter.

CHAPTER 7

The Radio Hobby Comes in from the Cold, 1945–1955

"OM's PSE Don't Forget the German Hams!"
German Ham: OM: "We want to go back on the air foor [sic] meeting again—working for international understanding"
Foreign ham: "They always come back!"[1]

We often refer to the immediate post-Second World War period in Germany as "Zero Hour" (Stunde Null). The term speaks to the complete desolation in which Germany found itself upon defeat, a physical and moral desolation the like of which few countries have ever experienced in modern times. Yet simultaneously, the term also refers to immense possibilities open to a Germany which was able to start from zero and build its future, as if the burdens of the past had disappeared. They hadn't. But there was a new start for the public sphere and private associations in general, and for the radio hobby in particular.

[1] Blank QSL card issued to commemorate the "First German postwar Ham Fest" in Stuttgart, June 7–8, 1947. The card was given to American military occupational authorities by German hams arguing for licensing. in: Wolfram Körner et al.: "Application for Amateur-Transmitting Licenses" United States National Archives (NARA) RG260 box 968 Records of United States Occupation Headquarters, WWII, Office of Military Government for Germany (US) OMGUS, Office of Military, Wuerttemberg-Baden. Records of the Education and Cultural Relations Division: Community Active Branch Chief: Corresp & Rel Recs. 1945–1949: Amateur Radio thru K9 American-French Zone Merger.

B. B. Campbell, *The Radio Hobby, Private Associations, and the Challenge of Modernity in Germany*, Palgrave Studies in the History of Science and Technology, https://doi.org/10.1007/978-3-030-26534-2_7

Despite the desperate living conditions after the end of the War, and even with the opposition of the Allied occupation forces, the "ham" radio portion of the radio hobby (and in the long term, only this subset) rapidly reestablished itself in Germany, albeit first on a local, then regional basis. This might be seen as speaking to the general health of the public sphere and its ability to sustain democratic structures, despite 12 years of dictatorship. This is true, but there is much more to the story, as we will see. And yet, the radio hobby did experience both reestablishment and resurgence after 1945, albeit with a limited focus on ham radio. This was helped by the sheer usefulness of radio and the fear of falling behind in science and innovation, but was ultimately the product of countless radio hobbyists insisting on the right to practice their hobby. This is an extraordinary development and bears careful thought. There is something about hobbies which make them remarkably resistant to war and oppression.

One major change after the Second World War was that the vibrant do-it-yourself community of the 1920s and 1930s did not revive. There was, to be sure, a brief renaissance of home-built radios in the immediate post-Second World War period due to the very low postwar standard of living, but this situation did not last. This was due not least to the fact that, by the point large numbers of people had the time, energy, and capital to engage in a hobby again, the rapid reestablishment of German industry with the currency reform of 1948 meant that manufactured radios became available at a price that made home-building unnecessary. In Western Germany, a culture of consumption rapidly spread, while in East Germany, political control limited radio do-it-yourselfers to a strictly circumscribed set of club members within tightly controlled, government-sponsored structures.[2] The rapid spread of transistors to replace the older vacuum tubes also made home-building more difficult, and thus commercially purchased radios more attractive. Moreover, there was less interest in the social aspect of radio clubs in the postwar era, with its retreat into domesticity.[3]

[2] There is something of a paradox here, which bears future research. A pervasive climate of shortage in the Soviet Zone of Occupation and later the German Democratic Republic led to a very widespread do-it-yourself culture in most areas of life, which did not decline until German Unification in the 1990s. Government suspicion of anyone able to build a radio out of simple parts goes part of the way toward explaining the lack of a widespread radio do-it-yourself community, but is likely insufficient by itself.

[3] On the notion of an increased emphasis on domesticity in the postwar era, see: Hannah Schissler, *The Miracle Years: A Cultural History of West Germany, 1949–1968* (Princeton: Princeton University Press, 2001).

The broadcast- and shortwave-listening portion of the radio hobby also soon declined as organized groups, for reasons beyond this present study, but which included structural change in German broadcasting, the rapid ubiquity of radios after the transistor revolution, and above all, the rise of TV.[4] Radio in the late 1940s was no longer the novelty it had been in the 1920s. Other technologies rapidly stole the limelight as symbols of modernity and cutting-edge science. And of course, after Auschwitz and Hiroshima, technology as a symbol of modernity and a better future was seriously called into question.[5] Technology had moved on, and radio was no longer the public's darling but rather a mundane workhorse—except for the transmitting amateurs.

The hams were forced by postwar conditions to cling to a tight organizational structure. This was partly in the nature of the hobby itself, but largely due to the realities of Allied occupation: obtaining permission to transmit again required amateur radio hobbyists to band together and engage in serious lobbying and negotiations on two fronts: with the Allied occupying powers, and with local and regional German governments and the German postal authorities once they were reestablished. But in fact, strong organization came fairly easy to hams in Germany after the war: they could build on the legacy of the old DASD (Deutsche Amateur Sende- und Empfangsdienst), and the disciplined structures, which had existed in the Third Reich. Indeed, the post-Second World War amateur radio hobby in (West) Germany was characterized by an uncomfortable inheritance from the old DASD. To begin with, the Nazis had been meticulous in purging the DASD (and all other associations) of known socialists and communists. The men who had belonged to and been trained by the DASD in the Third Reich did not have any such sympathies, or at the very least, had learned to hide them well. Very quickly after the end of the war, as the Cold War grew, the Western Allies increasingly tried to limit socialist influence in their zones. When German hams in the western zones began to organize again, there was never much question about whether they

[4] For the history of TV in postwar Germany, see: Knut Hickethier and Peter Hoff *Geschichte des deutschen Fernsehens* (Stuttgart, Weimar: J.B. Metzlar, 1998).

[5] If poetry was no longer possible after Auschwitz, as Theodor Adorns famously if somewhat erroneously is held to have said, then who could believe in the positive good of technology after Hiroshima? See, for example, Spencer R. Weart, *The Rise of Nuclear Fear* (Cambridge: Harvard University Press, 1988, 2012). See also, of course, Theodor Adorno, "An Essay on Cultural Criticism and Society", trans. Samuel and Shierry Weber, *Prisms* (Cambridge, Mass.: MIT Press, 1967): 34.

harbored socialist elements, and this made it much easier for them to be accepted by the Western Allies.[6] Conversely, it was harder for amateur radio to become reestablished in the Soviet zone because the DASD had been so closely linked to the Nazi regime and purged of socialist elements. (There had also been a socialist amateur radio tradition in the Weimar Republic which might have become an argument for amateurs in the Soviet zone but didn't.) Either way, the legacy of the Third Reich cast a long shadow, even over amateur radio.

When discussing the revival of amateur radio in Germany after the Second World War, postwar histories written by German amateurs stress the importance of the international brotherhood of amateur radio enthusiasts and the role of sympathetic amateurs within the Allied administrations—not least to avoid mentioning some of the messier realities of the postwar situation. They also foreground their own tenacity and commitment to the hobby, which was certainly considerable. But beyond this, one of the most important elements in the reestablishment of amateur radio in the Germanies was precisely the strong continuity with the Weimar- and Nazi-era DASD. A very large number of those Germans interested in amateur radio after the Second World War had already been involved in it before the war ended, and nearly all had been members of the DASD. The postwar West German *Deutscher Amateur Radio Club* (German Amateur Radio Club, DARC), in particular, demonstrated a remarkable continuity with the Nazi-era DASD both in organization and personnel. This is an uncomfortable fact and, as a result, is usually ignored or downplayed by post-Second World War writings by West German amateurs. Reorganization after 1945 was simply easier for (West) German ham radio enthusiasts because they could fall back on models and leadership from a pre-war organizational structure for their hobby which had continued under the Nazis, whereas all the other parts of the radio hobby had essentially been suppressed under National Socialism and forced to endure 12 years of inactivity. It is clear that the reestablishment of ham radio in the two Germanies after 1945 was very greatly helped by being able to rely on strong elements of continuity with the pre-war ham radio organization, the DASD—other parts of the radio hobby simply did not have this ability.

[6] To be fair, there was a resurgent working-class amateur radio movement in the Western zones in the late 1940s, but it quickly faded, and some postwar hams did have left-wing sympathies, but they were not at all characteristic of the new ham organizations as a whole. See below.

The DASD did not survive the end of the Third Reich. In May 1945 the Allies ordered the dissolution of all organizations with Nazified by-laws, and this, of course, included the DASD, both because the Minister of Popular Enlightenment and Propaganda named the chair of the DASD, but also because its by-laws contained the "Aryan-paragraph" banning membership by Jews.[7] Note that neither of these two points, necessarily, make the DASD a fully Nazified organization, nor does it mean that the later DARC was just a continuation of a Nazi organization or had any Nazi-sympathies. As we have seen in the previous chapter, there is, in fact, a convincing case to be made that the DASD was able to retain a certain autonomy in the Third Reich, and that it provided in some ways an escape from the all-pervasive Nazification of society. But it was certainly a Nazi-dominated and controlled organization, which served both state and Party interests, and many members had been demonstrably pro-Nazi. In the immediate postwar period, the Allies disbanded and banned all organizations which had been too closely tied to the NSDAP, including the DASD.

Anyone who joined the DASD as a young man very likely came into contact with it via the Hitler Youth (HJ), and lived, worked, played, and was thus presumably socialized in a thoroughly Nazi-dominated context in both peacetime and in war. While some of this socialization as National Socialists occurred within the DASD, most of it happened elsewhere: as we saw in the last chapter, the day-to-day activities of DASD members contained very little overt Nazi ideology. Nevertheless, the Nazis had rather successfully Nazified the German society. This was the inescapable background of most post-Second World War radio hobbyists (and indeed, of an entire generation), whatever their trajectory after the war.[8] While a

[7] History of Ortsverband Coburg of the DARC, accessed July 24, 2016. http://www.darc-coburg.de/modules/wiwimod/index.php?page=chronik

> Im Mai 1945 wird der DASD wegen §4 (Vorstandsernennung durch das Reichspropagandaministerium) und §6 (Mitglieder nur arische Deutsche) seiner letzten Satzung vom 20.10.1934 als Verein verboten. Es ist bemerkenswert für die damalige Zeit, der Löschungsvermerk im Vereinsregister des Amtsgerichts Berlin-Charlottenburg (No. VR 7691) wird erst nach acht Jahren am 16.11.1953 eingetragen.

There is some controversy about whether there was a specific ban on the organization, or simply a blanket ban on all Nazified structures. As stated in the quote, the DASD was not formally stricken from the legal registry of associations (*Vereinsregister*) until November 11, 1953.

[8] See the recent article by Glenn Cuomo on the Nazi past of German writers: "The NSDAP's Enduring Shadow: Putting in Perspective the Recent Outing of Brown Octogenarians", *German Studies Review* 35, No. 2, (May 2012): 265–288.

few postwar hams could legitimately lay claim to having been dissidents during the Third Reich (and while they certainly provided welcome galleon figures and go-betweens with the Allies for the postwar radio organizations), they were certainly not in the majority.[9] Nor were the majority necessarily died-in-the-wool Nazis, either, but the Allies were careful. They preferred to ban organizations which might become centers of resistance, and judge individuals individually.[10] The legacy of the DASD for amateur radio after the war was nevertheless not entirely negative. The technical training it offered was of a very high standard. Moreover, the organizational structure of the DASD, while rigid, was also very utilitarian, and the strict discipline to which its members were accustomed gave postwar German amateurs great cohesion—at least toward the outside.[11]

Grim Conditions at the End of the War

The context was grim at the end of the war. Germany had clearly lost, and lost badly. Her cities had been reduced to rubble. Foreign armies occupied all of German territory, and much of Germany had been the site of ground fighting. While casualty estimates vary widely, at a minimum, some five million German military personnel,[12] and roughly 850,000 to one million civilians had been killed in the war.[13] Most adult German men still alive

[9] W(olfgang) F(elix) Körner, DL1CU, *Geschichte des Amateurfunks 1909–1963* (Hamburg: FT-Verlag Rojan & Kraft, 1963). Körner mentions, for example, Hans Haberl, DA2DH, later DL1AX, who led negotiations between the postal authorities and German amateurs in 1948 and 1949. According to Körner, Haberl had been sentenced to death for illegal political transmission and high treason in 1942. Körner, *Amateurfunk*, p. 185. Körner himself had been arrested by the Nazi authorities and punished for illegal transmission.

[10] Some Nazi organizations were declared criminal organizations in their entirety, like the General-SS or the Gestapo, and their members arrested automatically, but the DASD did not fall into this category.

[11] This book cannot hope to address the ins and outs of internal divisions within the post-Second World War German amateur community, and within the DARC itself. For this information, see Körner, *Amateurfunk*, which, however, barely touches on the topic. The most reliable and objective rendering is found in the online "Amateurfunk in Deutschland" series prepared by the Vienna-based Dokumentationsarchiv Funk (Dokufunk) (http://dokufunk.org/amateur_radio/history_dl_2/ and following).

[12] There are a number of contradictory studies; the most reliable recent study is Rüdiger Overmans, *Deutsche militärische Verluste im Zweiten Weltkrieg*, Beiträge zur Militärgeschichte 46 (Oldenbourg: DeGruyter, 2004): 313–324.

[13] Roughly 350,000 were killed in the strategic bombing campaign. Richard Overy, *The Bombers and the Bombed: Allied Air War Over Europe 1940–1945* (London, New York: Penguin, 2014): 304–307. Some 500,000 civilians were killed in late- and postwar flight and

were in prisoner of war camps. A total of 11 million ethnic Germans had been or were in the process of being expelled from Poland, Bohemia, Romania, Yugoslavia, and other parts of Eastern Europe. Millions more had fled before the advancing Red Army. Millions of Germans and other Europeans on German soil were homeless, displaced, and dispossessed. Those still alive faced rape, theft, and massive shortages of the most basic necessities. It is no exaggeration even to speak in terms of starvation in the first year after the end of the fighting.[14] A sovereign German state had ceased to exist, and Germany was split into four zones of occupation, each administered by a different foreign power. For a time, each zone was a world unto itself, and communication between zones was deliberately made difficult, even beyond the massive destruction of the German rail system and communications infrastructure during the war. In this situation, who had time or energy to think about a hobby? Simple survival was the main preoccupation.

On the other hand, catastrophe breeds a hunger for information, and radio was a cheap and rapid way to still that hunger. It had become an important, even indispensable source of news during the war, as progressive shortages of paper (and then, the destruction of the urban printing presses themselves) caused German newspapers to shrink and then almost disappear, save for the Nazi Party's own organs. The Nazis had contributed to making radio a simple necessity through their massive campaign to flood the market with affordable and robust "People's Receivers". By 1945 Germans had adopted the radio as one of their main sources of information and entertainment.

For the Allies, radios were suspect during the early months of the occupation. Though the Allied occupation authorities sometimes confiscated radio receivers, they were less of a problem than private transmitters, which were completely banned.[15] Rapidly, the occupying authorities

expulsions. Hans Henning Hahn and Eva Hahn, *Die Vertreibung im deutschen Erinnern. Legenden, Mythos, Geschichte* (Paderborn: Verlag Ferdinand Schöningh, 2010): 659–726. Note that this last figure is particularly contested, with the German government still giving a figure just over two million. (Statistisches Bundesamt—Wiesbaden, *Die deutschen Vertreibungsverluste. Bevölkerungsbilanzen für die deutschen Vertreibungsgebiete 1939/50* (Stuttgart: Verlag W. Kohlhammer, 1958)).

[14] On the immediate postwar situation, see: Ian Kershaw, *The End: The Defiance and Destruction of Hitler's Germany, 1944–1945* (New York: Penguin, 2011).

[15] But of course, Germans had already learned how to hide radios and radio listening from their own authorities during the war.

appreciated the usefulness of broadcast radio as a means of "reeducating" and controlling the German population, and set up German-language broadcast stations under their own control, using either new equipment or quickly repairing existing transmitters. The first of these Allied stations were on the air already in the Spring of 1945.[16] With the destruction caused by the war, particularly in urban centers where radio ownership was concentrated, the confiscation of some radio receivers by the Allies, and the even more widespread theft practiced by Allied soldiers, (who appreciated the quality of German radios), there was, to be sure, a shortage of civilian receivers among some groups, particularly refugees and expellees.[17] There was also a shortage of money or other capitol to procure new radios to replace those stolen or destroyed. Yet given the difficulties of communication between zones and with the outside world, radio had an advantage over all other forms of media in that it was largely unaffected by physical and political boundaries.

Consequently, there was a short renaissance of home-built radios in the first few years after the end of the fighting. Commercial parts, and even electricity, were scarce, so that there was even a return to primitive crystal radios, which were cheap, relatively easy to build, and did not even require electricity. Magazines and newspapers, once they began to reappear, carried articles and plans to home-build radios, and even broadcast radio carried programs about basic radio skills.[18] Do-it-yourselfers formed local clubs after the war in some places,[19] and in the American zone, the "Württemberg-Badischer Radio-Club" (WBRC), founded in 1946, included radio hobbyists of all kinds throughout the entire US zone, including home builders.[20] Local groups in other zones were probably also mixed in the beginning, but within the first year or two, the "true hams"

[16] Hans Bausch, *Rundfunkpolitik nach 1945: Teil 1, 1945–1962, Rundfunk in Deutschland* Vol. III (Munich: Deutsche Taschenbuch Verlag, 1980).

[17] Malzahn, *Sourcebook*, p. 166, confirms a shortage of radios immediately after the war.

[18] So, for example, The hessische Rundfunk carried a weekly program "Funktechnik für Alle" beginning in March 19, 1948. The broadcast was prepared with the help of the "Hessischer Radio-Club". *CQDL* 4 (1997): 316.

[19] Interview with Gerhard Hoyer, DJ1GE, Hamburg, June 9, 2014. This interview recounted do-it-yourself activities and clubs after the founding of the Federal Republic of Germany, and not during the occupation period.

[20] Körner, *Amateurfunk*, 175–176; Note that by 1947, there were severe differences between the amateur radio contingent and the others in the club, which eventually led to a split. Ibid, 184, 192–193. Apparently, only around 500 of the 4000 members in the WBRC were shortwave amateurs or hams.

or transmitting enthusiasts split off from most radio associations.[21] Hams had been a major backbone of the radio hobby during the Weimar Republic, and their disappearance from the other radio clubs after the Second World War seriously weakened them. Sometimes the do-it-yourselfers were able to maintain separate clubs to aid people in their handiwork,[22] but there just wasn't the time, energy, or sustained interest in a widespread club culture. Most people built a working radio in a club and then disappeared. The do-it-yourselfers, separated from the hams in the postwar era, did not rebuild the vibrant organizational culture of the interwar years.

Interestingly, there was a real attempt to revive the old Weimar-era *Arbeiter-Radio-Bund* in West Germany in the early 1950s, and for a time, there was again a strong competition between middle-class and working-class radio clubs in some areas.[23] But in the context of the growing Cold War, neither the Western Allies nor the later West German government had any interest in supporting radio amateurs with a specifically socialist outlook. And most hams coming out of the former DASD also had no interest in supporting a socialist competitor. Moreover, the postwar tactic of unity among radio clubs to pressure the Allies and the German authorities was manifestly successful, and in any case, class ties slowly broke down in the postwar economic recovery.

Broadcast radio as a medium remained important into the 1950s and did not decline until the widespread introduction of TV.[24] The uncertainties of occupation, division, and the various crises of the Cold War kept the hunger for immediate sources of information alive in Germany well into the early 1950s. Radio also provided a cheap and welcome source of enter-

[21] For example, by late January the "Hessischer Radio-Club" had some 1200 members in 29 local groups, of which 400 were "true hams" interested in transmitting. There was sharp in-fighting between the various interests within the club, and by March 1949, most non-hams had left, leaving the club with only 300 members.

[22] Interview with Gerhard Hoyer, DJ1GE, Hamburg, June 9, 2014.

[23] Folder 43 Arbeiter-Radio-Bund Deutschland Verein Bremerhaven (Hauptgeschäftsstelle Hannover), NARA RG 260 BOX 590 Records of United States Occupation Headquarters, WWII, Office of Military Government for Germany (US) OMGUS; Records of Office of Military Government, Bremen; Rcs of Bremerhaven Liaison & Security Detachment: Records of Clubs and Youth Activities, 1947–1949; Volksbund "Deutsche Kriegsgräberfursorge E.V." Thru Y Viii. Playgrounds.

[24] Regular broadcast of television programs began separately in both German states in 1952, although TV ownership remained low until the 1960s. See Hickethier and Hoff *Fernsehen*.

tainment and culture, even after the reintroduction of the pre-war radio tax. As daily life stabilized and began to improve, radio remained important as a way of access to the wider outside world, both politically and personally. In a world where both Germanies remained largely the pawns of outside forces, access to the outside world was psychologically very important. But (broadcast) radio was now simply a practical tool, it had lost its glamour.

Amateur Transmitters Revive: Illegally

Things were a bit different for the much smaller group of amateur radio operators within the larger radio hobby. True, they were economically, politically, and psychologically in the same boat as everyone else, and even the most die-hard members of the ham community could not afford to think much about the hobby in the very early days of the occupation—which most of them spent, in any case, as prisoners of war (POW). But their knowledge of the practice and technology of radio also gave them some important advantages. First, the ability to build a working radio receiver out of very basic parts or repair a broken one were valuable skills. Second, the amateur radio community was tightly knit and had been tightly organized under the Nazis and even before. The old "ham spirit" of self-help was very welcome in the difficulties of the postwar world, and networks of former radio amateurs quickly formed on this basis. Moreover, if their preferred activity—transmitting over the airwaves—was potentially more dangerous under the occupation than simple listening, and also much harder to hide, the need to convince the local occupation authorities that it ought to be allowed once again put a premium on organization and teamwork. And within the US, French, and British occupational administrations, fellow radio amateurs could also often be found who might be persuaded to lend a hand or at least look the other way.[25]

The issue of getting back on the air was not a trivial one for German amateurs. Just as after the First World War, but in a much more intense

[25] Radio amateurs naturally served in radio-related communications and intelligence functions during and after the war in all armies. Moreover, amateurs within all three Western Allied armies were allowed to practice their hobby while in Germany as part of the occupation. Many German hams came into contact with Allied amateurs in this way, either by having to do with Allied communications and intelligence troops because of their radio activities, or by hearing—even contacting—Allied hams on the air. While the Soviet Army also contained many radio amateurs, it did not allow them to transmit while on active duty.

fashion, Germany was an international pariah after the Second World War and Holocaust. Because of this, the ability to participate again in the friendly, open worldwide community of amateur radio operators was a prize many German hams longed for after the war, for it meant precisely the kind of contact with the outside world and participation in the international community which was denied Germany as a whole. Whatever the personal responsibility of German hams for Nazi crimes may have been, practicing their hobby gave them the chance to again be accepted in and by the wider world. This was a powerful psychological motivation for resuming amateur radio activities whose importance should not be underestimated.[26]

Growing numbers of German amateurs found each other in the post-war period, and first local, then regional (zonal) organizations grew out of them. This was made easier by the fact that the structure of the old DASD was not forgotten, and lots of old-timers were around who remember how amateur radio had to be organized.[27] The old leaders still often held respect and were often available to take on new duties. Procedures and organizational models could be copied and adapted to new conditions instead of needing to be invented outright. Newly trained radio specialists were also around in even greater numbers after the war: thousands of men had been trained as radio and telegraph operators in the military, many times more than in the First World War. Many of these men first came into contact with old-time DASD members, and thus with the idea of amateur radio as a hobby during the war (many DASD-trained hams became instructors in the wartime radio schools), and got infected with the amateur radio bug even then.[28] Massive amounts of former military radio equipment from all sides were also available immediately after the war, and quickly found new uses in the hands of men who understood how to modify them for hobby use.

[26] The general hunger for trustworthy information from, and contact with the wider world is conveyed in the first issue of the new newspaper *die Zeit*, which sets out its task for the future. *Die Zeit* 1 No. 1 (February 21, 1947): 1. Quoted in Malzhan, *Sourcebook*, 167–168. Note the parallel with the situation after the First World War: in both cases, Germany was a pariah state, and German amateurs after both wars were often grateful—and a little surprised—when they were met with courtesy by foreign hams.

[27] It should also be acknowledged that the strong organizational structures developed by German hams also meant that they have left behind more documents with which to trace their history than did the much more ephemeral clubs of listeners and do-it-yourselfers. This may well skew the history written here.

[28] Letter from Dipl. Ing. Fritz Trömel to Dipl. Ing. Helmut Ahlborn of February 1, 2001, in: Hamburg DARC Archive.

Initial Allied Plans

As the Allies began to prepare for the occupation of Germany, and then to gradually occupy more and more German territory in 1944 and 1945, they had tremendous fear of organized partisan activities. The Nazis had promised as much, and fanatical resistance was expected. This is reflected in the immediate confiscation of all weapons, carrier pigeons, and radio transmitters by the Allies, and the automatic arrest of numerous categories of Germans, from high Nazi Party figures and officers down to SS officers and intelligence operatives.[29] In fact, with one or two isolated exceptions and to the great relief and surprise of the Allies, no such resistance appeared. This fact was decisive in the eventual Allied acceptance, even encouragement of amateur radio.

Germans living under Allied occupation were obviously very interested in divining what the occupational authorities were thinking, and they were often keenly aware of what was going on. As German hams began to organize and push for licensing, they were quite aware that there were differences of character and policy between the different Allies, and they quickly adopted policies which they reckoned would appeal to the particular occupying power where they lived. Though the evolution of the radio hobby was broadly similar across the entire occupied area which had once been Germany, hobbyists in each zone faced a particular situation dictated by the Allied power in charge. Speaking very broadly, it became easier for radio hobbyists (and for all aspects of German civil society) as time went on. Once the Allied powers were convinced that there was no longer a danger of armed German resistance, they were increasingly likely to view radio hobbyists with sympathy. And again, speaking very broadly, radio hobbyists faced the greatest sympathy in the US and British zones (BZs), a bit less in the French, and even less in the Soviet zone. Everywhere, though, there was a kind of cat and mouse game as some radio hobbyists in each of the four zones of occupation insisted on practicing their hobby even under conditions of great illegality and sometimes at significant personal risk. This is perhaps not surprising, given that German ham radio

[29] Arrest of top Nazi leaders was agreed upon by the Allies at the Potsdam Conference, and was reflected in specific regulations within each Allied zone. In the US zone, there were two categories of those subject to arrest: a "blacklist" of specific individuals, and a separate list of certain categories of individuals who were subject to automatic arrest based on the position or rank they held. See Elmer Plischke, "Denazification Law and Procedure", *The American Journal of International Law* 41, No. 4 (October 1947): 807–827, here 811–814.

operators in particular had rarely been allowed to transmit freely, and since their inception in the early 1920s had learned to cope with a restrictive legal climate. Given the long experience of semi-legality in the Weimar Republic and the very restrictive legality of the Third Reich, the illegal continuation of ham radio during the occupation period was rather close to business as usual, though publicly all German radio amateurs naturally stressed their complete conformity to whatever regulations applied—just as they had earlier, in the Weimar Republic and Third Reich.

In the immediate postwar phase of Allied occupation, Allied interests and therefore actions were quite similar in every zone. The Nazis had promised armed resistance, and ostensibly had created a widespread "Werewolf" organization to carry it out. Allied policy was therefore geared to making sure that such resistance was unsuccessful.[30] For example, in the state of Hessen, American forces immediately restricted all forms of travel and communication to a bare minimum. Among other things, US forces demanded:

> §8 Radio transmitters as well as any other means of transmission, firearms and other war materials as well as ammunition and explosives must be handed in to the military authorities. It is illegal to hold such things or to have access to them. Only the police may carry hand-held weapons and ammunition with a permit issued by the military authorities, in order to keep up law and order.
> §9 All cameras as well as binoculars will be handed in to the military authorities.
> §10 It is forbidden to release pigeons. They must either be killed, or have their wings cut back.
> §11 Any communication such as by mail, telephone, telegraph or radio will be stopped at once.[31]

Thus, from the beginning of the US occupation of Germany, and well before the Allied Control Council began to meet, radio transmitters were considered dangerous contraband. Legislation in the other three zones was similar. Interdiction of radio transmitters was codified for all zones of

[30] See: Volker Koop, *Himmlers letztes Aufgebot: die NS-Organisation "Werwolf"* (Cologne: Böhlau, 2008).

[31] From Wolf-Arno Kropat, ed., *Hessen in der Stunde Null 1945/1947: Politik, Wirtschaft und Bildungswesen in Dokumenten* (Wiesbaden: Selbstverlag der Historischen Kommission fur Nassau, 1979): 13–17; Original source given as: "Reg P 1 Gen 1945 Bl.2-5".

occupation in Control Council Law No. 76. The law included a long list of provisions, including the outlawing of weapons and military equipment, but also specifically and prominently banned the possession of radio transmitters by civilians under Allied occupation, and specifically allowed punishment up to and including the death penalty for violation of the law.[32] This certainly indicates Allied thinking in the early period of the occupation, and this and similar laws were also certainly enforced. However, as in all Allied regulations, there were difference from zone to zone, and also over time. Laws such as this were rigorously enforced early in the occupation, when resistance was still expected, but tended to be loosened later on in actual application.

The Foundation Myth[33] of Postwar Amateur Radio

Interestingly, "Control Council Law No. 76" is referred to prominently by all of the postwar histories written by amateur radio enthusiasts themselves,[34] and also by the few academic studies on the topic.[35] Beyond its actual application, it serves as a part of the (mythic) origin story of postwar ham radio. In actual practice, the Allied Powers all rapidly concluded that broadcast radio was an important tool for reeducating the German population and administering the occupation. While they all remained more skeptical about radio transmitters, punishment for illegal transmission was generally not as draconian as Control Council Law No. 76 might lead us to believe.

[32] This law certainly existed, but despite the many references to it and its almost mythic status in the German amateur radio community, it is devilishly hard to track down. It was referred to at the time, for example, in: "Die Gesetzgebung der Besatzungsmächte", *Süddeutsche Juristen-Zeitung* 1, No. 2 (May 1946): 46. But it does not appear at all in the "Official Gazette of the Control Council for Germany" (Berlin: Allied Secretariat, 1945–1948.), which otherwise lists all the laws, orders, and degrees issued by the Allied Control Council.

[33] I use myth here in an anthropological and value-neutral sense. I define it as an epistemological story or narrative, usually concerning origins, which is reflective of the values of a particular group. I am not suggesting that the "foundational myth" of post-Second World War German amateur radio I describe here is false or untrue, but I am stating that the historical reality was much more complex than the myth.

[34] For example, Körner, *Amateurfunk*, 174.

[35] For example, Christian Senne, *Rahmen und Organisationsbedingungen für Funkamateure in der SBZ und DDR (1945–1990). Zwischen Selbstzweck und gesellschaftlichem Auftrag*, Studien zur Zeitgeschichte 70 (Hamburg: Verlag Dr. Kovač, 2008): 87.

The later emphasis on Control Council Law No. 76, and on other measures against private transmission cited in chronicles written by German amateurs serves a different purpose. Immediately after mentioning Control Council Law No. 76, the postwar German chronicles all immediately go on to state how the died-in-the-wool German hams immediately got back on the air despite the ban and the risks involved. The chronicles describe the tricks used by plucky and clever amateurs to conceal their true identities and locations. They then usually go on to tell a version of this simple story: The clever and dedicated German hams burned to get back on the air, despite the potential danger. Seeking to end the ban on amateur transmission, they soon approached the local Allied military commander (whether American, British, French, or Soviet) to ask for permission to transmit, or at least to form a radio club. Lo and behold, said Allied commander (or his adjutant, or communications officer) turned out to be a fellow amateur radio operator. With his connivance, the German "OM's"[36] were soon back on the air, tolerated by the local Allied officials in spite of the official ban. Countless versions of this foundational myth exist.[37] The idea that German hams transmitted during the occupation, even in the face of brutal Allied suppression, is part of the founding myth of West German amateur radio. Like all such foundation myths, there is some, perhaps even much truth to it: certainly, many German amateurs transmitted illegally during the Allied occupation.[38] Many military personnel in all the Allied

[36] "O.M." originally stood for "Old Man", and in international ham radio jargon, means a ham radio operator, usually male (as opposed to a "YL", a Young Lady or female ham), and usually licensed.

[37] For example, see: Jürgen Bennöhr, DL1NP, "Geschichte des Ortsverbandes Schleswig". Accessed June 25, 2015. http://www.viehl-radio.de/homeda/chronik/chronik_m13.pdf; Heinz Schifferdecker, DL7AC, Horst Eligering DL9MH and Manfred May DJ1KF, "Chronik des Amateurfunks in Deutschland und im Bereich Köln-Aachen". ("von Heinz Schifferdecker, DL7AC und Horst Ellgering, DL9MH, © 1991; für das Internet aufbereitet von Manfred May, DJ1KF (neu bearbeitet von DO1KXL)"). Accessed June 25, 2015. http://www.datv-agaf.de/links/chronik.html#a7

[38] "Letter from Lt. Col. Harry H. Pretty, GSC S-2,S-2 Branch to Deputy Director of Intelligence, European Command, Frankfurt, APO757, US Army subject: Unlicensed Amateur Transmitters (Secret)", NARA Record Group 260 BOX 3 Records of United States Occupation Headquarters, WWII, Office of Military Government for Germany (US) OMGUS; Records of functional Offices and Divisions, The Office of the Director of Intelligence: General Corresp of the Analysis and Research Branch 1945–1949; 000.9-2 Scientists, German, IOA Evacuees thru 004. German Industrialists and Industry (1); FOLDER 000.77 RADIO INFORMATION OF INTELLIGENCE INTERESTS 7-20-1/28. The letter has two appendices, one listing the illegal German and Austrian transmitters

armies were also amateur radio enthusiasts.[39] Situations similar to the mythic story also seem to have actually occurred in some cases.[40] But even then, actual events were a bit more complicated than the mythic story of origins. First, it is highly unlikely that such tolerant Allied commanders existed everywhere, particularly in the Soviet and French zones. The story is just a little too good to be true, and draws just a bit too much on the myth (and reality) of "ham spirit" among the international brotherhood of radio amateurs. Where events approached the myth, reality was just a bit coarser. Moreover, though we have already seen that Allied occupation tended to be harsh in the beginning, the myth also undoubtedly exaggerates the threat involved in illegal transmission—at least outside of the Soviet zone. In many ways, the actual story is more interesting. The myth provides a heroic origin story, which emphasizes the obstacles faced by ancestors in reestablishing amateur radio in Germany, and thus aids modern cohesion among amateurs. But it also serves to obscure the complexity of the story, and even forget aspects which might not be seen in a positive light today.

The tasks faced by German radio hobbyists seeking to revive their hobby were huge, but also clear. There were economic difficulties to overcome, but this was nothing new—in fact, the radio hobby in Germany had first evolved to meet exactly this challenge in the first place. More difficult were the political difficulties presented by the need to persuade four different Allied occupational authorities that the radio hobby was not dan-

whose location had been identified, and a separate list of German and Austrian transmitters with unidentified locations. The letter states that the lists were based on British signals intercepts. Note that the location known was only down to the level of a particular city, in most cases. The number of unlocated transmitters was largest; together, the two lists contain roughly 300 illegal transmitters, with call sign. Along with being stated in numerous memoirs and local chronic of amateur radio clubs, there is also hard and fast proof of this activity found in numerous QSL cards sent to and from German amateurs during this period. A number of these QSL cards may be found in the Dokufunk Archive.

[39] See the list of all licensed Allied ham radio operators attached to the "letter from Schimmel (For the Chief) II E2-5332-2 to U.S. Military Government, Communications Branch, Berlin-Steglitz, RE: YOUR LETTER OF 27.7.49 CBC 216/49 (COPY) OF AUGUST 16, 1949", NARA RG 260 BOX 581 Records of United States Occupation Headquarters, WWII; Office of Military Government for Germany (US) OMGUS

Records of the Berlin Sector. Records of the Communications Branch: General Records, 1945–1949. Newspaper clippings 1949 (6) thru Action Sheets. There were enough US and British amateurs among the troops, for example, that the military authorities soon issued regulations and special call signs to allow them to operate.

[40] See below.

gerous, perhaps even helpful. Perhaps most daunting, however, was the task of then persuading local and national German authorities to go beyond Weimar and openly allow private radio transmission. The tradition of a governmental (postal) monopoly on communications media in Germany was still very strong.

In this quest, German hams were aided by Allied plans to democratize German society, and by their growing preoccupation with the Cold War rivalries between Allied powers. In particular, the British and Americans were both very unwilling to see the reestablishment of the centralized, government-controlled media landscape of the Weimar Republic and Third Reich. British and US occupational government authorities were therefore unwilling to see the German postal authorities regain their control over broadcast radio, and insisted on a more open and decentralized media regime. This had repercussions on the discussion over ham radio licensing as well, and was a major factor in the legalization and reestablishment of amateur radio in West Germany. It just doesn't fit the promethean myth.

Perhaps the strongest argument that German hams pushing for legalization of their hobby had was the promise that a well-organized, disciplined, and legal amateur radio hobby would be able to restore order on the airways and keep bad apples in check. This argument had been made before, both in the Weimar Republic (unsuccessfully) and Third Reich (with limited success). It proved stronger in the postwar period and is eventually what convinced US and British authorities. The British and Americans were open to this line of argument for two reasons. First, rapid redeployment and repatriation of troops in the immediate postwar period left the military governments of both nations increasingly understaffed. Moreover, both military governments simply lacked the technical means to locate illegal transmitters on a large scale. By this time, it was technically possible to do so, but the occupation authorities (at least in the British and American zones) simply lacked the manpower and equipment.[41] Ultimately, it was simply cheaper for the British and Americans to tolerate and moni-

[41] "Lt. Col. Robert Walker, Chief, European Command Headquarters, Army Security Agency Europe, APO757, to Deputy Director of Intelligence, Headquarters, European Command, APO757 U.S. Army, ASA-94/MJC/mp of 17 February 1948 (copy) (Confidential)", NARA Record Group 260 BOX 3, Records of United States Occupation Headquarters, WWII, Office of Military Government for Germany (US) OMGUS; Records of functional Offices and Divisions; The Office of the Director of Intelligence: General Corresp of the Analysis and Research Branch. 1945–1949; 000.9-2 Scientists, German, IOA

tor (disciplined) illegal transmission, and later to allow licensed amateurs, than it was to repress them. Once they had convinced themselves that the German hams were capable of making good on the promise to act in a disciplined way, the British and Americans were willing to pressure the German and French authorities to go along with licensing German hams. Paradoxically, the Americans and British were reassured by observing the disciplined conduct of German hams on the air—as they transmitted illegally! (Of course, the willingness of some German hams to cooperate with Western intelligence services helped. See below.) The strong discipline of German amateurs was, naturally, a reflection of the graveness of the postwar situation, but it was also part of the inheritance of the DASD. The willingness to both self-police and to aid the Allies and later the German government in policing the airways is another major element in the refounding of amateur radio in Germany.

Events in the US and British Zones: Ham Spirit and Civil Society

I will begin with the American zone of occupation, which included much of the southern part of Germany centered on Bavaria, but also a part of Berlin and the port city of Bremen. Given that the most important amateur history of ham radio in Germany was written by a direct participant in activities in the American zone, it is natural that we know most about events here, and that they sometimes tend to overshadow developments in the other zones.[42]

Certainly, preoccupation contacts between radio enthusiasts everywhere in occupied Germany played a major role in the earliest efforts to reestablish amateur radio. All Germans needed to re-find and re-create older friendship networks in order to survive the difficulties after the end of the war. As early as August 1945, a small group of ham radio enthusiasts, most freshly liberated from POW camps, began to meet privately in the home of one of their members, a certain Seiferheld, (ex D 4 MCN). Most had known each other before the war. By Fall 1945, a handful of south German transmitters were back on the air—illegally.[43] In April 1946,

Evacuees thru 004. German Industrialists and Industry (1); FOLDER 000.77 RADIO INFORMATION OF INTELLIGENCE INTERESTS 7-20-1/28.

[42] This is Körner, *Amateurfunk*.

[43] Körner, *Amateurfunk*, 174.

some of these illegal transmitters arranged to "meet" on the air every Saturday night—the origins of the so-called Saturday Night Club (SAC).[44] Their number grew to around 20 by May 1946, and soon they issued a regular Saturday evening on-air bulletin. Still fully illegal, pre-war hams were able to begin to organize based on pre-war radio connections and using the medium of radio to make anonymous links between individuals on the air.

In the summer of 1946, radio enthusiasts in Stuttgart (led mainly by hams, including Egon Koch, (DL1HM)), obtained permission from the American military authorities to found a radio club for the entire US zone. The "Württemberg-Badischer Radioklub" (WBRC) was formally founded on August 17, 1946. This was a club like old DFTV, which was for anyone interested in radio, whether ham, "do-it-yourselfer", or broadcast listener. Amateur transmissions were still illegal, so the club was ostensibly focused on radio listening and the do-it-yourself building of receivers, making the club palatable to the US authorities. Nevertheless, it did contain an informal separate "Shortwave Section" ("*Sektion-Kurzwelle*"), which became the foundation of the later ham radio movement in southern Germany.[45] As the first radio club to be founded in any zone, it rapidly attracted the interest of radio enthusiasts in other zones, not least because many of its members embarked on illegal transmission on an ever growing scale, and thus could be heard on the air. Not much later, the "Verband Bayerischer Kurzwellenamateure" also applied for club status.[46] While transmission remained illegal in 1946, the Americans were willing to allow the careful return of the public sphere in the form of some club activities, including radio hobby clubs.

Even by the end of 1946, ever larger numbers of German amateurs were willing to flout Allied laws and even risk arrest by transmitting illegally, but without some sort of a call sign, their ability to participate in the international hobby was severely limited. Remember that without a legal (or legal seeming) call sign, a ham is really without an identity, and cannot make radio contact with other hams. What to do? Any German ham wishing to transmit and make contact with like-minded enthusiasts around the world faced a major problem in 1945. International law stated that amateur radio operators were only allowed to make contact with other legally licensed amateurs, a fact vouched for by the possession and use of a unique

[44] Körner, *Amateurfunk*, 174–175.
[45] Körner, *Amateurfunk*, 175–176.
[46] Körner, *Amateurfunk*, 190.

personal "call sign", which identified each individual. The international transmission of contact reports ("QSL cards"), which were the basis for awards and recognition within the amateur community were also contingent on the use of legal (or apparently legal) call signs. But in 1945, the DASD had been dissolved and stricken from membership in the International Amateur Radio Union (IARU), a member entity of the International Telecommunications Union (ITU). This meant that there was no national body to issue call signs in Germany, and thus no internationally valid amateur call signs for Germans. If German amateurs wanted to regain admission to the international world of ham radio, they needed legal, internationally recognized call signs—or at least something, which might be taken as such and which would allow contact reports from foreign amateurs to reach them. Readers of this book will recognize that the same problem existed for the numerous illegal transmitters during the Weimar Republic. Given that recent history, which many post-Second World War amateurs had experienced firsthand, it will be of no surprise that German amateurs after 1945 found similar solutions.

Many former amateurs simply gritted their teeth and refrained from illegal transmission altogether, in the hope that respect for international law and good discipline would one day win them legality. This was the official strategy (if not always actual practice) of the radio enthusiasts in the British and French zones, as hams began to organize there. But within a few months after war's end, amateurs in Allied nations and neutral countries who had been prevented from transmitting during the war came back on the air. Worse, foreign amateurs (mostly soldiers) on German soil in the US, British, and French occupational governments also began to transmit, with special "German" call signs reserved for Allied staff. The amount of international amateur radio traffic around the world grew rapidly, and ever more German amateurs became tempted to jump back into the game.

In the long run, the only way to solve the call-sign dilemma was to gain permission from the Allied occupying power and the local German civil authorities (the postal administration, given prior German law) to transmit legally, and then to have a national radio club focused on ham radio which could apply for recognition by the ITU. Respect for the international laws governing radio, and strict discipline in using radio, were key elements of the amateur radio hobby worldwide. The experience of the war, and the pre-war emphasis on strict on-air discipline, which had characterized the DASD, only reinforced this characteristic among German amateurs. After all, legality had also been the public strategy of the DASD and DFTV already during the Weimar period. Waiting until legality could be achieved was the "right thing to do". The problem was, the wait might be very long, and some amateurs were impatient. The clash between the strategy of legality and the strategy of deliberate and widespread illegal transmission as a means of pressuring the authorities to grant legal transmitting licenses led to a great deal of conflict within German amateur circles in the immediate post-Second World War period.

In a sense, history repeated itself. A handful of German amateurs began illegal transmissions quite early after the end of the war, because the authorities refused to grant them legal transmission licenses; their numbers only increased as time went on. This recalls the situation in the Weimar Republic. Just as in the early 1920s, illegal German hams who dared transmit first had to use made-up call signs, in order to initiate a contact with a foreign ham. Many of those who insisted on transmitting illegally simply used their pre-war call sign, but they could be traced. Others misrepresented themselves by using a foreign call sign or later, the call sign of an Allied soldier. But all of these solutions were only temporary measures; they allowed a conversation to take place on the air but could not lead to the receipt of a QSL card to document the contact. Moreover, as more and more illegal German stations appeared on the air, chaos threatened if no order or system could be brought to the call signs used. Too much of the practice of the international amateur radio hobby depended on the orderly administration of the system of call signs and the transmission of QSL cards. A more stable and orderly solution had to be found. Then again, if there could not be legal licenses and call signs yet, that didn't mean that there could be no call signs at all. To give the appearance of legality, many illegal German transmitters began to systematize their illegal call signs more or less spontaneously, in the spirit of the existing international system used for the allocation of national identifiers in call signs. Thus, by May 1946, many illegal German amateurs began to use (self-assigned) call signs beginning with D4 or D3, followed by a number, two other letters, and ending in an "X".[47] By late in the year, many had begun to make up call signs ending in "A".[48] This looked legitimate, and introduced a bit of order. It was also not a complete solution.

On July 7 and 8, 1947, the WBRC held the first postwar "Shortwave Conference" (*Kurzwellentagung*), which attracted some 500 amateurs

[47] Körner, *Amateurfunk*, 175, cites D4GAX, D4FUX, D4AHX, D3DQX, as examples.

[48] Körner cites D3 ADA, D4 ATA as examples. Körner, *Amateurfunk*, 176.

from all over Germany, not just the American zone. It was a major milestone in the development of amateur radio organizations in the postwar period.[49] Concerned hams from all over Germany met together to discuss common problems. With illegal German transmission increasing, pressure mounted to address the call sign problem once and for all, yet the Allies were still unwilling to allow legal transmission. Something had to be done to bring order to the airwaves, facilitate the transmission of OSL cards, and perhaps put pressure on the reluctant Allies. The southern German hams felt that massive illegal transmissions would pressure the Americans into legalizing amateur transmission, if only to avoid embarrassment, though they also hoped to demonstrate that their hobby was harmless, and that in fact, German hams could become partners in policing the airways.

A June 1947 editorial in *QRV*, the journal of the WBRC gives a good impression of the thinking of the leaders of the WBRC in 1947 as they sought to gain legalization of amateur transmission and to convince German hams elsewhere of the correct path to follow in preparation for the July 1947 Shortwave Conference. It is worth looking at the text in detail:

> The amateur's natural counterpart is inevitably whatever political power holder is currently in place. In the countries that won the war in 1918 the state was little or not at all interested [in controlling radio], in Germany a police state was set in motion from all sorts of office holders, who themselves were either brought to a fall or chased out again by somebody else after a few months. The more decent a man was, the more suspicious he seemed—it is a miracle that there were even [radio] receivers in Germany.
>
> There is now a once-in-a-lifetime opportunity for us in the American and British zones, compared to the post-WWI years. The real power-holders come from countries that themselves possess sizable amateur organizations and that approach the question in an absolutely well-disposed manner. It's up to us to create a licensing regime which can meet this situation. Moreover, it is also up to us to demonstrate that we have absolutely no political ties to the outside, either openly or covertly—and we should be able to succeed in giving proof of this fact. In fact, for the occupying power—and it is only from them that we will get licenses—we are thus more transparent than any political agency, we see ourselves as being simply more credit-worthy than a random agglomerate of random politicians. Moreover, we spent the "thou-

[49] Körner, *Amateurfunk*, 176–177.

> sand years" [a sarcastic reference to the "Thousand-Year Reich"] neither in a safe foreign country, nor as nattering nabobs in the style of Kasimir Edschmid.[50] Instead, we fought for our very lives many times over with rifles and hand grenades—we are impregnated through and through with mistrust of everything and everybody. Recently, political rat-catchers seem to flee our company for this very reason, so that our credibility seems assured for the foreseeable future.
>
> So let's just write a [amateur radio] licensing law right now, that we can take to the occupational power, and [successfully] request the authorization to start amateur radio operations.[51]

The arguments presented are both insightful and bizarre in turn, but contain a basic truth: the ham radio operators in the WBRC were not interested in politics, and certainly did belong to a generation shorn of all illusions. They shrewdly understood that the main reservation of the British and Americans was political, and their historical argument that they

[50] Edschmid (pseud) was a (precious) pre-expressionist writer, who went into internal emigration during war, and published a self-serving book in 1947 describing the conflict between a "good" German writer and his family living in a remote mountain village and Nazi supporters who were quartered in the same house, as if there was this moral conflict within all Germany. The point is that German hams were all very practical men without illusions.

[51] *"Der natürliche Gegenpart des Amateurs ist zwangsläufig der jeweilige politische Machthaber. In den Siegerländern von 1918 war der Staat wenig oder gar nicht interessiert, in Deutschland wurde ein Polizeistaatapparat von allerlei Sesselchen aus getätigt, deren Inhaber ihrerseits wieder alle paar Monate von irgendwem gestürtzt oder weggejagt wurden. Je anständiger ein Mensch, desto verdächtigtiger—ein Wunder, dass es in Deutschland überhaupt Empfänger gab.*

In den amerikanischen und britischen Zone besteht nun im Vergleich zur Nachweltkriegszeit die einmalige Möglichkeit für uns, daß der wirkliche Machthaber aus Ländern stammt, die selbst große Amateurorganisationen besitzen und der Sache absolut freundlich gegenüber stehen. Es liegt an uns, eine Lizenzordnung aufzustellen, die dem Rechnung trägt. Es liegt ferner an uns, zu beweisen, daß wir keine parteipolitischen Bindungen haben, weder offene noch getarnte—ein Beweis, das uns gelingen dürfte. Wir sind damit für die Besatzungsmacht, von der allein eine Lizenz zu bekommen ist, überschaubarer als jede politische Behörde, wir halten uns ganz einfach für kreditwürdiger als ein zufälliges Agglomerat zufälliger Politiker. Ferner haben wir die tausand Jahre weder im sicheren Ausland noch als brabbelnde Bürger Edschmidschen Stils verbracht, sondern uns mit Karabiner und Handgranate etliche Male unseres Lebens gewehrt—wir sind durch und durch imprägniert mit Mißtrauen gegen alle und jeden. Politische Rattenfänger scheinen deshalb neuerdings unsere Nähe zu fliehen, so daß auch für die Zukunft die kreditwürdigkeit gesichert scheint.

Stellen wir also zunächst einmal eine Lizenzordnung hin, mit der wir zur Besatzungsmacht gehen können, um die Genehmigung eines Amateurbetriebs zu erbitten". From QRV, No. 6 (June 1947): 97–99 (here p. 99).

had lost their illusions in 12 years of dictatorship and war, while risky, had a solid ring of truth. That it also betrayed resentment of exiles and intellectuals, and contempt for Weimar democracy, seems shocking today, but would probably not have been noticed by most American or British soldiers in 1947. The German hams were men who had been formed by dictatorship and war, more so than they themselves perhaps realized, but they also understood that the fundamental traditions of British and American government, as well as the presence of thousands of amateurs in the British and American ranks,[52] predisposed the British and Americans favorably toward radio hobbyists—if they could make a convincing case that they were not "political", meaning neither Nazi nor communist. They also understood that they had nothing to hope for from the German postal authorities, who had fought amateur licensing and transmission since the 1920s. The German hams in Baden-Württemberg calculated that their best chance was with the British and Americans, and they were determined to seize their chance.

Not addressed in the public editorial was another aspect of the German strategy. On the one hand, the south German amateurs really encouraged illegal transmission, despite public protests to the contrary, in order to pressure the Americans into conceding legal licenses. Meanwhile, the north German hams in the British zone held strictly to a policy of following the law, and publicly criticized the WBRC for its illegality (see below). The illegal transmitters sought to place the Allies before a fait accompli. Yet by imposing discipline on the illegal transmitters, as described below, the WBRC also made the implicit argument to the Allies that they would be good partners in ending the chaos on the airways. These were very

[52] By July 1949, there were some 455 hams within the American administration alone who were licensed to transmit from Germany. "Letter from E.T. Martin, Chief, Telecommunications Branch, to Office of Military Government, Berlin Sector, Chief, Communications Branch, A.P.O. 742-A U.S. Army, Attention Mr. John H. Gayer, A.G. 311.23 (EA) of July 21, 1949, Subject: List of U.S. Amateur Radio Operators", in: NARA Record Group 260, BOX 581; Records of United States Occupation Headquarters, WWII; Office of Military Government for Germany (US) OMGUS; Records of the Berlin Sector. Records of the Communications Branch: General Records, 1945–1949. Newspaper clippings 1949 (6) thru Action Sheets; FOLDER 4-81-1/17 "LICENSES FOR AMATEURS (RADIO) 51A-". Attached to the letter is a 20-page list of US hams currently transmitting from the US Sector. By this time, a West German amateur radio law had been passed, so all of these American amateurs had been issued special German call signs for the occupation forces, but the list gives a good indication of the numbers of hams in the US administration earlier.

clever arguments. They worked well with the Americans and British, but less well with the French, and not at all with the Soviets.

To facilitate and organize the growing illegal radio activity of German amateurs, a group of South German hams grouped around the previously mentioned "Saturday Evening Club" and with the help and support of the leadership of the amateurs within the WBRC, decided to issue their own systematic—but still resolutely illegal—call signs for German hams in all four Allied zones and Berlin.

The idea was to create order and uniformity in the call signs used by illegal German hams, but also to create a kind of underground license system: not anyone would be awarded one of the new call signs, the recipient should have a good command of the technology and know the rules for good on-air communications, something which was to everyone's advantage. Significantly, the system they came up with was—except for being illegal—in outward conformity with the worldwide call-sign system voted at the most recent ITU meeting in Atlantic city in 1947. Thus, a ham outside Germany encountering a German ham on the air would have no reason to think that the German was operating illegally with a non-official call sign. The new system went into operation on December 31, 1947. Within two months, some 400 of the unofficial call signs had been issued. The system grew steadily, and soon there were some 600 transmitting hams registered, with over 10,000 names in the card file. Significantly, this was always conceived as a system for all of Germany, irrespective of Allied zone; it specifically reserved whole blocks of call signs for German hams in the Soviet and French zones and the Saarland.[53]

To put a public face on what they were doing (and in alignment with pre-war practice), the same group of hams within the WBRC also began to issue the so-called DE-Numbers. These were not call signs per se, but were instead a reappearance of the Weimar-era invention later maintained within the DASD which amounted to a shortwave listeners' license as a first step to subsequently obtaining a full call sign, which would authorize transmission. It demonstrated that one had passed a test in the basics of radio and Morse code, and intended to take part in the amateur radio

[53] Körner, *Amateurfunk*, 178–181. The unofficial call signs began with the letters DE, followed by a number indicating the district or region: DA1=Württemberg & Baden, DA2=Bavaria, DA3=Hessen, DA4=Nordrhein-Westfalen, DA5=Niedersachsen, DA6=Schleswig-Holstein, DA/DK7=Berlin, DA/DK8="Ostzone" [Soviet Zone], DK9="Französische Zone", EZ=Saargebiet. This was then followed by two more letters, the first of which often pointed to a particular town.

hobby. They were a first step in the direction of true call signs, although they had no legal standing.[54] But they were acceptable to the Americans, since they ostensibly only concerned passive listening. They were thus a good way to move toward legal transmission, by bringing radio hobbyists together, and by having a legal cover for the still illegal transmitting activities. By the end of the year, the American authorities had even authorized the sending of reception reports ("QSL cards") to DE-license holders, although they still explicitly forbade transmission by Germans.[55] This was important, because it meant that being seen to receive QSL cards in the mail would not necessarily mean the recipient was using an illegal transmitter.

Simply having a legal cover for receiving QSL cards did not fully solve the problem, particularly as the number of illegal German transmitters increased. A way had to be found to deal with the growing volume of reception reports concerning illegal transmissions, which were being sent to and from German amateurs. This was not a trivial problem, since Allied or German authorities could quite easily identify those illegally transmitting by intercepting a QSL card addressed to them, which mentioned a two-way contact. In most national amateur radio clubs a central forwarding system for QSL cards was used, not least to save money, but also to avoid the exchange of full addresses. While German hams found this a desirable next step, it was still fraught with danger, since by seizing the records either of the central "call sign" file, or of the QSL transmission office, which naturally had to have a list of the addresses of anyone to whom it wanted to send a QSL card, the Allied or German authorities could identify illegal transmitters. There needed to be a secure way of both keeping the central call-sign register, and of handling the forwarding of QSL cards.

This led to a rather curious development. According to the postwar ham lore, a long-time Stuttgart amateur, Günter Frech, (DE7215 N, DL1CP, after 1945 EK2ND, D3OPA, D4GNX, DA1AP) had an old school friend in the US military administration who had immigrated to

[54] Körner, *Amateurfunk*, 176.

[55] Licensed amateurs (those allowed to legally transmit) usually sent each other QSL cards to document a two-way radio contact, but simple listeners could send them as well. Thus, one could send QSL cards even if one were not allowed to transmit. See Chap. 1 for more on QSL cards.

the US before the war.[56] As the story goes, Frech explained to his friend what amateur radio was all about, and explained the problem with not having a secure way to forward QSL cards. This friend ostensibly came up with the solution: why not let the cards all be sent to an anonymous address, "P.O. Box 585 in Stuttgart".[57] This became the new location of the (illegal) central clearinghouse for German QSLs. What made the new address secure was the fact that it was located within the US Military Administration in Stuttgart, and was thus beyond the reach of the German postal authorities. As Körner coyly puts it, the "friend" belonged to a "very important American military office in Stuttgart" ("*sehr wichtige Dienststelle der Amerikaner in Stuttgart*"), who intervened with the US Military government to tolerate this illegal activity. Naturally, transmission was still illegal, and if other parts of the US Military administration asked too many questions, it could lead to problems, but one could always contact the friend again for help.[58] US authorities subsequently held a meeting in Stuttgart with several holders of the new unofficial call signs, and came to a "gentlemen's agreement": the illegal call-sign system, the transmitting of QSL cards, and even illegal transmission would be secretly tolerated by the Americans, and in exchange, they would be given the list of call-sign holders, and weekly updates on new additions.[59] The organizers of the unofficial system rapidly agreed.

[56] Though it cannot be said with 100% certainty, it is likely that Frech's contact in the American administration was T4 William J. Sailer. Sailer was a Stuttgart native who immigrated to the US before the war, and had been a "Richie boy" trained as an intelligence officer at Camp Richie, Maryland. During the war, he served in a propaganda unit that attempted to persuade German soldiers to surrender. He later served in the American Military government in Stuttgart, first as a soldier, and then as a civilian. I am indebted to Dr. Thomas Boghardt and his colleagues, Dr. Nick Schlosser and Dr. Kathy Nawyn and to researcher Mr. Dan Gross, for tracking down Sailer and providing information about his career. See also: Bruce Henderson, *Sons and Soldiers. The Untold Story of the Jews Who Escaped the Nazis and Returned with the U.S. Army to Fight Hitler* (New York: William Morrow/ Harper Collins, 2017). Sailer's name appears on p. 404.

[57] Though participants refer to "P.O. Box 585", in 1946 this was probably "APO 585" (Army Post Office box 585).

[58] Körner, *Amateurfunk*, 177–178, p. 181.

[59] The document prepared by the WBRC for this meeting survives in the US National Archives NARA Record Group 260, box 968, Records of United States Occupation Headquarters, WWII; Office of Military Government for Germany (US) OMGUS; Office of Military Government, Wuerttemberg-Baden; Records of the Education and Cultural Relations Division: Community Active Branch Chief: Corresp & Rel Recs. 1945–1949; Amateur Radio thru K9 American-French Zone Merger. Note particularly the "memo from 'Committee for the treatment of questions relating to amateur-broadcasting-licenses of the Radio Club

How did this come about? It would be tempting to dismiss this as just another version of the "foundation myth" I discussed earlier, save that we have a record of contacts between the American Military government in Stuttgart and members of the WBRC. Similarly, we also have records documenting that the US and British Military governments[60] considered the issue carefully.[61] Within this file, there is further the recommendation of Lt. Col. Robert Walker, the head of the Army Security Agency Europe within the US European Command Headquarters, admitting that the US Army lacked the technical means to track down large numbers of illegal transmitters, and suggesting instead, that the US should instead cooperate with the German hams, and in this way obtain not only a full list of names and addresses of those transmitting illegally, but also perhaps valuable intelligence gained from monitoring their logbooks and transmissions.[62] This is exactly what was done, beginning locally in Stuttgart, then more widely in the US zone, and later, in the American-British combined zone. It was an elegant solution to the problem, which recognized US shortcomings and secured German cooperation in exchange for information, self-policing, and potentially, usable intelligence. Of course, it broke US regulations in the American zone and violated common Allied policy agreed upon in the Control Council, but that could conveniently be used to shut down the whole experiment if it didn't work out.[63]

Württemberg-Baden, Chief Delegate Wolfram Körner, Licensed Publisher, Stuttgart, Tagblatt-Turmhaus, 8th Floor' to: Federal Communications Commission, [and] Chief Communications Branch; Subject: Granting of Transmissions Licenses to German Shortwave Amateurs in Württemberg-Baden of July 8, 1947", and "Memoir" Stuttgart, July 1947.

[60] See the section on the British Zone below.

[61] Not least, see NARA Record Group 260 BOX 3, Records of United States Occupation Headquarters, WWII, Office of Military Government for Germany (US) OMGUS; Records of functional Offices and Divisions; The Office of the Director of Intelligence: General Corresp of the Analysis and Research Branch. 1945–1949; 000.9-2 Scientists, German, IOA Evacuees thru 004. German Industrialists and Industry (1); FOLDER 000.77 RADIO INFORMATION OF INTELLIGENCE INTERESTS 7-20-1/28.

[62] "Lt. Col. Robert Walker, Chief, European Command Headquarters, Army Security Agency Europe, APO757, to Deputy Director of Intelligence, Headquarters, European Command, APO757 U.S. Army, ASA-94/MJC/mp of 17 February 1948 (copy) (Confidential)", in: NARA Record Group 260 BOX 3 Records of United States Occupation Headquarters, WWII, Office of Military Government for Germany (US) OMGUS; Records of functional Offices and Divisions; The Office of the Director of Intelligence: General Corresp of the Analysis and Research Branch. 1945–1949; 000.9-2 Scientists, German, IOA Evacuees thru 004. German Industrialists and Industry (1); FOLDER 000.77 RADIO INFORMATION OF ILTELLIGENCE INTERESTS 7-20-1/28.

[63] Ibid.

Though it is never mentioned by Körner (a direct participant, after all), and only hinted at by some other chronicles of amateur radio in German,[64] there is a direct connection here to US Army Intelligence. Walker, an intelligence officer, mentions in his memo that hams might produce usable intelligence, though he does so only in passing.[65] While the Americans and British naturally wanted intelligence on the activities of the (half-) illegal German amateurs, that was not what Walker meant when he spoke of "useful intelligence". There was a much bigger intelligence problem for the Western Allies, who had very little knowledge about what was going on in the Soviet Zone of Occupation. As tensions mounted with the Soviets and the Cold War developed, the US and British had a growing interest in obtaining information from the Soviet zone, and feared that Soviet spies might use amateur transmissions as cover for spying in their zones.[66] They wanted to use German amateurs as a way of learning what might be happening in the Soviet zone. US intelligence circles realized that there was much to gain from tolerating the illegality of the German amateurs, provided they had easy access to the addresses of all those involved. What they said and did on air was already closely monitored by the Americans and all other intelligence services in Germany. With access to the central address file of the (illegal) German license scheme, they could better monitor who was transmitting, and could make use of any intelligence-worthy information a QSL card might contain. In this way, the nascent amateur movement in postwar West Germany became a willing tool in the Cold War, because the prize was the ability to fully engage in their hobby. In other words, rather than being motivated by international "ham spirit", there is good evidence that the toleration illegal German amateurs found in the

[64] Note that the first public mention of the association between German amateurs and US Intelligence probably came in an article in the March 1948 issue of journal of the DARC/BZ, "*CQ*". It was a fairly open secret.

[65] Ibid.

[66] See NARA Record Group 260 BOX 3 Records of United States Occupation Headquarters, WWII, Office of Military Government for Germany (US) OMGUS; Records of functional Offices and Divisions; The Office of the Director of Intelligence: General Corresp of the Analysis and Research Branch. 1945–1949; 000.9-2 Scientists, German, IOA Evacuees thru 004. German Industrialists and Industry (1); FOLDER 000.77 RADIO INFORMATION OF INTELLIGENCE INTERESTS 7-20-1/28 and also NARA Record Group 260 BOX 167 Records of United States Occupation Headquarters, WWII; Office of Military Government for Germany (US) OMGUS; Records of the Office of Military Government, Bavaria; Records of the Intelligence Division: General Intelligence Recs. 1946–1948 Research Control THRU youth, general folder 10/84-2 26 Radio, General for documentation of this fear.

US zone may well actually come from the fact that US Army Intelligence officers (and later the CIA and its precursor, the "Strategic Service Unit") had decided to make use of amateurs in their zone to try to find out what was actually going on in the Soviet zone.[67] The expansion of US intelligence efforts in the Soviet zone is described in detail by recently declassified CIA documents, and the interest of the CIA in amateur radio is well documented, making the slightly earlier connection with German hams quite plausible.[68] Of course, the Soviets had similar intelligence operations in the Western zones.[69] It is not known if they also sought to make use of German amateur radio operators in their own zone, but it would not be surprising. While postwar mythology attributes this and later US goodwill toward the growing amateur movement as due to the fundamental good humor of the Americans, combined with "ham spirit", in fact, it is clear that the German amateur movement in the US zone (and indeed, in the other three Allied zones of occupation as well) owed their rebirth in part to Allied calculation that it was better (and cheaper) to cooperate with German hams than to try to repress them, and in part to the usefulness of German amateurs as intelligence assets and pawns in the Cold War.[70]

[67] Thomas Boghardt, "America's Secret Vanguard: U.S. Army Intelligence Operations in Germany, 1944–47", *Studies in Intelligence* 57, No. 2 (June 2013): 1–18. Accessed September 5, 2015. https://www.cia.gov/library/center-for-the-study-of-intelligence/csi-publications/csi-studies/studies/vol-57-no-2/pdfs/BookStudies%2057%20No%20 2-June2013.pdf

[68] See, for example, Memorandum by Richard Helms of January 10, 1947 "Targets of German Mission" (Confidential), accessed June 25, 2015. https://www.cia.gov/library/center-for-the-study-of-intelligence/csi-publications/books-and-monographs/on-the-front-lines-of-the-cold-war-documents-on-the-intelligence-war-in-berlin-1946-to-1961/1-4.pdf; "Report on Berlin Operations Base", MADOC-B226 from Chief of Station Karlsruhe Dana Durand to Chief of Foreign Branch M (eyes-only) of April 8, 1948, (Secret), accessed June 25, 2015. https://www.cia.gov/library/center-for-the-study-of-intelligence/csi-publications/books-and-monographs/on-the-front-lines-of-the-cold-war-documents-on-the-intelligence-war-in-berlin-1946-to-1961/1-2.pdf. Later CIA interest in amateur radio is documented in the following article, which draws on newly declassified documents: Rick Lindquist, WW1ME, "The CIA, the Cold War, and Amateur Radio", *QST* 102, No. 2 (February 2018): 64–68.

[69] "Reorganization of the RIS in Germany", Report from "Information Control, OSO", September 11, 1947 (Control), accessed June 15, 2015. https://www.cia.gov/library/center-for-the-study-of-intelligence/csi-publications/books-and-monographs/on-the-front-lines-of-the-cold-war-documents-on-the-intelligence-war-in-berlin-1946-to-1961/1-10.pdf

[70] Körner, *Amateurfunk*, 182. Note the outward similarity between this agreement and the 1933 cooperation of the DASD. At both occasions, German amateurs made unpleasant compromises in exchange for freedom to transmit.

Amateur Radio in the British Zone of Occupation

Developments on the British zone of occupation were broadly similar to those in the US zone. Immediately after the end of the war, the British were, if anything, even more suspicious of German amateurs than the Americans were: searches, interrogation, even preventive detentions were the norm. But just as in the US zone, amateurs in the British zone (BZ) began to come together around former DASD members. Almost from the beginning, an amateur and former DASD member in Kiel, Alfred Müller (DE1885, D4VJV, DL1FL) began spreading word of international news concerning amateurs and pushing/planning for a new German amateur radio club. Unlike the choice in the US zone, the amateurs in the North of Germany saw this new club from the beginning as being exclusively for amateur radio enthusiasts.[71]

As early as February 1946, amateurs in Hamburg around Rudi Rapcke (DE0356, D4WJ, DL1WA) applied to the local British authorities for permission to found a radio club—unsuccessfully, at that time. Meanwhile, other amateurs attempted to influence the military authorities by seeking help from foreign amateur organizations. In March 1946, Otfried Lührs (DE6220, DL1KV, HK3AH), with support of Göttingen hams, wrote via a "middleman" to the ARRL in the US for help; it eventually replied that it would allow individual Germans to join, but could not send its journal, *QST* to Germany.[72] With this, Alfred Müller then contacted the Radio Society of Great Britain (RSGB) via a British national he knew (German access to international mail, telegraph, and telephone was severely limited at this time). Müller eventually heard back that the RSGB already had four applications for support from the BZ; it was going to wait and discuss all of them at a meeting, and then coordinate with the ARRL; but for now, things were "perhaps a few months too early". Soon, in July 1946, the RSGB announced publicly that it knew of no proof that the German hams had worked with the Nazis, but then, a bit later, it announced that it would not allow German members into the RSGB because a state of war still existed between the two countries. Though frustrating, the replies by the ARRL and RSGB at least seemed well meaning and gave hope that future support might still come.[73]

[71] Körner, *Amateurfunk*, 188–189.
[72] Körner, *Amateurfunk*, 188–189.
[73] Körner, *Amateurfunk*, 189–190.

Meanwhile, German amateurs had made contact with British members of the "Radio and Research Section" of the Allied Control Council in Hamburg. The officer in charge of that office, Col. Kidd, and most of his section heads were radio amateurs who were already transmitting under the D2 call signs reserved for members of the Allied forces in Germany. They were all quite open to the concerns of German amateurs. This is especially true of the secretary of the amateur club for British servicemen, Maj. R.D. Shears (D2KW/G8 KW), who became a long-standing friend and supporter of the German amateurs.[74]

Once the US began to allow radio clubs in its zone, the amateurs in the British zone redoubled their efforts. In September 1946, amateurs in Hamburg under Rudi Rapcke, together with amateur representatives from other parts of the British zone, met to draw up a new set of draft by-laws for a new club, and to plan a second application to the British authorities. The Hamburg Military government again told them to wait, but also said that their application had been sent on to a higher authority. Apparently, the British were at least willing to consider the founding of a radio club.[75] The existence of a precedent in the US zone likely helped. The economic unification of the US and British zones on January 1, 1947, made cooperation between the two allies even more important, even though they each retained their own military governments.

For communication and coordination, the amateur movement in the British zone depended on a series of circulars first typed and copied by hand and sent from person to person. By the beginning of 1947, amateurs in the British zone were able to send out a proper journal, the "*QRV*". This was thanks, in part, to Col. Kidd, who intervened with the British authorities to obtain a special allocation of paper, which was still tightly rationed. This was a great help to the amateurs in the BZ, and is a sign that their efforts toward obtaining permission to found a club were heading in the right direction.

In April 1947, the British Military government in Kiel approved the founding of the "Kiel Radio-Amateur Club" ("Kieler Radio-Amateur Klub", KRAK), the first legal radio club in the British zone. This was an

[74] Though not mentioned, it is at least plausible that the British radio intelligence personnel would be not only supportive of fellow hams, but also interested, like the Americans, in using German amateurs to learn about conditions in other zones. Körner, *Amateurfunk*, 190.

[75] Körner, *Amateurfunk*, 190–191. See also Jochen Hindrichs, "60 Jahre DARC-Distrikt Hamburg", *Hamburger Funkerblatt*, (October 19, 2007), accessed August 4, 2015. http://www.df3xz.de/Hamburger%20Funkerblatt.pdf

important precedent, and the amateurs in Hamburg could again approach the British Military government about forming a club for the entire British zone. On May 17, 1947, the leaders of the Hamburg amateurs and those of other parts of the British zone were invited to British headquarters to discuss their proposals. After careful discussion of the proposed by-laws, and after careful warnings against transmission, which remained illegal, Col. Kidd gave oral permission to found the new "German Amateur Radio Club/British Zone" (DARC/BZ). Col. Kidd, Major Chalk and Maj. Shears assisted at the "*feierliche Zeremonie*" (solemn ceremony) which followed. Copies of the British certificate allowing the founding were sent out to amateurs throughout the British zone, so that they could found local clubs (and prove to the local British authorities that they were allowed to do so.) Note that, just as in the US zone, the chapter and regional boundaries of the new club were aligned with the official postal boundaries, and all the first "District Managers" (regional club representatives) were former DASD members, as were the new presidents Rapcke and Lürs, and their office manager (*Geschäftsführer*) Müller.[76]

The first big general meeting of the newly formed DARC/BZ was held at the Stuttgart "Shortwave Conference" hosted by the WBRC from June 7–8, 1947.[77] Even though the South Germans were the hosts of this conference, and even though they managed the illegal call-sign system, their influence was on the decline. At its founding, the WBRC had included all sorts of radio hobbyists, not just hams. By 1948, dissention between the shortwave amateurs (hams) and the other types of radio enthusiasts in the WBRC broke out into open warfare and paralyzed the club, which gradually collapsed.[78] Henceforth, the DARC/BZ rapidly took over leadership among German amateurs.

Relations between the south- and north German amateurs were complex. Publicly, the north German amateurs completely rejected the south German tactic of mass illegal transmission. In private, many individual north German amateurs had also obtained an unofficial call sign from the south German group, and were also transmitting illegally. There were some tense moments, as members of one group publicly criticized the

[76] Körner, *Amateurfunk*, 190–192. These were Senior Postal Direction (Ober Post Direktion OPD) Zone Düsseldorf/Köln—Pazem (DL 1 KB), OPD Zone Bremen—Goldmann (DL1 KH), OPD Zone Hamburg—Rapcke (DL1 WA) and OPD Zone Kiel—Müller (DL1 FL).

[77] Körner, *Amateurfunk*, 192.

[78] Körner, *Amateurfunk*, 192.

policy of other. Yet, on the other hand, there was quite a bit of cooperation. The leadership of both groups communicated, and the north German amateurs attended the first "Shortwave Conference" held by the WBRC in Stuttgart, June 7–8, 1947, in large numbers.[79] Nevertheless, the public exchange of opinion in the circulars and journals of the respective clubs, and between those who favored illegal transmission and those who favored a policy of strict legality was often very heated. A real split threatened to divide German hams by the Spring of 1948. In April 1948, a meeting in Munich of the leaders of the north and south German clubs came to an agreement that both sides would work together as much as possible, and a full schism was avoided. Despite often contentious exchanges at meetings and in their respective journals, the leaders of the amateur movement in the two zones well understood that they had to work together to be able to get what they wanted from both the Allies and their own German postal authorities. Cooperation eventually won out.

By the Spring of 1948, events began to move rapidly in a favorable direction for the amateurs in the US and British zones. At the end of April negotiations with the bizonal German Central Postal Administration (*Hauptpostverwaltung*) in Frankfurt began in earnest. Hans Haberl (DA2 DH, later DL1AX) led the negotiations with the German postal authorities.[80] At this time (Spring 1948), General Clay informed the Saturday Afternoon Club that the report of his intelligence organs on ham radio was positive, and that he was going to speak with Gen. Robertson, his British opposite number, about a formal licensing of hams in both zones.[81]

Despite the connivance of some US authorities, German postal authorities (and even, occasionally, US and British military authorities), continued to seize illegal radio transmitters—the ties to US intelligence and to the Military government were secret, as was the US toleration of illegal transmitting. So, for example, the German postal authorities in the US zone launched a wave of seizures of illegal transmitters in April 1948. Several stations were seized by the postal authorities, but other amateurs were immediately warned on the air that there was a crackdown in process, so that numbers of arrests and seizures were fairly low.[82]

[79] Körner, *Amateurfunk*, 176.

[80] Körner, *Amateurfunk*, 184–185. According to Körner, Haberl was arrested by the Nazis for "politischen Schwarzsenden" and "Landesverrat" and sentenced to death 1942, and only escaped execution by "a miracle".

[81] Körner, *Amateurfunk*, 185.

[82] Körner, *Amateurfunk*, 182–184.

In fact, the stance of the German postal authorities in both the US and British zones was contradictory. On the one hand, many, in the long tradition of the postal authorities in the German Empire, Weimar Republic and Third Reich, felt that the Post should have a monopoly over all forms of telecommunications. This is the current of thought responsible for so severely restricting the access of German amateurs to the airwaves in the Weimar Republic. These officials were on the whole opposed to private transmission and unwilling to treat amateur organizations as negotiating partners. But already during the war, more and more hams had obtained positions within the Post Office, and this trend was only intensified under Allied occupation, as the bureaucracy was de-Nazified. At least some among the postal authorities after the Second World War were open to granting German amateurs more access to the airways. Whereas in the Weimar Republic and Third Reich, the Post Office and amateurs had had a largely adversarial relationship, by 1948, this had changed, and it would not be too much of an exaggeration to discuss the relationship as more of a partnership, albeit one still marked by periodic reversion to a more adversarial relationship on occasion. Certainly by 1948, when it became clear in the US and British zones of occupation that German amateurs would not be stopped by the sporadic and largely ineffective searches and seizures by the postal authorities, those within the German postal apparatus who wisely decided that the best policy would be to work with amateurs toward creating a legal framework within German law for amateurs wishing to transmit gained greater influence. While the focus here has been on the US and British zones (united into "Bizonia" in 1947),[83] a similar dynamic was true in principle even in the French and Soviet zones, as we will see below. Taken together, this led to a real sea-change in the attitude of the postal authorities. The postal leadership in the US and British zones, and later in Bizonia, officially took a strong position against amateur licensing, but the presence of some officials who supported amateur radio was crucial when it came to respond to Allied pressure in favor of legal licenses. To be fair, the German authorities under Allied military rule did not have too much leverage, but it was to the advantage of everyone concerned that relations be good and cooperative. As we have already seen, neither the British nor the Americans trusted the German postal authorities, and were determined to create a new media landscape in

[83] Körner *Amateurfunk*, 184–185.

Germany for the future. Allied intervention in favor of amateur radio, as we will see below, was crucial.

On May 8–9, 1948, a second big Shortwave Conference held in Bad Lauterbach led to founding of the bizone "German Amateur Radio Club" or DARC, the fusion of the separate radio clubs in the British and US zones. With satisfactory progress in the negotiations with the German amateurs, who were able to propose an acceptable draft licensing scheme (as they had planned a year earlier), the Bizone postal authorities announced that the first legal license tests would be held in May.[84] As it turned out, this was a few months too soon, but barely.

Events were helped along by some pressure from the British and US Military governments. On July 1, 1948, the independent German amateur radio journal *QRV* published a joint declaration of Generals Clay and Robertson announcing that as far as the British and Americans were concerned, hams could be licensed in the Joint Economic Area (the Bizone or Bizonia), but that the actual licensing arrangements would be the job of the German authorities. Hoping that a resolution to the licensing problem was neigh, the DARC then ordered a transmission moratorium for all German amateurs in preparation for legal licensing.[85] To back up the joint declaration by Clay and Robertson, on July 27, 1948, the Allied Bipartite Control Office recommended to the Administrative Council of the *Wirtschaftsrat* (literally, "Economic Council", but in fast, a kind of West German proto-parliament) that the *Wirtschaftsrat* immediately pass legislation governing all non-broadcast radio transmission in Bizonia, including, but not limited to amateur radio.[86] It is hard not to see this recommendation as a massive intervention in favor or amateur radio. A "recommendation" by the British and US Military government was not just an idle request, particularly since they reserved the right to review and veto all German legislation. Without this clear Allied position in favor of amateur radio, postwar West German amateur radio would have been much more restricted, just as it had been in the Weimar Republic.

[84] Körner, *Amateurfunk*, 184–185.

[85] Körner, *Amateurfunk*, 185.

[86] Fritz Kirchner, DJ2NL, "Einige Daten und Fakten zur Entstehung des Amateurfunkgesetzes (AFuG) vom 14. März 1949", *CQDL* (October 1987): 632–634. The text of the letter in German is reprinted on p. 263. Note that the Bipartite Control Office specifically asked for amateur radio licenses for Allied personnel, as well as for German amateurs.

The DARC and the bizonal Central Postal and Telecommunications Administration (*Hauptverwaltung für Post und Fernmeldewesen*) in Frankfurt began to negotiate a new licensing law, which would then need to be submitted to the *Wirtschaftsrat* of Bizonia for approval.[87] But the *Wirtschaftsrat* was overworked, and some politicians, as well as many in the German Postal Authority, were worried about losing the traditional postal telegraph monopoly, and about the wisdom of authorizing private transmitters. The Bizone postal authorities wanted to base amateur licensing on the old 1928 "Law Concerning Telecommunications Installations" (*Gesetz über Fernmeldeanlagen vom 14. Januar 1928*) and the 1939 "Decree on Transmitters for Amateurs" (*Verordnung über Sender für Funkfreunde vom 9. Januar 1939*). Moreover, they preferred that an omnibus Postal and Telecommunications law be written first, under which any amateur regulations would fall. They preferred to release any future amateur radio regulations not as a law, but rather as a decree (*Verordnung*) by the Director of Post and Communications (*Direktor für das Post- und Fernmeldewesen*) of the United Economic Zone (Bizonia). From the point of view of the British and Americans, but also from the standpoint of West German politicians seeking democratic change, there were problems with this approach. It was cumbersome, and would delay negotiations with the hams, who by this point had gained a measure of Allied trust. It also would lead to too much continuity with the Weimar Republic and Third Reich, in which the government—in the form of the Postal Ministry—retained too much power over the media landscape. In a similar fashion, if the amateur rules were issued in the form of a ministerial decree rather than a formal law, they could be rescinded at any future time without parliamentary control. In other words, the bizonal German postal authorities would practically inherit the powers of the old Postal Ministry of the German Empire, and retain considerable legal autonomy.[88] Many (not all) German political leaders and the British and US Military governments simply did not want this much continuity with the old empire, Weimar Republic and Third Reich, and argued for something new. Moreover, the German amateurs did not want to wait for a larger and more general law over the post, telephone, and telegraph to be written first, which might take years. The German amateurs wanted a separate amateur radio law immediately.[89]

[87] Körner, *Amateurfunk*, 185–186.
[88] Kirchner, "AfuG" 632.
[89] Kirchner "AfuG", 632.

This eventually led the South German amateurs to make a show of force: on October 16, 1948, the SAC (now just an unofficial group within the DARC) recommended the resumption of illegal transmission—some 700 amateur stations immediately went back on the air. With no progress by early 1949, the DARC came up with a vivid means of pressuring the deputies to pass the law: they proposed that all hams send a (genuine) brick to the Chairman of the *Wirtschaftsrat* around January 15, 1949, "to provide a good foundation for the new law on Amateur radio" ("*zur Untermauerung des Amateurfunkgesetzes*"), and also as a "complaint" ("*Beschwerung*", a pun on the German words for "complaint" and "heavy" *Beschwerde* and *schwer*) about the slow process of the law. This was the fabled "Brick Action" ("*Backsteinaktion*"): hundreds of hams from all over Bizonia sent them in, and it was fully covered by the press. The post office of the *Wirtschaftsrat* was overwhelmed, and it was temporarily forced to cease accepting any packages at all.[90]

This act of creative protest plays an important role in the origin story told among German hams today. Körner adds another part to the story: according to him, around January 10, 1949, a new, unknown amateur station appeared on the air, UA1KAA, purporting to be the legal and official station of the amateurs in the Soviet Zone of Occupation. This allowed West German hams to put pressure on the West German government at a very sensitive point, by claiming that the Soviets had already officially licensed amateurs and permitted them to transmit. In fact, this was absolutely untrue, and the station was actually a fake, and was located in Württemberg in Bizonia.[91] It isn't known if the "*Backsteinaktion*" or the appearance of a (fake) legal amateur station purporting to be from the Soviet zone actually had any real influence on the negotiations, despite their media-impact. Yet these (true) stories certainly fit the postwar narrative of a group of doughty hobbyists meeting and besting the bureaucracy with humor and tenacity, until their goals were met, and amateur transmission once again became legal. Pressure from the US and British, and those German leaders seeking a break with the past, certainly had much more impact.

In any case, the amateurs finally got their separate amateur radio law. In early December, a first draft law was finally produced, and presented to the *Wirtschaftsrat*. After discussion and amendment in the Postal Committee,

[90] Körner, *Amateurfunk*, 185–186

[91] Körner, *Amateurfunk*, 186.

the draft was sent back to the *Wirtschaftsrat* in early January 1949 for a vote. Once it passed vote in the 31 Plenary Session of the *Wirtschaftsrat* on January 19, 1949, it was then accepted by the *Länderrat* (Council of States, the proto-parliamentary upper house of Bizonia) on February 1, 1949.[92] At that point, the law was then submitted to the British and US Military government for final approval. The US and British accepted the law, on the condition that language in the new law limiting amateur transmission to research purposes be struck. The draft law went back to the committee on Post and Telecommunications, then back to the full *Wirtschaftsrat*, and then to the *Länderrat*. The Allied conditions were accepted in all cases, and the final law was passed on March 14, 1949.[93] For the first time in history, [West] German hams enjoyed a generous, widespread legal licensing policy.

On January 21, 1949 DA1AB, the transmitter used by the illegally transmitting south German amateurs to spread information, again ordered a cessation of (illegal) transmissions by hams in preparation for the issuing of real licenses. At long last, the West German postal authorities began to issue the hotly awaited legal licenses to transmit on March 13, 1949.[94] Amateur transmission was legal again—in Bizonia. German amateurs elsewhere were still not included. Still, this was a major step, and created the legal and organizational foundations of the German amateur radio movement today.

The March 14, 1949, law remained in effect after the creation of the German Federal Republic and the promulgation of the Basic Law (*Grundgesetz*) on May 23, 1949.[95] At first, the new law only applied in the British and American zones, even after the establishment of the Federal Republic; it was only extended to the French zone of occupation in May

[92] Kirchner, "AfuG", 632–633.

[93] Gesetzblatt der Verwaltung des Vereinigten Wirtschaftsgebietes No. 7 vom March 22, 1949. Kirchner, "AfuG", 633.

[94] Körner, *Amateurfunk*, 186.

[95] At the same time, the old *Fernmeldeanlagegesetz* of 1928 was technically restored, but was automatically superseded by the "Gesetz über den Amateurfunk" of March 14, 1949. In any case, the German Federal Republic did not regain sovereignty over telecommunications until May 1955. The "Gesetz über den Amateurfunk" of March 14, 1949, like all laws passes by the Wirtschaftsrat, remained in force thanks to Articles 123I and 124 of the Basic Law. See: Thomas Neufeld (DARC), "Geschichte des Amateurfunk und die rechtlichen Grundlagen" (2017), accessed May 28, 2018. https://www.darc.de/der-club/distrikte/q/ortsverbaende/11/chronik/geschichte-des-amateurfunks/

1955. West Berlin had its own Amateur Radio law from 1950 to 1967,[96] and of course, laws in the Soviet zone and later German Democratic Republic (GDR/DDR) were completely different.[97]

With the founding of the German Federal Republic, a new amateur radio law firmly in place and transmission legal again, there was no longer any reason for a separation between North and South German hams. On September 8, 1950, the German-Amateur-Radio-Club (Deutsche Amateur-Radio-Club e.V.; DARC) was founded in Bad Homburg. On July 24, 1951, the DARC was admitted as a member club in the International Amateur Radio Union.[98] West German Hams were once again legally part of the international hobby of ham radio.

The Soviet Zone: Return to the Past

Legalization of transmission elsewhere in Germany lagged far behind what had transpired in Bizonia. Conditions were not at all good in the Soviet zone. It might have been different: The Soviets appeared to move fast to restore the public sphere in their zone of occupation. The Soviets were the first of the Allied powers to allow Germans to form political parties in their zone. As early as June 10, 1945, Order No. 2 of the Soviet Military Administration in Germany (SMAD) allowed the formation of anti-fascist/democratic political parties. The German Communist Party was duly founded a day later. On the 13th, the first trade union was also formed in the Soviet zone. It was not until June, 17, 1945, that the Christian Democratic Union formed in Cologne in the British zone.[99] Yet despite the quick founding of political parties, unions, and newspapers in the Soviet zone, it cannot be said that the Soviets were particularly encouraging to clubs in general, and certainly not to radio in particular. Whatever the attitude of individual Soviet commanders, the system itself had no tolerance for disobedience, and great suspicion of espionage, none of which decreased as the Cold War began to develop. In this context, the

[96] Thomas Neufeld "Amateurfunk". Neufeld states that the introduction of the law in the French Zone was delayed so long due to attempts to negate it in favor of the 1928 law.

[97] See below.

[98] Thomas Neufeld "Amateurfunk". According to Neufeld, there were already 2000 German amateurs when the DARC was founded. He states that ca. 85% of all German amateurs in 2017 belonged to the DARC.

[99] Wolfgang Benz, ed., *Deutschland seit 1945. Entwicklungen in der Bundesrepublik und in der DDR. Chronik, Dokumente, Bilder* (Munich: Moos & Partner, 1990): 10.

SMAD and later the dictatorial government of the German Democratic Republic (GDR) both felt threatened from the outside and were deeply suspicious of potentially dangerous amateur radio transmitters. Thus, paradoxically, amateur radio in the German Democratic Republic developed in ways which closely paralleled the historical experience of amateurs in the Weimar Republic and the Third Reich.

Of course, German amateur radio operators who happened to live in the Soviet Zone of Occupation immediately after the War were identical to their comrades in the other zones: they had the same training in the DASD, the same set of life experiences, and they wanted very much to get back on the air. Many did, in strict illegality and at very great personal risk.[100] Just like their fellow amateurs in the US and British zones, German amateurs in the Soviet zone tried to obtain the right to organize radio clubs by approaching the SMAD and asking for permission. Transmission in the Soviet zone was just as illegal as in the other Allied zones, so hams had to be circumspect. Just as in the West, East German radio enthusiasts hoped that general radio clubs, without an open emphasis on shortwave work or transmission, might be the best way to begin.

The reformation of the radio hobby in the Soviet zone initially took place in the same ways it did in the US and British zones, only it lagged developments there by a considerable amount. Amateurs who knew each other from earlier were in contact, and at the very least, those who were active with illegally transmitting (or even simply listening in on the amateur bands) learned of each other on the air, though this sort of contact rarely led to a face-to-face meeting due to the dangers involved, unless the amateurs already knew each other. The fact that more and more amateurs in the US and British zones began to transmit and could be heard on the air must have motivated those in the Soviet zone to attempt to follow suit. The problem was that the Soviet authorities were much more suspicious of spontaneously organized associations than their British and American counterparts.

Radio hobbyists in the Soviet zone and later in the German Democratic Republic were also not helped by the fact that the pre-war DASD to which most had belonged had been a state-sponsored organization under strict state control, while the working-class and socialist Workers Radio Union (ARB) and Communist Free Radio Union (FRB) had been rigorously suppressed, and their members persecuted. By itself this was not necessar-

[100] Senne, *SBZ und DDR*, 88–92.

ily a major problem, for many former members of Nazi-dominated state organizations were inevitably active in postwar society even in the GDR. Moreover, the historical precedent of the ARB and FRB as autonomous, left-wing organizations, which "resisted" Hitler might have caused the Soviet and East German authorities to look more favorably on all facets of the radio hobby—certainly, former members of the ARB and FRB were available to act as leadership cadres in any future radio organization. We will never have a definitive answer to the "what if?" question, but from their own perspectives, Soviet and East German officials had good reason be careful about the legalization of amateur radio. Chief among them, of course, was a deeply ingrained suspicion of spontaneous associations. The Soviets and the East German Communists, given their own histories in the 1930s and 1940s, were deeply afraid of espionage and subversion, even before the Cold War started. The notion of illegal radio transmitters was absolutely intolerable, even if there had not been the efforts by US, French, and British intelligence services to make use of the radio amateur community.[101] Moreover, they shared with other totalitarian regimes a suspicion of the very idea of a hobby, even one such as radio with such clear ties to the world of work and science. Private passions were suspect, because they might lead an individual to deviate from the larger political goal. In short, Soviet and East German authorities reacted much as those of previous German regimes.

Some rather loose and informal radio hobby groups probably existed from 1947 on,[102] but they had trouble organizing officially. On August 26, 1947, the first legal radio club, the "Bautzen Working Group of Radio Do-It-Yourselfers" ("*Arbeitsgemeinschaft Radiobastler Bautzen*") was formed in that East German city with the express permission of the city authorities and the Soviet Military Administration. Originally an independent club, in 1949, they were attached to the East German "Free German Association of Trade Unions" (*Freier Deutscher Gewerkschaftsbund*, FDGB) and allowed only local activities. The members objected, and attempted to argue for the creation of a radio club for the entire Soviet zone within the FDGB. This failed, and the organizer of the Working Group, Rudolf Berneis, was arrested.[103]

[101] See above, pp. 281–283.

[102] Senne, *SBZ und DDR*, 88.

[103] Senne, *SBZ und DDR*, 90–91.

Similarly, in May 1948, two amateurs, Werner Schütz and Edgar Krause, approached a local official of the East German "Cultural Union for the Democratic Renewal of Germany" ("*Kulturbund zur Demokratischen Erneuerung Deutschlands*"), a socialist cultural organization in the Soviet zone.[104] They wished to discuss licensing radio amateurs to transmit, but this was rejected out of hand, even though the authorities knew by then that radio enthusiasts were active illegally without any form of state supervision.[105] If these efforts were uninspiring but not completely useless, other attempts to create a legal framework for amateur radio in the Soviet zone led to worse results: in the Spring of 1948, the request of a group of radio amateurs in Thuringia to found a radio club for the entire Soviet zone was not only turned down, but led to a series of house searches and arrests. On the other hand, at the same time, the SMAD began itself to operate what was ostensibly an amateur transmitter for propaganda purposes.[106] Another similar attempt by two Dresden amateurs, Heinz Morawa and Fritz Trömel, in 1949, also led nowhere. They visited the "Chamber of Technology" (*Kammer der Technik*) in East Berlin to argue that it should create a "shortwave group" within the chamber.[107] The Chamber did not act, but the two were permitted to create a shortwave

[104] This was originally an organization of intellectuals dedicated to the cultural renewal of Germany in a humanistic and anti-fascist sense. By 1947, it had become an unofficial organ of the Socialist Unity Party (SED). It gradually collected a number of hobby groups under its umbrella, such as stamp collectors and hobby photographers, likely because the regime didn't know where else to place them. See: Gerd Dietrich, "Kulturbund". In: Gerd-Rüdiger Stephan et al., ed., *Die Parteien und Organisationen der DDR. Ein Handbuch* (Berlin: Dietz, 2002); Magdalena Heider, *Politik—Kultur—Kulturbund. Zur Gründungs- und Frühgeschichte des Kulturbundes zur demokratischen Erneuerung Deutschlands 1945–1954 in der SBZ/DDR*, Bibliothek Wissenschaft und Politik 51 (Cologne: Verlag Wissenschaft und Politik Claus-Peter von Nottbeck, 1993); Magdelena Haider, "Der Kulturbund zur Demokratischen Erneuerung Deutschlands (KB)", in *SBZ-Handbuch: Staatliche Verwaltungen, Parteien, gesellschaftliche Organisationen und ihre Führungskräfte in der Sowjetischen Besatzungszone Deutschlands 1945–1949*, edited by Martin Broszat and Hermann Weber, 714–733 (Berlin: De Gruyter/Oldenbourg, 1990, Second Edition 1993).

[105] Senne, *SBZ und DDR*, 88–92.

[106] In other words, a SMAD transmitter pretending to be an amateur. Senne, p. 90.

[107] The KdT was an official corporative professional organization in East Germany for technicians, engineers, and scientists, first in the Soviet zone, and later in the GDR. See Dolores L. Augustine, *Red Prometheus: Engineering and Dictatorship in East Germany, 1945–1990*, Transformations: Studies in the History of Science and Technology (Cambridge, MA: MIT Press, 2007). see also Günter Fischhold, *Kammer der Technik. Ein Beitrag zur Geschichtlichen Aufarbeitung des Ingenieurverbandes* (Norderstedt: Books on Demand GmbH., 2011).

group within the Dresden chapter of the "Cultural Union for the Democratic Renewal of Germany" ("*Kulturbund zur Demokratischen Erneuerung Deutschlands*").[108] The Shortwave Group did not last more than a few years, not least because most amateurs were afraid to attend the meetings lest they attract the attention of the Soviet MVD (Ministry of Internal Affairs). The group folded by the Fall of 1952.[109] The growing tensions in the Cold War did not help the case of the amateurs in the Soviet zone and, later, in the German Democratic Republic, and led to a vigorous crackdown in 1950 and 1951.[110]

The Soviets and the East German authorities knew, of course, that there were radio enthusiasts who were engaged in illegal transmission, and did not like it. In a year-end report in 1950, the Ministry for State Security (the feared East German "Stasi") estimated that, based on their monitoring, there were some 120 radio enthusiasts engaged in illegal transmissions in the German Democratic Republic and East Berlin. The report quite rightly stated that most of the East German amateurs used call signs based on the guidelines of international radio law and following the call signs issued in West Germany by the DARC, though some East German amateurs, for whatever reasons, were still using fantasy call signs of their own design. The report also rightly noted the influence of the old DASD and its former members in the new West German DARC and the closeness of the DARC to the Western Allies. The Stasi report further mentioned that an illegal organization of East German amateurs had been formed in 1950 during a radio exposition in Berlin-Witzleben, and what is most significant, blamed this on the lack of an official East German radio organization.[111]

Meanwhile, some amateurs in the German Democratic Republic continued to transmit illegally. Worse, they generally obtained call signs from the West German DARC. Remember that illegal amateur radio operators in the Soviet zone, along with their political problems, had a problem of identity. (They shared this problem with German amateurs in another occupied region, the Saar region, which was then still not only under

[108] The Culture Union might not seem to be the most obvious place for amateur radio, but in fact, it included sections for other hobbies, such as photography.

[109] Letter from Dipl. Ing. Fritz Trömel to dip. Ing. Helmut Ahlborn of February 1, 2001, DASD Archive.

[110] Not least, the Korean War took place from June 1950 to July 1953.

[111] Senne, *SBZ und DDR*, 98. The document in question is cited here as BStU MfS/ Zentralarchiv Allg. S 31/54 Band I, Bl. 253.

French occupation, but which France then sought to annex.) Since amateur radio was not yet legal in the Soviet zone, nor later in the German Democratic Republic, illegal East German hams could not obtain an official (or official-sounding) call sign, and thus had trouble making contacts internationally. Naturally, when transmission was illegal, using a call sign which could be traced to an individual was very dangerous. But on the other hand, hams wanted to make contacts on the air and possibly receive QSL cards. What to do? Just as in the Western zones, some East German hams simply continued to use their pre-1945 call signs (if they had had one), while others simply made-up a call sign.

But there was a way to get a real (or real-looking) call sign: East German amateurs could make contact with the representatives of the WBRC, and later, the DARC. Though based in Stuttgart, remember that from the beginning, the South German amateurs issued call signs to German amateurs irrespective of their zone of residence. This policy was continued by the DARC for the subsequent legal call signs and was ultimately a reflection of the "One Germany" policy adopted by the German Federal Republic. After the founding of the German Federal Republic, a block of call signs in the DK8 and DL8 range was specifically set aside for amateurs in the Soviet zone by the West German authorities. This lasted until 1953, when amateur radio was finally legalized in the German Democratic Republic and amateurs there could be issued legal East German call signs.[112] Obtaining a new call sign from the West German amateurs solved the "identity problem", and allowed illegal East German amateurs to make contacts on the air and receive QSL cards.

The DARC chapter in West Berlin functioned as the contact point for amateurs in the Soviet zone. In the late 1940s and early 1950s, it was fairly easy to move from one part of Berlin to another. East German amateurs could make a legal and normal-seeming visit to the Soviet-controlled part of Berlin, and then cross over into the American zone and meet with DARC officials. DARC District Chairman Rudi Hammer and the DARC Berlin office manager (*Geschäftsstellenleiter*) Bruno Garnatz (both former DASD officials) ran the program. They administered tests, issued licenses (call signs), and apparently, served to weed out potential Soviet agents

[112] See below. "Verordnung über den Amateurfunk" vom February 6, 1953, Gesetzblatt DDR No. 21 vom February 17, 1953.

who might try to obtain information about the amateur movement.[113] Naturally, neither the Soviets nor the East German authorities looked favorably on this arrangement, which seemed (and was) easy for Western intelligence agencies to exploit.

There were thus two contradictory alternatives for the authorities in East Germany by early 1951: one was to continue to ban amateur radio and stamp out illegal transmitters and the organizations supporting them as quickly as possible. Yet the Ministry for State Security recognized that the postal authorities did not then have the equipment necessary to locate illegal transmitters on a broad scale, and was not likely to get it in the near future. Thus, the second alternative was to respond to the clear public interest in radio and radio transmission by creating an approved organization in East Germany, which could bring order and clarity to the current situation, and which would make use of the skills of East German amateurs for spreading knowledge of radio science, for helping to police the airways themselves, and in preparing future soldiers for national defense.[114] Eventually, this second alternative won out, but there was much hesitation before this came to pass.

The "Chamber of Technology" (*Kammer der Technik*, KdT)[115] seemed to be the most likely vehicle for creating a national radio organization for both the amateurs themselves and the authorities of the East German Interior Ministry. Already in Thuringia, this had happened on a local level.[116] "Radio-Technical Working Groups" (*Arbeitsausschuß Rundfunktechnik*) existed within the KdT in Jena, Weimar, and Erfurt by 1950. These were seen by the KdT as primarily industry-related groups, but since they applied for permission to transmit for "experiments", it is clear that amateurs were also involved.[117] On May 19 and 20, 1951, a public meeting was held by the "Radio-Technical Working Group" of the Jena KdT. According to a participant, some 100–150 amateur radio operators

[113] Letter from Dipl. Ing. Fritz Trömel to dip. Ing. Helmut Ahlborn (then Head of the DASD Archive) of February 1, 2001, Dokufunk Archive.

[114] Senne, *SBZ und DDR*, 99. Here he cites two documents: 1. Letter from State Secretary Warnke of the Ministry of the Interior to the Minister for State Security, Zaiser of 3.3.1951, BStU MfS/Zentralarchiv Allg. S 31/54 Band I, Bl. 345 and 2. BStu MfS/Zentralarchiv Allg. S 31/54 Band I, Bl. 345 f.

[115] See note 108 above.

[116] Senne, *SBZ und DDR*, 100–105.

[117] Senne, *SBZ und DDR*, 100. Of course, in East Germany just as nearly everywhere else, many radio scientists and professionals were also amateur radio operators.

attended[118] from all over the GDR. The authorities were quite discouraging and stated that there could not be a return to the conditions or organizations of the Weimar Republic, and that there was no place today for "playing around" with hobbies. However, they did also mention participation in scientific work might be possible, and that the one area where transmitting amateurs might be able to play a role was in spreading the word abroad of the good conditions in the GDR and thus countering anti-socialist propaganda.[119] The irony of this last remark, and its parallel to the role Goebbels envisioned for amateur radio in 1933, was surely not lost on at least some of the participants who had lived through those earlier events. East German Postal authorities were supportive of the amateurs, though they cited the Allied Control Council law banning transmitters as a major hindrance. The amateurs left the meeting frustrated. Even a telegram of support from the Moscow Central Radio Club of the Soviet Union did not sway the East German authorities. But at least the amateurs were able to voice their wishes without major repercussions. No further progress was made in 1952, though the KdT did continue to seem like the proper location for discussion.[120]

All of the arguments for and against amateur radio in East Germany arguments came to discussion at an open meeting of radio enthusiasts with government representatives held in Jena in May 1951. Ostensibly, the purpose of the meeting was to discuss the reestablishment of amateur radio in East Germany through the KdT. Yet some amateurs felt that the real reason was to set a trap, and draw out into the open radio hobbyists.[121] Given prior experience, this was a logical reaction, and in fact, the idea for the meeting originated in a suggestion from the Ministry for State Security (*Ministerium für Staatssicherheit*, MfS, the dreaded "*Stasi*").[122] Despite fear of a trap, the meeting was strongly attended. The meeting was chaired by Dr. Heinze of the "RFT Funkwerk Leipzig" (a state-owned company) who was also First Chairman of the Trade Committee for Broadcast Technology in the KdT (*Erster Vorsitzender des Fachausschusses Rundfunktechnik*). Though there was quite a bit of discussion, Heinze stuck to message, and what he had to say was not very encouraging.

[118] Senne, *SBZ und DDR*, 100.

[119] Ibid, 100–105. Senne was able to use a protocol of the meeting, which he obtained from a participant.

[120] Ibid.

[121] Senne, *SBZ und DDR*, 100–102.

[122] Ibid.

Though he cited all the reasons given above in favor of legalizing amateur radio, he very firmly told the disappointed amateurs that licenses to transmit were out of the question. Instead, he pushed for a scientifically oriented interest group within the KdT which would educate its members about the science—but not the practice—of radio. Though such a radio group within the KdT was set up after the meeting, the expectant hams were deeply disappointed. Nevertheless, the very fact that the meeting was held shows that the authorities were toying with the legalization of amateur radio. By 1952, there was also growing Soviet support for the move. Despite all this, the East German authorities simply pulled back at the last minute.[123]

What seems to have turned the tide was the creation of an organization which could suitable and sensibly chaperon amateur radio activities. This was the "Sport and Technology Association" (*Gesellschaft für Sport und Technik*, GST), a new mass state paramilitary organization which served as an umbrella organization for the supervision and encouragement of a range of sports with a paramilitary character. It included shooting sports, parachuting, motorcycle, and automobile sports—and eventually, amateur radio. It carried out pre- and post-military training in schools, factories, and clubs, working closely with the National People's Army (NVA).[124] The GST was founded in August 1952, in the midst of a wave of militarization in the German Democratic Republic in the wake of the Korean War and the signing of the "German Treaty" (*Deutschlandvertrag*) between the German Federal Republic and the Western Allies. Amateur radio was included in the list of paramilitary sports it was designed to support from the beginning, even before amateur radio was legalized.[125] With the organizational structure secured and a socialist purpose defined for the hobby, amateur radio was finally legalized in the German Democratic Republic in February 1953.[126] Amateur radio thus became a central part of the milita-

[123] Senne, *SBZ und DDR*, 100–105.

[124] On the GST, see: Paul Heider, "Die Gesellschaft für Sport und Technik (1952–1990)". In *Handbuch der bewaffneten Organe der DDR*, edited by Torsten Diedrich, Hans Ehret and Rüdiger Wenke, 169–199. Berlin: Christian Links Verlag, 1998; Paul Heider, *die Gesellschaft für Sport und Technik. Vom Wehrsport zur "Schule des Soldaten von Morgen"* (Berlin: Fides Verlag, 2002); and Ringo Wagner, *Der Vergessene Sportverband der DDR. Die Gesellschaft für Sport und Technik in sporthistorischer Perspekektive.* (Aachen: Meyer & Meyer Sport, 2006).

[125] Senne, *SBZ und DDR*, 109–110.

[126] Senne, *SBZ und DDR*, 113. "Verordnung über den Amateurfunk" vom February, 6, 1953, Gesetzblatt DDR No. 21 vom February 17, 1953.

rization of East German society in the name of national defense as the suborganization of the GST known as the Radio Sport Union of the German Democratic Republic (*Radiosportverband der DDR*, RSV).

When amateur radio was finally legalized in East Germany in early 1953, it was less due to pressure from amateurs themselves, and more to the Cold War context. With West German amateurs on the airwaves legally since 1949, the East German authorities could see the propaganda value of having their own amateur radio operators as a means of demonstrating the values of German socialism to the world. The international contacts of amateur radio enthusiasts might play an important role in the Cold War propaganda war, and whatever the real usefulness of ham radio to this end, the West Germans were already doing it, which was a powerful argument. Certainly, the contribution of amateur radio to science and radio technology could also not be denied, and there was concern in industry, the scientific community, and parts of the government that failure to reestablish amateur radio would have a negative impact on the economy. Finally, amateurs had important skills which might be harnessed for national defense, and for the training of future generations of radiomen. Radio remained an important strategic technology. All of these arguments will certainly be familiar by now. In the light of the Cold War, any competitive edge the East German state could gain was welcome. There was, to be sure, a certain amount of pressure from would-be amateurs in the GDR. The resumption of the international amateur radio traffic after the Second World War, and in particular, the legalization of amateur radio in West Germany by 1949 made East German amateurs increasingly desperate to get back on the air. Legalization might help to regulate and control the situation, which could potentially lead to acts of rebellion. But in the East German case, the strongest arguments for legalizing amateur radio came from Cold War competition with West Germany.

Conclusion: Why the Rebirth of Amateur Radio After 1945 Matters

In retrospect, it is clear that German civil society in all its forms made a remarkable comeback after the end of the Second World War and the Third Reich. Despite Allied control and acute economic deprivation, club activities rapidly resumed. This is true of radio clubs, but also of most other forms of clubs (*Vereine*). Self-repair and construction of radio receivers was one impetus for the revival of the radio hobby, ham radio (transmission) was another, more lasting one. Both were very closely linked to a compelling desire to re-create civil society and the public sphere.

The revival of the public sphere after 1945 occurred in an originally tightly supervised and fundamentally de-Nazified context under Allied control and supervision, but it did not occur in a vacuum. Germans who joined clubs after 1945 had mostly been in similar clubs before then, and quite naturally the new clubs were built on the foundations of the old ones. In some cases, there was near total continuity, in others there was less, but rarely was there a full and total break with the past. That would have required the destruction of the historical memory of an entire people, which was a clear impossibility. Despite all the damage they caused, the Nazis were unable to completely destroy or Nazify civil society and the public sphere.[127]

The energy and drive with which German amateurs fought to reestablish their hobby is both remarkable and admirable. At a time when most Germans were focused on simply having a warm place to sleep and enough to eat, German hams began to organize, and had the courage to even get back on the airways, despite the possibility of draconian punishment. Certainly, those who survived the war must have come out of it with a great deal of determination. Certainly, too, the tight organization of the DASD, and the experience of living through dictatorship provided important foundations for postwar reorganization. And as we have seen, earlier experience prepared both West and East German hams in several different ways for cooperation with intelligence agencies if that was the price of continuing their hobby. It never hurts to have a little help from your friends. Nevertheless, the drive for the reestablishment of amateur radio in all four zones of occupation came from German hobbyists. It did not come easily, and took much time and sacrifice. This is an extraordinary fact. It is proof that hobbies are not simple pastimes.

Much as hunger for contact with outside world helped drive the radio revolution of the 1920s, the desire to return to the community of nations after the Second World War surely played a role in motivating the activities of German hams after the Second World War. Having a (legal) call sign, receiving and sending QSL cards with hams around the world, this was a normality, which surely allowed many German hams in the immediate

[127] In this regard, the fact that the Third Reich only lasted 12 years—less than a generation—is certainly of central importance.

postwar years to temporarily forget the misery of the immediate postwar period. Returning to an identity as just another ham also allowed many to abandon an earlier identity as soldier, "party comrade" or even perpetrator. There is a strong similarity here with the intense (East and West) German desire for travel after the war.[128]

The German radio hobby in the post-Second World War period contained many elements of continuity with earlier periods of German history, including both the Weimar Republic and the Third Reich.[129] The experience of dictatorship in the Third Reich was most directly relevant, of course, in the Soviet Zone of Occupation, and later the German Democratic Republic. The German state always considered the amateur community deeply suspect, and only reluctantly tolerated it because of its importance for militarization. This is one major continuity. West Germany is ultimately the great outlier: only with Allied pressure in Bizonia did state suspicion of amateur radio really change, though one can certainly find a continued current of hostility to it even long after the foundation of the German Federal Republic. During both the Third Reich and the Cold War, amateur radio became an important propaganda tool in a larger, global ideological conflict, another element of continuity. And finally, there was great personal and organizational continuity between amateur radio organizations during the Weimar Republic, Third Reich, and postwar West Germany.[130] These are uncomfortable facts for German amateurs today. But there were changes, too, over time. In the German Federal Republic, there was originally a very Weimar-like competition between different radio hobby clubs, but they did eventually coalesce to form the DARC. Similarly, the old rivalry and separation between middle-class and working-class radio clubs, present at the beginning of the post-Second World War period, also ended quickly. Even the old opposition of the postal authorities to the radio hobby also ended after 1945, and interestingly, in both Germanies.

[128] Christine Keitz, *Reisen als Leitbild. Die Entstehung des modernen Massentourismus in Deutschland* (Munich: Deutsche Taschenbuch Verlag, 1997). Unfortunately, Keitz does not go into the reasons for the postwar German love of travel at much length.

[129] This is **not** in any way to make the superficial and stupid argument of essential equivalency between the Third Reich and the German Democratic Republic.

[130] There was certainly some continuity between the DASD and the radio hobby in East Germany as well, but it was much less pronounced, and certainly did not extend to its leadership to any great extent.

Somewhat surprisingly, for a country whose legitimization was drawn from the memory of the working-class culture of the 1920s and the subsequent anti-fascist struggle, the German Democratic Republic did not seek to instrumentalize the history of the radio hobby (the legacy of the ARB and FRB) in this same way. In East Germany, despite the acceptance of amateur radio by the postal authorities, the old Weimar-era and Nazi suspicion of a public with uncontrolled access to the airwaves dominated, as did the instrumentalization of the radio hobby as an adjunct to military preparedness. But in the end, there still was a radio hobby in East Germany, as stunted and controlled as it may have been. There was a public sphere in the form of hobby activities in the GDR, it was just limited by the authoritarian hand of the state and its organs. The ability of East German radio hobbyists to be easily incorporated into the DARC after unification in 1990 is proof of the underlying existence of civil society, even after 40 years of dictatorship.

For all the instrumentalization of the radio hobby community in both Germanies (but particularly in the GDR), its members still participated, even at some personal risk, because they were dedicated to the *hobby*. This is a surprising fact: there is something about hobbies, which is intimately connected to the public sphere and civil society and to individual self-realization. It is so important to some people, that it is worth great effort, even sacrifice and risk of arrest and serious punishment. Despite the repression of the Third Reich, despite the careful control under Allied occupation, and despite the separation of Germany into two separate and opposing states, the radio hobby (at least for the small group of ham radio operators) survived and continued to assert itself. It is still there in a now united Germany, while the Third Reich, Allied occupation, and the German Democratic Republic are not.

CHAPTER 8

Conclusions and Questions

This book has been looking at how radio became integrated into the public sphere in Germany in the first half of the turbulent twentieth century, and how ordinary people interacted with it in the context of hobby clubs. Radio is one of the most iconic media of the twentieth century, which has been as important as a symbol of modernity as it has been as a tool for entertainment, business, government, and warfare. To turn Freud on his head, ordinary people took the *unheimlich* of unfamiliar radio technology, and made it *heimlich* by bringing it into clubs and domestic spaces.[1] To look at how people interacted with radio is to look at how they came to terms with technological modernity, whose vast changes characterize the twentieth and, now, twenty-first centuries. My thesis has been that people "domesticated" radio technology by bringing it into their homes in the form of a hobby. In this way, a mysterious and even frightening new thing could be understood, and the ways in which it might be useful explored. This is supremely important, because mass media, in order to be important, depend on masses of ordinary people to find room for it in their daily lives.

The importance of hobbies, hobbyists, tinkerers, and "makers" has been a constant subtext in this book. While "makers" have become fashionable in modern America as an emulation of the successful and

[1] Siegmund Freud, "The 'Uncanny' ", in *The Complete Psychological Works of Sigmund Freud*, Vol. 17, Transl. and ed. James Strachey (London, Hogarth Press, 1964: 217–256).

B. B. Campbell, *The Radio Hobby, Private Associations, and the Challenge of Modernity in Germany*, Palgrave Studies in the History of Science and Technology, https://doi.org/10.1007/978-3-030-26534-2_8

wealthy Silicon Valley culture of tech entrepreneurs, we still do not really give enough credit to the "lowly" tinkerer and hobbyist. Without hobbyists to "domesticate" the technology of the early twentieth century, it would never have become mass technology and would never occupy such an important place in people's daily lives. Though we at least acknowledge the "nerd" today as a social type,[2] we still treat them with distain and amusement—at least until they can buy their first Lamborghini. We rarely think of the crucial role (technological) hobbyists play in the spread, adoption, and even shape a given technology takes in society. Nor do we usually credit the importance of tinkerers and do-it-yourselfers in this process, and in the further development and refinement of technology. Not all tinkerers are Steve Jobs, but certainly in the development of radio in its early years, hobbyists helped drive technological progress. The same was, of course, true in the early days of the PC revolution, too.

If hobby culture is important in coping with/coming to terms with modernity in the form of the flood of technology, which hit people's lives in twentieth century, it was also a key aspect in the development of that same modernity: much of the mass technology so characteristic of the twentieth century was impossible to imagine without public—hobby—engagement with science. The twentieth-century inventor was increasingly a professional scientist, even part of a team, but this is only half the story: hobby scientists helped to provide part of the context within which scientific discovery was possible, and helped create the markets which drove those same discoveries. Hobbyists are, to be sure, only a small portion of the public, but they were and are often leaders—early adopters—of modern technologies, alongside the government and the military/industrial complex itself. Not least, technological hobbies often created a context in which young people gained motivation to become scientists and inventors. Moreover, particularly in the first half of the twentieth century, hobbyists fed into science itself, making new discoveries in basic science, confirming theories, and above all, finding new uses for inventions which were not thought of by their inventors. Here is the point: to create a culture of technical invention and entrepreneurship, the hobby culture is essential. There would have been no Apple without the hobby of phone phishing, no Dell or Gateway without the electronics hobby. The clubs, cafes, and garage workshops matter more than we are often willing to acknowledge.

[2] See the popular television series "The Big Bang Theory" on the US network CBS, which first premiered in 2007.

In fact, the power of hobbyists is enormous: the "citizen scientist" and "maker culture" are only the most modern manifestation of a much larger phenomenon of modern society—particularly in liberal democracies, but also in dictatorship: individual citizens not only aware of technology and science, but certain that they need a personal engagement with technology and science as a way of mastering the present and guaranteeing the future (if only to get a good job for themselves or a better job for their children). In order to succeed, these technological hobbyists must be educated enough to master specialized knowledge, and rich enough in time, money, and other resources to get personally involved. But most of all, they must be willing to get into technology with their own hands. It is time we ceased using adjective "amateur" in a negative way and began to be more aware of the wide variety of skill and experience which the term contains. While most scientific and technological discoveries today are made by specialists, they are absorbed, digested, and "produced" in a social sense by non-specialists. Hobbyists, "super-users" are the ones who lead the way in this process. Hobbyists matter a lot.

Associations and clubs matter, too. Hobbies, even or perhaps especially technological ones, are social activities. While (male) hobbyists can and do isolate themselves in their basements and garages to pursue their hobbies, even they also depend on hobby-related institutions and networks. The culture(s) of the hobby created around radio is/are an important part of the story of how technologies like radio became accepted and spread. Associations are networks of knowledge and knowledge seekers. They provide the social space within which individuals can engage with new technologies.

In Germany, the characteristic form of voluntary, civil society is the club or *Verein*. Clubs are an important part of the public sphere in nearly all developed countries, but *Vereine* are characteristic and particularly important institutions in the German-speaking world. As radio became a hobby in Germany, it naturally found expression within the existing framework of *Vereine*. This is an important part of the story. The radio hobby clubs were given an early role by the Postal Ministry when (state-run) broadcast radio was first introduced in Germany. They were bought in as tools of the government, to create a market for broadcast listening, to share the regulatory and administrative burden, and to teach people how to use the somewhat complex early radio sets in a way which would minimize both radio-frequency interference and government costs. But the clubs refused to become mere tools of the state-industrial complex, and their importance in the spread of radio as a pastime or hobby goes far beyond this original role. It was easy for Germans to found and run clubs, because there was a great store of both law and practical experience already in existence which taught people how to do it. Most Germans had experience in one or more *Vereine* before radio was invented. Thus, by packaging the involvement of the public in the new medium of radio in the well-known form of the *Verein*, a great deal of the uncertainty surrounding the new medium was eliminated. The familiar social form made the unfamiliar technology easier to absorb.

Clubs provided many things to their members. On a basic level, the clubs taught the technology behind radio. They gave classes, held talks, lectures and demonstrations, and brought people together with others to share knowledge. In short, they provided a social space within which people could learn about radio on their own terms. Not least, they provided work or "maker" spaces within which people could build their own radios, a financial necessity for most new radio listeners in the 1920s and 1930s. The clubs also provided a wide range of social activities, many of which had little to do with the technology as such, but which were a very important part of bringing people to the medium of radio. In a nutshell, people came for the parties, and stayed for the technology.[3]

The form of private clubs, and the social space allotted to them in society, varied over time with the vicissitudes of German history. The exuberant club life of the 1920s took on a much more politically polarized and serious character in the context of the Great Depression and the rise of fascism in Germany. When the Nazis seized or were handed power in Germany, most of the radio clubs were shut down. Nazi dictatorship abhorred the anarchic independence of club life, though it must also be said that industrial and technological progress by the early to mid-1930s was increasingly making the do-it-yourself part of the radio hobby unnecessary. Only the relatively small part of the radio hobby known today as "ham" radio survived the Nazi takeover, and then only because the Nazis felt it had some utility to them as a propaganda tool and agent in rearmament. Ham radio survived under the Nazis, and even thrived to a certain extent, but only at a great price. After the end of the Third Reich, club life again flourished in Germany. Radio as a hobby, after a short upswing in the immediate postwar period due to economic constraints, quickly focused again on the relatively small group of amateur radio operators or "hams" who wanted to transmit as well as receive. But again, political

[3] A heretical thought, but would the internet be as popular today without the pornography?

context was important. While a flourishing ham radio culture grew rapidly in West Germany, it was much more slow to take off in East Germany, due to government suspicion of citizens with the skills to transmit. And in both Germanies, ham radio was greatly influenced by what had happened in the Third Reich, and by the Cold War. Much had changed, but much had stayed the same.

If technology matters, and the amateur matters, then so does the context: politics and economics set real barriers to all individuals. One reason why a study of hobbyists in twentieth-century Germany can be so rich is because the variance in political and economic context within a single (German) culture was so very great. Economically, and in terms of standards of education, Germany was and is a major world industrial economy. Germany was, and is, a country with a high standard of education and a well-developed industrial economy. It is relatively rich in both money and resources. This makes Germany comparable to other major industrial economies. And yet, much of the development of radio in Germany occurred in a context of relative poverty and scarcity: the economic and social disruption caused by two world wars, an international status as a political pariah, a period of kinetic hyperinflation, and then the Great Depression all took their toll on German wealth, power, and confidence. Not unique among industrialized nations, yet quite characteristic of Germany is the country's very highly developed engineering and crafts system, which works as a bridge between pure science and the shop floor. Equally important for the study of how technology was absorbed in Germany is the highly developed culture of clubs and private civic associations, which helped bridge the public and the private realms and ease the introduction of radio into each. None of this makes Germany completely different from all the other modern industrial societies, but like the great instability of Germany politically in the twentieth century, it does make the story of the growth of radio in Germany much less linear than in the US, Britain, or France. Context matters a lot, even in an essentially globalized world.

Certainly, the experience of losing two world wars, and of being considered an international pariah state long after each of them played a role in the popularity of radio in Germany, in ways related to, but different from other industrialized countries. Politically, the great instability of German politics in the twentieth century, that is, the persistent imperial legacy within Weimar democracy, the rise of National Socialism and dictatorship, the experience of foreign occupation, the separation of Germany into a competing liberal democratic German Federal Republic, and a socialist German Democratic Republic all affected the public sphere and people's ability to have a hobby in profound ways. Current debates about the future of "the internet" (or any other tech complex) may miss the point: local political and economic context still matter, even in a heavily globalized economy/world.

Nevertheless, the rise and mutation of the radio hobby in Germany cannot simply be reduced to a mirror of German political history. Larger economic and technological trends at both the national and global levels also matter in the story. While the experience of dictatorship under National Socialism profoundly warped the radio hobby, the Nazis by themselves were not solely to blame. In the long run, what killed the broader radio hobby of organized listeners and do-it-yourselfers,[4] which was so vibrant in the first third of the twentieth century, was as much due to technological changes as it was produced by political repression. Affordable commercial receivers by the mid-1930s made building one's own receiver unnecessary, particularly because rising incomes and prosperity in the postwar years (after the immediate poverty of the late 1940s) led to enough disposable income to purchase commercial radios. In any case, the growing complexity of radio receivers (not least with the introduction of FM broadcasting and particularly, the transistor) made it increasingly hard (though still possible) to home-build the equivalent. The authoritarianism in East Germany—as in the Third Reich—was a factor in limiting the radio hobby there to only a handful of carefully watched ham radio operators. But what really killed the radio hobby as it had existed in the 1920s was prosperity—and television.[5] In the end, this proved to be a much more tenacious enemy than either Communism or Fascism, though there still is a vibrant, if much smaller, ham radio hobby in united Germany today. Radio simply lost its glamour, its "buzz", and people moved on.

The big question we are left to answer is why hobby culture has proven to be so resilient despite political and technological threat. I have demonstrated here that this was the case in Germany, but I can only hint at answers. Certainly, in seeking an answer, the social aspect of hobbies must

[4] Note that there remain small organized groups of hobbyists who engage in both of these activities, but their numbers are quite small compared to the large and vibrant radio hobby organizations of the 1920s and early 1930s.

[5] See Knut Hickethier and Peter Hoff *Geschichte des deutschen Fernsehens* (Stuttgart, Weimar: J.B. Metzlar, 1998).

be more important than the purely technical, though both also surely work together. This is a tantalizing question for new research. The answer is likely to be more important than we realize. It is certainly extraordinary that people would risk severe punishment, perhaps even death, in order to pursue a hobby. As a society, we are comfortable with the notions that a person might make big sacrifices and take big risks for money, love, family, or political commitment. But for a hobby? Perhaps we have underestimated the hobbyists, or the engagement that a hobby can bring. There is much we still need to know about how and why people engage and identify with a hobby, particularly tech-related ones such as radio or computers. Certainly, the neat dichotomy between "work" and "play" which is so much a part of our basic assumptions in our modern liberal-capitalist world is demonstrably false, even harmful when we try to gauge the impact of makers, hobbyists, and tinkerers.

Archives, Libraries, and Private Collections

Berlin Museum für Kommunikation

Berliner Staatsbibliothek

Bundesarchiv (German Federal Archives, BArch)

NS19/3917 Persönlicher Stab, Reichsführer-SS, Havelinstitut
NS19/3918 Pers. Stab Reichsführer-SS, Amateurfunkwesen
NS23/vorl.194.Sturmabteilung der NSDAP
R/4701/10858 Reichspostministerium Z 1/17936 Akten Betr. Funkgesetznovelle 1922
R/4701/8673 Reichspostministerium Sitzungsberichte R.F.K. Geh. Reistratur Z, Band 1 1919–1925
R/4701/8673 Reichspostministerium, Sitzungsberichte R.F.K. Geh. Registratur Z, Band 1 1919–1925
R4701/8994 Reichs=Postamt, Akten betreffend Reichsfunkkommission 1918–1924, Geh. Registratur Z, Band 2
R/1501/20400 St10 RMdI KPD-Radio-Propaganda Bd. 2 Mai1931–Juli1933
R/1501/20401 St10 RMdI KPD-Radio-Propaganda Bd. 3 Dez. 1929–Febr. 1931 St10/62, Bd. 3
R/1501/20402 St10 RMdI KPD-Radio-Propaganda Bd. 4a March 1931–February 1933
R/1501/20403 St10 RMdI KPD-Radio-Propaganda Bd. 4b March 1931–February 1933

B. B. Campbell, *The Radio Hobby, Private Associations, and the Challenge of Modernity in Germany*, Palgrave Studies in the History of Science and Technology, https://doi.org/10.1007/978-3-030-26534-2

R/1501/20404 St10 RMdI KPD-Radio-Propaganda Bd. 5 March 1933–March 1934.

R/1501/20062 Reichsministerium des Innern KPD Radio Propaganda, 1930–1934

R55/229 Reichsministerium für Volksaufklärung und Propaganda, DASD

Sammlung Schumacher 405

Bundesarchiv-Militärarchiv in Freiburg

MSG/3/3668 "Mitglieder-Verzeichnis des Offiziersbundes der Nachrichtentruppe Berlin e.V.", Berlin, 1927

PH2/204 Kriegsministerium Nr.731/18 g.Anch. (Geheim) Berlin W.66, den April 8, 1918, "Etatisierung von Funkerformationen"

PH3/80 "Gliederung und Stärkenachweisungen der Nachrichtentruppen im Felde 1917–1918"

PH 16/12 "Etatsstärke Schw. Funken Stat. 1 v. 18.8.18"

PH21/31 Heeres-Nachrichtenschule Namur. Lehrgang vom February, 21–23, 1918

PH21/32 Materials from the Heeres Nachrichtenschule Namur (Geheim!) January 1918

RH1/75 "Chef der Heeresleitung Beilagenheft zu den Bemerkungen des Chefs der Heeresleitung auf Grund seiner Besichtigung im Jahre 1921"

RH 1/8 Heeresleitung/Oberkommando des Heeres (OKH) – Oberbefehlshaber des Heeres

BAMA RH12/7 vol. 8 "Bericht über die Funkübung am 24.6.1933"

RH/12/7 vol. 16 OKH Inspektion der Nachrichtentruppen (In.7) Handakte Gen.Lt. Thiele Chef d.St.d.Insp. d. Nachr.-Truppen 1919–1934

RH12/7 vol. 17 OKL-Inspekteur der Nachrichtentruppen (In7) "Gen. Lt. Thiele: Materialzusammenstellungen und Vorarbeiten zum II. Teil von Th.'s Geschichte der Nachrichtentruppe" 1924

RH12/7 vol. 18 O.H.L Inspekteur der Nachrichtentruppen (In.7) "Gen. Lt. Thiele: Materialzusammenstellungen und Vorarbeiten zum II. Teil von Th's "Geschichte der Nachrichtentruppe" 1924–1930"

RH12/7 vol. 19 OHL-Inspekteur der Nachrichtentruppen (In7) "Die Nachrichtentruppen im Osten und Südosten des 1.Weltkrieges"

RH12/7 vol. 20 OHL Inspektion der Nachrichtentruppen (In7). "Gen.Lt. Thiele: Manuskripte und andere verschiedene Ausarbeiten 1936–1937"

RH12/7 vol. 21 OKH-Inspektion der Nachrichtentruppen (In7) "Lehr- und Vortragsmaterial aus dem Wehrmacht-Manöver 1937 in Mecklenburg-Vorpommern"

RH12/7 vol. 23 OKH-Inspektion der Nachrichtentruppen (In7)

RH17 (Heeresschulen) /184 Sammelmappe "Ausnutzung des Festnetzes"

RM20/1978 Oberbefehlshaber der Kriegsmarine, Deutsche Amateur- Sendedienst (D.A.S.D.)
RM20/1983 Oberkommando der Kriegsmarine, Freiwillige Wehrfunkgruppe (Marine)
RM20/1975 Oberbefehlshaber der Kriegsmarine, Sonderdienst M
RM20/1979 Oberkommando der Marine, Sondergruppe M (DASD)
RM20/2975 Oberkommando der Marine, Marinekommandoamt der Reichsmarine und Kriegsmarine
RW42/23 OKW / Sonderbeauftragter für Überprüfung des zweckmäßigen Kriegseinsatzes (OKW/Stab z.b.V.) DASD

DARC, BRAUNATHAL

DARC, DISTRIKT E ARCHIVE HAMBURG
DEUTSCHES RUNDFUNKARCHIV, BABELSBERG
DEUTSCHES RUNDFUNKARCHIV, FRANKFURT
DEUTSCHES MUSEUM, MUNICH (LIBRARY)
HAMBURG MUSEUM FÜR KOMMUNIKATION (NOW CLOSED)
SWEM LIBRARY AT THE COLLEGE OF WILLIAM & MARY
UNITED STATES NATIONAL ARCHIVES, ARCHIVES II, COLLEGE PARK, MD.
Captured German Records Microfilmed at Alexandria, Virginia, Records of the Reich Leader of the SS and Chief of the German Police [RF-SS]. Microfilm Publication T175
Microfilmed Records Received from the Berlin Document Center:
NSDAP Ortsgruppenkartei. Microfilm Publication A3340, Series MFOK.
NSDAP Zentralkartei. Microfilm Publication A3340, Series MFKL.
NSDAP Antraege. Microfilm Publication A3340, Series NSDAP-A.
NSDAP Party Census (July 1939). Microfilm Publication A3340, Series PC.
NSDAP Gaukorrespondenz. Microfilm Publication A3340, Series NS-MIS.
NSDAP Oberstes Parteigericht Akten. Microfilm Publication A3340, Series OPG.
NSDAP OPG Nachsortierung. Microfilm Publication A3340, Series OPGNA.
NSAP Partei Korrespondenz. Microfilm Publication A3340, Series PK.
SA Personnel Files. Microfilm Publication A3341, Series SA-Kartei.
SA Personal & Process Akten. Microfilm Publication A3341, Series SA.
Other SA Collections. Microfilm Publication A3341, Series SA.
Einwandererzentrale. Microfilm Publication A3342, Series EWZ.
SS Officer Personnel Files. Microfilm Publication A3343, Series SSO.
Record Group 226, Box 4, Office of Strategic Services Entry 119A; London X-2 PTS Files; London X-2 PTS-3 Files Bovensiepen PF 602.627
Record Group 226, Box 24 Office of Strategic Services Entry 119A, London X-2 PTS Files

LONDON X-2-PTS FILES WRE/18 CSDIC Miscellaneous Reports THRU W.R.E Monthly Reports

Record Group 226, Box 33 Office of Strategic Services Entry 119A; London X-2 PTS Files; London X-2 PTS-3 Files Thran, Friedrich THRU Ulrich, Wolfgang

Record Group 226, Box 48 Office of Strategic Services Entry 119A, London X-2 PTS Files, LONDON X-2-PTS FILES (ILLEG) THRU Medele

Record Group 226, Box 83 Office of Strategic Services Entry 119A, London X-2 PTS Files

London X-2 PTS-3 Files Hardouin, Roger THRU 2012-SA

Record Group 226, Box 87 Office of Strategic Services Entry 119A, London X-2 PTS Files, LONDON X-2-PTS-69 Folder 64 THRU London X-2-Pts69 Folder 70

Record Group 260 BOX 3 Records of United States Occupation Headquarters, WWII, Office of Military Government for Germany (US) OMGUS; Records of functional Offices and Divisions; The Office of the Director of Intelligence: General Corresp of the Analysis and Research Branch. 1945a–1949; 000.9-2 Scientists, German, IOA Evacuees thru 004. German Industrialists and Industry (1); FOLDER 000.77 RADIO INFORMATION OF INTELLIGENCE INTERESTS 7-20-1/28

Record Group 260 BOX 11 Records of United States Occupation Headquarters, WWII, Office of Military Government for Germany (US) OMGUS, Records of the Executive Office, The Office of the Adjutant General: General Correspondence & OTR RECS ('Decimal File'). 1945–1949. AG 010.6 Military Government Regulation (General) THRU AG010.8 Traffic Regulations: Civil and Military

Record Group 260 BOX 123 Records of United States Occupation Headquarters, WWII, Office of Military Government for Germany (US) OMGUS, Records of Functional Offices and Divisions, The Office of the Director of Intelligence: Alphabetical Reference Subject File, A&R Branch, 1947–1949. SCC Minutes 1947–1949 Thru Russian Zone

Record Group 260 BOX 163 Records of United States Occupation Headquarters, WWII, Office of Military Government for Germany (US) OMGUS, Records of the Office of Military Government, Bavaria, Records of the Intelligence Division: General Intelligence Recs. 1946–1948 Ammunition Dumps THRU Jewish

Record Group 260 BOX 581 Records of United States Occupation Headquarters, WWII; Office of Military Government for Germany (US) OMGUS Records of the Berlin Sector. Records of the Communications Branch: General Records, 1945b–1949. Newspaper clippings 1949 (6) thru Action Sheets.

Record Group 260 BOX 590 Records of United States Occupation Headquarters, WWII, Office of Military Government for Germany (US) OMGUS; Records of Office of Military Government, Bremen; Rcs of Bremerhaven Liaison &

Security Detachment: Records of Clubs And Youth Activities, 1947–1949; Volksbund "Deutsche Kriegsgräberfursorge E.V." Thru Y Viii. Playgrounds

Record Group 260 BOX 795 Records of United States Occupation Headquarters, WWII, Office of Military Government for Germany (US) OMGUS, Records of the Executive Office, The Office of the Adjutant General: formerly Security Classified Intelligence Reports, 1945–1949

Record Group 260 BOX 968 Records of United States Occupation Headquarters, WWII, Office of Military Government for Germany (US) OMGUS, Office of Military Government, Wuerttemberg-Baden. Records of the Education and Cultural Relations Division: Community Active Branch Chief: Corresp & Rel Recs. 1945–1949: Amateur Radio thru K9 American-French Zone Merger.

Record Group 260 BOX 1067 Records of United States Occupation Headquarters, WWII, Records of the U.S. Element of Inter-Allied Organizations, The Communications Group

Record Group 260 BOX 1297 Records of United States Occupation Headquarters, WWII, Office of Military Government for Germany (US) OMGUS, Records of the Office of Military Government, Wuerttemberg-Baden, Records of the Intelligence Division: Intelligence Reports, 1945-1949, INTELLIGENCE AND CONFIDENTIL ANNEXES thru WEEKLY INTELLIGENCE REPORT NUMBER 7

WIENER DOKUMENTATIONSARCHIV ZUR ERFORSCHUNG DER GESCHICHTE DES FUNKWESENS UND DER ELEKTRONISCHEN MEDIEN – INTERNATIONALES KURITORIUM QSL COLLECTION (Dokufunk)

List of Journals Consulted

Arbeiterfunk

CQ Mitteilungen des Deutschen Sende- und Empfangsdienstes

CQMB

DASD Verordnungsblatt

Der Deutsche Sender

Der Funk-Amateur

Der Radio-Amateur. Zeitschrift für Freunde der Drahtlosen Telephonie und Telegraphie. Organ des Deutschen Radio-Clubs

Funk

Funk und Bewegung

Funk Bastler. Fachblatt des Deutschen Funktechnischen Verbandes E.V.

Funkpost

Funktelegramm

QRV

QST

Volksfunk

Bibliography

Absolon, Rudolf. *Die Wehrmacht im Dritten Reich* (4 Vols.). Schriften des Bundesarchivs 16, vol.I. Boppard a/R: Harald Boldt Verlag, 1969–1979.

Absolon, Rudolf. *Die Wehrmacht im Dritten Reich*, Vol. I: *30. January 1933 bis 2. August 1934*. Boppard am Rhein: Harald Boldt Verlag, 1969.

Adorno, Theodor. "An Essay on Cultural Criticism and Society," *Prisms*, trans. Samuel and Shierry Weber. Cambridge, MA: MIT Press, 1967.

Aitken, Hugh G.J. *Syntony and Spark: the Origins of Radio*. New York: Wiley, 1976.

Aitken, Hugh G.J. *The Continuous Wave: Technology and American Radio, 1900–1932*. Princeton: Princeton University Press, 1985.

Allierter Kontrollrat, Sekretariät, ed. *Amtsblatt des Alliierten Kontrollrats in Deutschland* (Berlin 1945–1948).

Aly, Götz, and Wolf Gruner, Susanne Heim, Ulrich Herbert, Hans-Dieter Kreikamp, Horst Möller, Dieter Pohl, and Hartmut Weber, eds. (Im Auftrag des Bundesarchivs, des Instutits für Zeitgeschichte und des Lehrstuhls für Neuere und Neueste Geschichte an der Albert-Ludwigs-Universität Freiburg) *Die Verfolgung und Ermordung der europäischen Juden durch das nationalsozialistische Deutschland 1933–1945*. Band 1: Wolf Gruner (ed.) *Deutsches Reich 1933–1937*. (Munich: R. Oldenburg Verlag, 2008).

Antoine, Herbert. "Hier und Dort. Schwarzhörerstatitik." *Der Deutsche Rundfunk* 9, No. 40 (1931): 9.

Applegate, Celia. *The Necessity of Music: Variations on a German Theme*. German and European Studies. Toronto, Buffalo, London: University of Toronto Press, 2017.

B. B. Campbell, *The Radio Hobby, Private Associations, and the Challenge of Modernity in Germany*, Palgrave Studies in the History of Science and Technology, https://doi.org/10.1007/978-3-030-26534-2

Applegate, Celia. *Bach in Berlin: Nation and Culture in Mendelssohn's Revival of the Saint Matthew Passion*. Cornell: Cornell University Press, 2014.

Applegate, Celia. *A Nation of Provincials: The German Idea of Heimat*. University of California Press, 1990.

Augustine, Dolores L. *Red Prometheus: Engineering and Dictatorship in East Germany, 1945–1990*, Transformations: Studies in the History of Science and Technology. Cambridge, MA: M.I.T. Press, 2007.

Bacht, Niklaus, ed. *Music, Theatre and Politics in Germany: 1848 to the Third Reich*. Farnham: Ashgate, 2006.

Bakhtin, Mikhail. *Rabelais and His World*. Bloomington: Indiana University Press, 2009.

Balderston, Theo. *Economics and Politics in the Weimar Republic*. New Studies in Economic and Social History, 45. Cambridge: Cambridge University Press, 2002.

Bartlett, Richard A. *The World of Ham Radio, 1901–1950, a Social History*. Jefferson, North Carolina and London: McFarland, 2015.

Bathrick, David, "Making a National Family with the Radio: The Nazi Wunschkonzert." *Modernism/Modernity* 4, No.1 (January 1997): 115–27.

Bauer, Thomas. *Deutsche Programmpresse 1923 bis 1941. Entstehung, Entwicklung und Kontinuität der Rundfunkzeitschriften*. Rundfunkstudien 6. Munich: Saur, 1993).

Baum, Vicki. *Menschen im Hotel. Berliner Illustrierte Zeitung* 38, No. 13–26 (April 30 to June 29, 1929).

Bausch, Hans. *Rundfunkpolitik nach 1945: Teil 1, 1945–196*. Rundfunk in Deutschland Vol.III. Munich: Deutsche Taschenbuch Verlag, 1980a.

Bausch, Hans. *Rundfunkpolitik nach 1945: Teil 2, 1963–1980*. Rundfunk in Deutschland Vol.IV. Munich: Deutsche Taschenbuch Verlag, 1980b.

Bausch, Hans, ed. *Rundfunk in Deutschland*. Munich: Deutsche Taschenbuch Verlag, 1980c.

Bausch, Hans. *Der Rundfunk im politischen Kräftespiel der Weimarer Republik: 1923–1933*. Tübingen: Mohr, 1956.

Beauchamp, Ken. *History of Telegraphy, Its Technology and Application*. IET History of Technology Series 26. London: the Institution of Engineering and Technology, 1994, 2006.

Beck, Hermann. *The Fateful Alliance: German Conservatives and Nazis in 1933. The Machtergreifung in a New Light*. New York, Oxford: Berghahn Books, 2008.

Becker, C. " '…dass sich in den Reihen des Vereins keine Juden oder Jüdisch-Versippte befinden.' Der Ausschluß der jüdischen Mitglieder aus den Turn- und Sportvereinen Hannovers in den Jahren 1933/34." *"Elf Freunde müßt ihr sein!" Einwürfe und Anstöße zur deutschen Fußballgeschichte*, edited by Dirk Schindelbeck and Andreas Weber Geschichtswerkstatt, 38. (Freiburg: Haug, 1995a):140–144.

Becker, C. "Zwischen (Selbst-)Gleichschaltung, kommunaler Sportpolitik und sporttreibenden NS-Gleiderungen. Die bürgerlichen Turn- und Sportvereine der Stadt Hannover in der Zeit des Nationalsozialismus." *Sozial- und Zeitgeschichte des Sportes* 9, No. 2 (1995b): 24–41.

Becker, Günther. "Esperanto im Rundfunk." *Esperanto Aktuell* 18 No. 4–5 (1999): 26–27.

Benjamin, Walter. *Das Passagenwerk*, in: *Gesammelte Schriften* V, No. 1–2. Frankfurt: Suhrkamp Verlag, 1991.

Bennett, Edward W. *German Rearmament and the West 1932–1933.* Princeton N.J.: Princeton University Press, 1979.

Benninghaus, Christina and Cohen, Deborah Laurie. "Mothers' Toil and Daughters' Leisure: Working-Class Girls and Time in 1920s Germany." *History Workshop Journal,* No. 50 (Autumn, 2000): 45–72.

Bennöhr, Jürgen, DL1NP. "Geschichte des Ortsverbandes Schleswig." Accessed June 25, 2015) http://www.viehl-radio.de/homeda/chronik/chronik_m13.pdf.

Benz, Wolfgang, (ed). *Deutschland seit 1945. Entwicklungen in der Bundesrepublik und in der DDR. Chronik, Dokumente, Bilder.* Munich: Moos & Partner, 1990.

Berger, Ullrich. *Frust und Freude. Die zwei Gesichter der Gesellschaft für Sport und Technik.* (Schkeuditz: GNN Verlag, 2002).

Bergerson, Andrew Stuart, *Ordinary Germans in Extraordinary Times: The Nazi Revolution in Hildesheim.* Bloomington: Indiana University Press, 2004.

Bergerson, Andrew Stuart. "Listening to the Radio in Hildesheim, 1923–53." *German Studies Review* 24, No. 1 (Feb., 2001): 83–113.

Bergien, Rüdiger. "Staat im Staate. Zur Kooperation von Reichswehr und Republik in der Frage des Grenz- und Landesschutzes." *Vierteljahreshefte für Zeitgeschichte* 56, No. 5 (2008): 643–678.

Bergmeier, Horst J. P. and Lotz, Rainer E. *Hitler's Airwaves: The inside Story of Nazi Radio Broadcasting and Propaganda Swing* (New Haven: Yale University Press, 1997).

Bernett, Hajo. *Der Weg des Sports in die nationalsozialistische Diktitur. Die Entstehung des deutschen (nationalsozialistischen) Reichsbundes für Leibesübungen.* Beiträge zur Lehre und Forschung im Sport 87. Schondorf: Hoffman, 1983a.

Bernett, Hajo. "die Zerschlagung des deutschen Arbeitersports durch die nationalsozialistische Revolution," *Sportwissenschaft* 13 (1983b): 349–373.

Bernett, Hajo. "*Der deutsche Sport im Jahre 1933,*" *Stadion* 7 (1981): 225–283.

Bertoni, Henry L. *Radio Propagation for Modern Wireless Systems.* (Upper Saddle River, NJ: Prentice Hall, 2000).

Bessel, Richard. "The Nazi Capture of Power." *Journal of Contemporary History* 39, No. 2, *Understanding Nazi Germany.* (Apr., 2004): 169–188.

Bessel, Richard. *Political Violence and the Rise of Nazism. The Storm Troopers in Eastern Germany 1925–1934.* (New Haven, Yale University Press, 1984).

Bessler, Hansjörg. *Hörer- und Zuschauerforschung* (*Rundfunk in Deutschland* Vol.V). Munich: Deutsche Taschenbuch Verlag, 1980.

Bijker, Wiebe E., Thomas P. Hughes and Trevor Pinch, eds. *The Social Construction of Technological Systems. New Directions in the Sociology and History of Technology.* Anniversary Edition. Cambridge, MA: M.I.T. University Press, 2012, 1987.

Binz, Rudolf, DL3SO "Mein Funker-Leben. (Privates und Unprivates)." Part 1. Accessed May 22, 2018. https://z37.vfdb.org/wp-content/uploads/DL0SO_Teil1.pdf. Part 2. Accessed May 22, 2018. https://z37.vfdb.org/wp-content/uploads/DL0SO_Teil2.pdf.

Blackbourn, David and Geoff Eley. *The Peculiarities of German History: Bourgeois Society and Politics in Nineteenth Century Germany.* Oxford: Oxford University Press, 1984.

Bley, Wulf (S.A.R. Mann Sta 9). *Deutsche Nationalerziehung und Rundfunk-Neubau! Was wollen die Nationalsozialisten im Rundfunk?* Hrsg. von der Verbandsgruppe Nationalsozialisten im Reichsverband Deutscher Rundfunkteilnehmer. Mit einer Einführung von Eugen Hadamovsky. Berlin: Verlag H. v Fehrentheil, 1932.

Blumtritt, Oskar. "Nesper, Eugen Heinrich Josef." *Neue Deutsche Biographie* 19 (1998): 70–71. Accessed July 16, 2014. http://www.deutsche-biographie.de/pnd129294594.html.

Bock, Gisela. *Frauen in der europäischen Geschichte: Vom Mittelalter bis zur Gegenwart.* Munich: C.H. Beck, 2005.

Bodek, Richard. "The Not-So-Golden Twenties: Everyday Life and Communist Agitprop in Weimar-Era Berlin." *Journal of Social History* 30, No. 1 (Autumn, 1996): 55–78.

Bodek, Richard. "Red Song: Social Democratic Music and Radicalism at the End of the Weimar Republic." *Central European History* 28, No. 2 (1995): 209–227.

Boghardt, Thomas. "America's Secret Vanguard: US Army Intelligence Operations in Germany, 1944–47." *Studies in Intelligence* 57, No. 2 (June 2013):1–18.

Bonacker, Max. *Goebbels' Mann beim Radio: der NS-Propagandist Hans Fritzsche (1900–1953).* Schriftenreihe der Vierteljahreshefte für Zeitgeschichte 94. Munich: Oldenbourg, 2007.

Bordieu, Pierre. *Outline of a Theory of Practice* (Cambridge Studies in Social and Cultural Anthropology 16). Cambridge: Cambridge University Press, 1977, 1972.

Bracher, Karl Dietrich, Wolfgang Sauer and Gerhard Schulz. *Die Nationalsozialistische Machtergreifung. Studien zur Errichtung des Totalitären Herrschaftssystem in Deutschland 1933/1934.* Wiesbaden: Springer, 1960.

Braun, Jutta and Wiese, René (Hg.): *Sport und Diktatur. Zur politischen und sozialen Rolle des Sports in den deutschen Diktaturen des 20. Jahrhunderts* (Köln: Zentrun für Historische Sozialforschung, 2007).

Brecht, Bertolt. "Der Rundfunk als Kommunikationsapparat." *Blätter des Hessischen Landestheaters Darmstadt* 16 (July 1932).

Bredow, Hans. *Vier Jahre Deutscher Rundfunk*. Berlin: Reichsdruckerei, 1927.

Broszat, Martin and Hermann Weber, eds. *SBZ-Handbuch: Staatliche Verwaltungen, Parteien, gesellschaftliche Organisationen und ihre Führungskräfte in der Sowjetischen Besatzungszone Deutschlands 1945–1949*. Berlin: De Gruyter/Oldenbourg, 1990, Second Edition 1993.

Broszat, Martin. *Die Machtergreifung. Der Aufstieg der NSDAP und die Zerstörung der Weimarer Republik*. Munich: Deutscher Taschenbuch Verlag, 1984.

Busch, Hans and Otto Reuter, "*Erfahrungen im Luftschiff-Funkdienst,*", *Arbeitskreis Seefunk - Erfahrungen - Probleme – Berichte* 4 (1967), http://www.seefunknetz.de/lzzep.htm. Accessed August 9, 2018.

Campbell, Bruce B. "The SA in the *Gleichschaltung*: Stress, Organization and Violence", in: Larry Eugene Jones and Hermann Beck (eds.). *From Weimar to Hitler: Studies on the Dissolution of Weimar Democracy and the Establishment of the Third Reich, 1932–34*. New York, Berghahn, 2018.

Campbell, Bruce B., Alison Guenther-Pal and Vibeke Petersen (eds.). *Detectives, Dystopias and Poplit: Studies in Modern German Genre Fiction*. Elizabethtown, NY: Camden House, 2013a.

Campbell, Bruce B. "The Detective Was the Killer: the Memory of the Nazi Past in Modern German Detective Fiction" in *Detectives, Dystopias and Poplit: Studies in Modern German Genre Fiction,* edited by Campbell, Bruce B., Alison Guenther-Pal and Vibeke Petersen, 133–151. Elizabethtown, NY: Camden House, 2013b.

Campbell, Bruce B. "Gewalt bis in die obersten Ränge. Die Höheren SA-Führer der SA-Gruppe Berlin-Brandenburg." Stefan Hördler (ed.) *Der SA-Terror als Herrschaftssicherung. "Köpinecker Blutwoche" und Öffentliche Gewalt im frühen Nationalsozialismus* (Berlin: Metropol Verlag, 2013c): 62–82.

Campbell, Bruce B. "Autobiographies of Violence: The SA in its own Words." *Central European History*, special issue on the SA, 46, No. 2 (June 2013d): 217–237.

Campbell, Bruce B. "The Radio Hobby, Government, and the Discourse of Catastrophe." in *Communities of the Air: Radio Century, Radio Culture,* edited by Susan Merrill Squier, 63–75. Durham: Duke University Press, 2003.

Campbell, Bruce B. " '*Kein schöner Land...*': The Spielschar Ekkehard and the Struggle to Define German National identity in the Weimar Republic." in *Music and German National Identity* edited by Pamela Potter and Celia Applegate (eds.), 128–139. Chicago: University of Chicago Press, 2002.

Campbell, Bruce B. and Arthur Brenner (eds.) *Death Squads in Global Perspective: Murder with Deniability*. New York: St. Martin's Press, 2000, 2002.

Campbell, Bruce B. *The SA Generals and the Rise of Nazism*. Lexington: University Press of Kentucky, 1998, 2004.

Campbell, Bruce B. “Gerhard Roßbach, the Spielschar Ekkehard, and the Cultural Attack on the Weimar Republic.” *Weimar 1930. Politik und Kultur im Vorfeld der NS-Diktatur* edited by Lothar Ehrlich and Juergen John, 243–259. Weimar: Böhlau, 1997.

Campbell, Bruce B. “The SA After the Röhm Purge and the Institutionalized State of Emergency in the Third Reich” in: *Journal of Contemporary History* 28, No. 4 (October 1993a), pp. 659–674.

Campbell, Bruce B. “The Schilljugend from Wehrjugend to Luftschutz: Bündisch Youth in the Weimar Republic in Search of a Political Voice.” in *Politische Jugend in der Weimarer Republik*, edited by Wolfgang R. Krabbe, 183–201. Bochum: Universitätsverlag Dr. Brockmeyer, 1993b.

Carey, James W. “Technology and Ideology: The Case of the Telegraph.” *Prospects* 8 (1983): 303–25.

Certeau, Michel de. *The Practice of Everyday Life*. Berkeley and Los Angeles: University of California Press, 1984.

Childes, Harwood L. and John B. Whitton, eds. *Propaganda by Shortwave*. Princeton: Princeton University Press, 1942.

Christie, Agatha. *Murder on the Orient Express*. Glasgow: William Collins & Sons, 1934.

Cohen, Deborah. *The War Come Home: Disabled Veterans in Britain and Germany, 1914 to 1939*. Berkeley: University of California Press, 2001.

Conan Doyle, Sir Arthur, “The Adventure of Silver Blaze”, first published in *The Strand Magazine*, December 1892.

Connor, Roger. “Wolfgang von Gronau and his Greenland ‘Whales’ ”, August 27, 2013, https://airandspace.si.edu/stories/editorial/wolfgang-von-gronau-and-his-greenland-%E2%80%9Cwhales%E2%80%9D. Accessed January 18, 2019.

Conway, J.S. “ ‘Machtergreifung’ or ‘Due Process of History’: The Historiography of Hitler’s Rise to Power. *The Historical Journal* 8, No. 3 (1965): 399–413.

Cooper, Merian C. „King Kong,“ RKO Radio Pictures, 1933.

Craig, Gordon A. *Germany, 1866–1945*. (The Oxford History of Modern Europe). Oxford: Oxford University Press, 1978.

Craig, Gordon R. *The Politics of the Prussian Army*. Oxford: Oxford University Press, 1956.

Crisell, Andrew. *An Introductory History of British Broadcasting* (2nd ed.). New York: Routledge, 2002.

Cuomo, Glenn. “The NSDAP’s Enduring Shadow: Putting in Perspective the Recent Outing of Brown Octogenarians,” *German Studies Review* 35, No. 2, (May 2012): 265–288.

Cuomo, Glenn. *Career at the Cost of Compromise: Günter Eich’s Life and Work in the Years 1933–1945*. Amsterdamer Publikationen zur Sprache und Literatur 82. Amsterdam and Atlanta: Rodopi, 1989.

Dahl, Peter. *Arbeitersender und Volksempfänger: proletarische Radio-Bewegung und bürgerlicher Runkfunk bis1945*. Frankfurt/aM.: Syndikat, 1978.

Dahm, Volker, "Nationale Einheit und partikulare Vielfalt. Zur Frage der kulturpolitischen Gleichschaltung im Dritten Reich", in: Vierteljahrshefte für Zeitgeschichte, Vol. 43, Nr. 2. (April., 1995), pp. 221–265.

David Welch, David "Nazi Propaganda and the Volksgemeinschaft: Constructing a People's Community" in: *Journal of Contemporary History*, Vol. 39, No. 2, *Understanding Nazi Germany* (Apr., 2004), pp. 213–238.

Dean, Alexandra. "Bombshell: the Hedy Lamarr Story." Reframed Pictures, 2017.

Desoto, Clinton B. *200 Meters and Down: The Story of Amateur Radio*. Newington CT: American Radio Relay League, 1985.

Deutsche Welle, *Mit 8 KW rund um die Welt. : Deutscher Weltrundfunk in der Weimarer Zeit. Geschichte des Kurzwellenrundfunks in Deutschland 1929–1932*. Berlin: Haude & Spener, 1969.

Deutscher Alpenverein, Österreichischer Alpenverein and Alpenverein Südtirol. *"Berg Heil!" Alpenverein und Bergsteigen, 1918 bis 1945*. Vienna: Böhlau Verlag, 2011.

Deutscher Amateur-Sende- und Empfangsdienst, Hauptverkehrsleitung, Technische Abteilung. *Kurzwellentechnik, Ein Leitfaden für den Amateur*. Berlin: Rothgiesser & Diesing, 1931, 2nd. edited edition 1935.

Dietrich, Gerd. "Kulturbund." In *Die Parteien und Organisationen der DDR. Ein Handbuch,* edited by Gerd-Rüdiger Stephan et al. Berlin: Dietz, 2002.

Diedrich, Torsten, Hans Ehret, and Rüdiger Wenke, eds. *Handbuch der bewaffneten Organe der DDR*. Berlin: Christian Links Verlag, 1998.

Diller, Ansgar. "Der Volksempfänger–Propaganda-und Wirtschaftsfaktor." *Mitteilungen des Studienkreises Rundfunk und Geschichte* 9 (1983): 140–157.

Diller, Ansgar. *Rundfunkpolitik im Dritten Reich*. Munich: DTV, 1980.

Diller, Ansgar. "Aus den Kinderjahren des Rundfunks. Wie Post und Rundfunk vor fünfzig Jahren auf Schwarzhörerfang gingen." *epd/Kirche und Rundfunk* 56 (August 24, 1974).

Dinçkal, Noyan, Christof Dipper and Detlev Mares, (eds.) *Selbstmobilisierung der Wissenschaft: Technische Hochschulen im "Dritten Reich"*. Darmstadt: Wissenschaftliche Buchgesellschaft, 2010.

Douglas, Susan J. *Listening In: Radio and the American Imagination*. Minneapolis: University of Minnesota Press, 2004.

Douglas, Susan J. *Inventing American Broadcasting, 1899-1922*. Johns Hopkins Studies in the History of Technology. Baltimore: Johns Hopkins University Press, 1987.

Doyle, Sir Arthur Conan. "The Adventure of Silver Blaze." *The Memoirs of Sherlock Holme. New York, London: Harper & Brothers, 1894.*

Dussel, Konrad. *Deutsche Rundfunkgeschichte*. Stuttgart: UTB, 2004.

Edwards, John Carver. *Berlin Calling: American Broadcasters in Service to the Third Reich.* Westport, CT: Praeger, 1991.

Eisenberg, Christiane "Medienfußball. Entstehung und Entwicklung einer transnationalen Kultur." *Geschichte und Gesellschaft* 31, No. 4 *Globalisierung.* (October–December 2005),: 586–609.

Erb, Dirk, ed. *Gleichgeschaltet. Der Nazi-Terror gegen Gewerkschaften und Berufsverbände 1930 bis 1933. Eine Dokumentation.* Göttingen: Steidl, 2001.

Fahnert, Friedrich and Ferdinand Nibler. *Mein Leben und meine Zeit. Die Lebenserinnerungen von General der Luftnachrichtentruppe Friedrich Fahnert. 18.01.1879–10.06.1964.* Munich: Universität der Bundeswehr, 1997.

Fayard, Anne-Laure and John Weeks, "Photocopiers and Water-coolers: The Affordances of Informal Interaction." *Organization Studies* 28, No. 5, (May 1, 2007): 605–634.

Fehrentheil, Heinz von. "Aus Geschichte und Organisation des RDR." In *Rundfunk im Aufbruch. Handbuch des deutschen Rundfunks 1934 mit Funkkalender.* Schauenburg: Lahr, 1934.

Feiereisen, Florence and Hill, Alexandra Merley (eds), *Germany in the Loud Twentieth Century* (Oxford: Oxford University Press, 2011).

Feldman, Gerald D. *The Great Disorder: Politics, Economics, and Society in the German Inflation, 1914–1924.* Oxford and New York: Oxford Univ. Press, 1996.

Fendler, Ernst (DL1JK) and Günther Noack (DL7AY). *Amateurfunk im Wandel der Zeit.* Baunatal: DARC Verlag, 1986.

Fergusson, Adam. *When Money Dies: the Nightmare of Deficit Spending, Devaluation, and Hyperinflation in Weimar Germany.* New York: Public Affairs, 2010.

Field, Andrew David. *Shanghai's Dancing World: Cabaret Culture and Urban Politics, 1919–1954.* Hong Kong: The Chinese University Press, 2010.

Fitzgerald, F. Scott. *The Great Gatsby.* New York: Charles Scribner's and Sons, 1925.

Flachowsky, Sören *Von der Notgemeinschaft zum Reichsforschungsrat: Wissenschaftspolitik im Kontext von Autarkie, Aufrüstung, und Krieg* (Studien zur Geschichte der Deutschen Forschungsgemeinschaft, vol. 3) Edited by Rüdiger vom Bruch, Ulrich Herbert, and Patrick Wagner. (Stuttgart: Franz Steiner, 2008).

Frank, Robert H. "Hitler and the National Socialist Coalition 1924–1932.", PhD. Diss. Johns Hopkins University, 1969.

Frauengruppe Faschismusforschung. *Arbeitsbuch und Mutterkreuz. Zur Geschichte er Frauen in der Weimarer Republik und im Nationalsozialismus.* Frankfurt/M.: Fischer, 1981.

Freud, Siegmund. "The 'Uncanny' ". In *The Complete Psychological Works of Sigmund Freud*, Vol. 17, Transl. and ed. James Strachey 217–256. London: Hogarth Press, 1964.

Frevert, Ute. *Frauen-Geschichte. Zwischen Bürgerlicher Verbesserung und Neuer Weiblichkeit.* Frankfurt: Suhrkamp, 1986.

Friedewald, Michael. *Die "Tönenden Funken". Geschichte eines frühen drahtlosen Kommunikationssystems, 1905–1914.* Aachener Beitrage zur Wissenschafts- und Technikgeschichte des 20. Jahrhunderts 2. Berlin: Diepholz/GNT Verlag, 1999.

Friedrich, Jörg. *Der Brand. Deutschland im Bombenkrieg –1945.* Hamburg: Spiegel-Verlag, 2007.

Fritzsche, Peter. "Machine Dreams: Airmindedness and the Reinvention of Germany." *The American Historical Review* 98, No. 3 (June 1993): 685–709.

Fritzsche, Peter. *A Nation of Fliers: German Aviation and the Popular Imagination.* Cambridge: Harvard University Press, 1992.

Fröhlich, Elke, ed. *Die Tagebücher von Josef Goebbels* (Im Auftrag des Instituts für Zeitgeschichte und mit Unterstützung des Staatlichen Archivdienstes Rußlands (18 vols.) (Munich: Sauer, 1998–2004).

Fuge, Janina. *An den Funkpranger gestellt und mit dem Wellendetektiv gejagt. Die Schwarzhörer als Konstante der Rundfunkgeschichte.* Nordwestdeutsche Hefte zur Rundfunkgeschichte 7, edited by Hans-Ulrich Wagner. Hamburg: Forschungsstelle Geschichte des Rundfunks in Norddeutschland, 2009.

Führer, Karl Christian. "German Cultural Life and the Crisis of National Identity during the Depression, 1929-1933." *German Studies Review* 24, No. 3 (October 2001): 461–486.

Führer, Karl Christian. "A medium of modernity? Broadcasting in Weimar Germany, 1923-1932." *The Journal of Modern History* 69. No. 4 (December 1997): 722–733.

Führer, Karl Christian. "Auf dem Weg zur "Massenkultur"? Kino und Rundfunk in der Weimarer Republik." *Historische Zeitschrift* 262, No. 3 (June 1996): 739–781.

Führer, Karl Christian. *Mieter, Hausbesitzer, Staat und Wohnungsmarkt. Wohnungsmangel und Wohnungszwangswirtschaft in Deutschland 1914–1960.* Vierteljahrschrift für Sozial- und Wirtschaftsgeschichte, Beiheft 119. Stuttgart: Franz Steiner, 1995.

Garratt, G.R.M. *The Early History of Radio, From Faraday to Marconi.* IET History of Technology Series 20. London: The Institution of Engineering and Technology, 1994, 2006.

Giles, Geoffrey. *Students and National Socialism in Germany.* Princeton: Princeton University Press, 1985.

Gerhard, Gesine. *Nazi Hunger Politics: A History of Food in the Third Reich.* London: Rowman & Littlefield, 2015.

Gersdorff, Ursula von. *Frauen im Kriegsdienst 1914–1945.* Beiträge zur Militär- und Kriegsgeschichte 11. Stuttgart: Deutsche Verlagsanstalt, 1969.

Gispen, Kees. *Poems in Steel: National Socialism and the Politics of Inventing from Weimar to Bonn*. Monographs in German History 6. Cambridge and New York: Berghahn Books, 2002a.

Gispen, Kees. *New Profession, Old Order: Engineers and German Society, 1815–1914*. Cambridge: Cambridge University Press, 2002b.

Goebbels, Joseph. *Signale der neuen Zeit. 25 ausgewählte Reden von Dr. Joseph Goebbels*. Munich: Eher Verlag, 1934.

Goebel, Gerhart. "Der Deutsche Rundfunk biz zum Inkrafttreten des Kopenhagener Wellenplans." Sonderdruck aus dem *Archiv für das Post- und Fernmeldewesen* 2, No. 6 (August 1950).

Golücke, Friedhelm, ed. *Korporationen und Nationalsozialismus*. Abhandlungen zum Studenten- und Hochschulwesen 2. Schernfeld: SH-Verlag, 1989.

Goodale, Greg. *Sonic Persuasion: Reading Sound in the Recorded Age*. Studies in Sensory History. Champaign: University of Illinois Press, 2011.

Grebing, Helga. *Der "deutsche Sonderweg" in Europa 1806–1945: Eine Kritik*. Stuttgart: W. Kohlhammer, 1986.

Greenblatt, Stephen *Renaissance Self-Fashioning: From More to Shakespeare*. Chicago: Chicago University Press, 2005.

Gruchmann, Lothar. *Justiz im Dritten Reich, 1933–1940. Anpassung und Unterwerfung in der Ära Gürtner*. Munich: Oldenburg, 3rd. improved edition 2001.

Guarnieri, Massimo. "The Age of Vacuum Tubes: Early Devices and the Rise of Radio Communications [Historical]." *IEEE Industrial Electronics Magazine* 6, No. 1 (March 2012): 40–43.

Günter, Hanns & C. Culatti. *Wer Gibt? Die Funkstationen der Welt, ihre Rufzeichen, Reichweiten u. Wellenlängen einschließl. d. europäischen Amateursender*. Stuttgart: Franckh'sche Verlagshandlung, 1925.

Habermas, Jürgen. *Strukturwandel der Öffentlichkeit: Untersuchungen zu einer Kategorie der bürgerlichen Gesellschaft*. Frankfurt: Suhrkamp, 1990.

Habermas, Jürgen. *The Structural Transformation of the Public Sphere*. Cambridge, MA: MIT Press, 1989.

Hahn, Hans Henning and Eva Hahn. *Die Vertreibung im deutschen Erinnern. Legenden, Mythos, Geschichte*. Paderborn: Verlag Ferdinand Schöningh, 2010.

Hampton, Stephanie E. and John N. Parker. "Collaboration and Productivity in Scientific Synthesis." *BioScience*, 61, No. 11 (1 November 2011): 900–910.

Hancock, Eleanor. *Ernst Röhm: Hitler's SA Chief of Staff*. New York: Palgrave Macmillan, 2008.

Haring, Kristen. *Ham Radio's Technical Culture*. Cambridge: MIT University Press, 2007.

Havemann, Nils. *Fußball unterm Hakenkreuz: der DFB zwischen Sport, Politik und Kommerz*. Bonn: Bundeszentrale für Politische Bildung, 2005.

Headrick, Danial J. *Invisible Weapon: Telecommunications and International Politics, 1851–1945*. New York: Oxford University Press, 1991.

Heidenreich, Bernd, and Sönke Neitzkel, eds. *Medien im Nationalsozialismus*. Paderborn: Ferdinand Schöningh, Wilhelm Fink, 2010.

Heider, Magdalena. "Der Kulturbund zur Demokratischen Erneuerung Deutschlands (KB)." In *SBZ-Handbuch: Staatliche Verwaltungen, Parteien, gesellschaftliche Organisationen und ihre Führungskräfte in der Sowjetischen Besatzungszone Deutschlands 1945–1949*, edited by Martin Broszat and Hermann Weber, 714–733. Berlin: De Gruyter/Oldenbourg, 1990, Second Edition 1993a.

Heider, Magdalena. *Politik - Kultur - Kulturbund. Zur Gründungs- und Frühgeschichte des Kulturbundes zur demokratischen Erneuerung Deutschlands 1945–1954 in der SBZ/DDR*. Bibliothek Wissenschaft und Politik 51. Cologne: Verlag Wissenschaft und Politik Claus-Peter von Nottbeck, 1993b.

Heider, Paul. *Die Gesellschaft für Sport und Technik. Vom Wehrsport zur "Schule des Soldaten von Morgen"*. Berlin: Fides Verlag, 2002.

Heider, Paul. "Die Gesellschaft für Sport und Technik (1952-1990)." In *Handbuch der bewaffneten Organe der DDR*, edited by Torsten Diedrich, Hans Ehret and Rüdiger Wenke, 169–199. Berlin: Christian Links Verlag, 1998.

Heinemann, Isabel, et al. *Wissenschaft, Planung, Vertreibung. Der Generalplan Ost der Nationalsozialisten. Katalog zur Ausstellung der Deutschen Forschungsgemeinschaft*. Deutsche Forschungsgemeinschaft, 2006. Accessed 15.May.2012 http://www.dfg.de/pub/generalplan/downloads/dfg_wissenschaft_planung_vertreibung_katalog.pdf.

Heinrichs, Jürgen Wilhelm Walter. " 'Blackness in Weimar': 1920s German Art Practice and American Jazz & Dance." Ph.D. Diss. Yale University, 1998.

Helling, Wilfried. *Gleichschaltung und Ausgrenzung. Der Weg der bremischen Anwaltschaft ins Dritte Reich*. Veröffentlichungen der Hanseatischen Rechtsanwaltskammer Bremen. Bremen: Hanseatischen Rechtsanwaltskammer, 1990.

Henderson, Bruce. *Sons and Soldiers. The Untold Story of the Jews Who Escaped the Nazis and Returned with the U.S. Army to Fight Hitler*. New York: William Morrow/Harper Collins, 2017.

Henke, Klaus-Dietmar: *Die amerikanische Besetzung Deutschlands (Quellen und Darstellungen zur Zeitgeschichte, Band 27)* (Oldenbourg: deGruyter, 1995).

Hensle, Michael P. *Rundfunkverbrechen. Das Hören von "Feindsendern" im Nationalsozialismus*. Berlin: Metropol Verlag, 2003.

Herzfeld, Andy. *Revolution in the Valley: the Insanely Great Story of How the Mac was Made*. Sebastopol, CA: O'Reilly Media, 2004.

Herzog, Markwart. *Der "Betze" unterm Hakenkreuz. Der 1.FC Kaiserslautern in der Zeit des Nationalsozialismus*. Göttingen: Verlag die Werkstatt, 2006.

Hickethier, Knut and Peter Hoff. *Geschichte des deutschen Fernsehens*. Stuttgart, Weimar: J.B. Metzlar, 1998.

Hiden, John "Germany, Home and Away" in: *The Historical Journal* Vol. 30, No. 2 (Jun., 1987), pp. 463–482.

Hilmes, Michelle, *Network Nations: A Transnational History of British and American Broadcasting*. New York: Routledge 2012.

Hilmes, Michele. "Is There a Field Called Sound Culture Studies? And Does It Matter?." *American Quarterly* 57, No. 1 (March 2005): 249–259.

Hilmes, Michelle, *Radio Voices: American Broadcasting 1922–1952*. Minneapolis: University of Minnesota Press, 1997.

Hindrichs, Jochen. "60 Jahre DARC-Distrikt Hamburg." *Hamburger Funkerblatt* (October 19, 2007). Accessed August 4, 2015. http://www.df3xz.de/Hamburger%20Funkerblatt.pdf.

Hochfelder, David. *The Telegraph in America, 1832–1920*. Johns Hopkins Studies in the History of Technology. Baltimore: Johns Hopkins University Press, 2012.

Hoffmann, Herbert. *In Gleichschritt in die Diktatur? Die Nationalsozialistische "Machtergreifung" in Heidelberg und Mannheim 1930 bis 1935*. Frankfurt, Bern, New York: Peter Lang, 1985.

Hoffmann, Karl Otto. *Die Geschichte der Luftnachrichtentruppe*. 2 vols. Neckargemünd: Vowinckel, 1965-1973.

Höhne, Heinz. *Die Zeit der Illusionen. Hitler und die Anfänge des Dritten Reiches 1933–1936*. Düsseldorf, Wien, New York: ECON Verlag, 1991.

Höhne, Heinz. *The Order Under the Death's Head. The Story of Hitler's SS*. New York: Ballantine Books, 1966, 1977.

Hördler, Stefan, ed. *Der SA-Terror als Herrschaftssicherung. "Köpinecker Blutwoche" und Öffentliche Gewalt im frühen Nationalsozialismus*. Berlin: Metropol Verlag, 2013.

Hörisch, Jochen. *Eine Geschichte der Medien. Von der Oblate zum Internet*. Frankfurt a/M.: Suhrkamp, 2001, 2004.

Hörner, Fernand. "Wenn das Original ins Wohnzimmer kommt. Musik und Reproduktion im Zeitalter von Walter Benjamin." *Lied und populäre Kultur / Song and Popular Culture* 56 (2011): 13–41.

Hopwood, Nick. "Producing a Socialist Popular Science in the Weimar Republic." *History Workshop Journal* 41 (Spring, 1996):117–153

Hugill, Peter J. *Global Communications Since 1844: Geopolitics and Technology*. Baltimore: Johns Hopkins University Press, 1999.

Innes, Harold. *Empire and Communications*. Toronto: University of Toronto, 1972.

International Telecommunication Union. *Radio Regulations*. Geneva: ITU, 2016. Accessed January 17, 2019. http://www.itu.int/pub/R-REG-RR-2016.

Jamin, Mathilde. *Zwischen den Klassen. Zur Sozialstruktur der SA-Führerschaft.* Wuppertal, Peter Hammer Verlag, 1984.
Jander, Thomas and Veit Didczuneit. *Netze des Krieges. Kommunikation 1914–18.* Museumsstiftung für Post und Telekommunikation/Brandenburgisches Verlagshaus, 2014.
Jarausch, Konrad H. *The Unfree Professions: German Lawyers, Teachers and Engineers.* Oxford: Oxford University Press, 1990.
Jessen, Ralf & Jakob Vogel, eds. *Wissenschaft und Nation in der Europäischen Geschichte.* Frankfurt, New York: Campus Verlag, 2003.
John, Richard. *Network Nation: Inventing American Telecommunications.* Cambridge, MA: Harvard University Press, 2010.
Jung, Leo H., DH4IAB. "QSL's erzählen deutsche Amateurfunkgeschichte." No. 9 "Beim DASD 1933-1945." *Funk-Telegramm* No. 12 (2003): 9–13.
Jung, Leo H., DH4IAB. "QSL's Erzählen deutsche Amateurfunkgeschichte", No 8 "Der DASD bis 1933." *Funk-Telegramm* No. 11 (2003): 28–29.
Kahn, David. *Hitler's Spies: German Military Intelligence in World War II.* New York: Macmillan, 1978.
Kallinikos, Jannis, Hans Hasselbladh and Attila Marton "Governing social practice: Technology and institutional change." *Theory and Society* 42, No. 4 (July 2013): 395–421.
Kampe, Hans Georg. *Nachrichtentruppe des Heeres und Deutsche Reichspost. Militärisches und staatliches Nachrichtenwesen in Deutschland, 1830 bis 1945.* Wandesruh bei Berlin: Project+Verlag Dr. Erwin Meißler, 1999.
Kater, Michael H. *Hitler Youth.* Cambridge: Harvard University Press, 2004.
Kater, Michael H. *Different Drummers: Jazz and the Culture of Nazi Germany.* New York: Oxford University Press, 1992.
Kater, Michael H. "The Jazz Experience in Weimar Germany." *German History* VI, No. 2 (April 1988): 145–158.
Keitz Christine. *Reisen als Leitbild. Die Entstehung des modernen Massentourismus in Deutschland.* Munich: Deutsche Taschenbuch Verlag, 1997.
Kennan, George F. *The Decline of Bismarck's European Order. Franco-Russian Relations, 1875–1890.* Princeton: Princeton UP, 1979.
Kerer, Hilde. *Ich war ein Blitzmädel.Frauenkamaradschaft in der Wehrmacht. Aufgezeichnet von Thomas Hanifle.* Bozen: Edition Raetie, 2014.
Kershaw, Ian. *The End: The Defiance and Destruction of Hitler's Germany, 1944–1945.* New York: Penguin, 2011.
Kirchner, Fritz, DJ2NL. "Einige Daten und Fakten zur Entstehung des Amateurfunkgesetzes (AFuG) vom 14. März 1949." *CQ-DL* (October, 1987): 632–634.

Kittler, Friedrich A. *Gramophone, Film, Typewriter* (Translated, with an Introduction by Geoffrey Winthrop-Young and Michael Wutz). Stanford: Stanford University Press, 1999.

Klein, Dolf. *Köln im Dritten Reich. Stadtgeschichte der Jahre 1933–1945*. Köln: Greven, 1983.

Kleinsteuber, Hans J. *Radio. Eine Einführung*. Wiesbaden:VS Verlag 2011.

Kliem, Konstantin. *Sport in der Zeit des Nationalsozialismus: Entwicklung und Zielsetzung im Höheren Schulwesen und in der Hitlerjugend*. Saarbrücken: Müller, 2007.

Klönne, Arno. *Jugend im Dritten Reich*. Cologne: Diederichs, 1982.

Koch, Hans-Jörg. *Das Wunschkonzert im NS-Rundfunk*. Medien in Geschichte und Gegenwart 20. Cologne, Weimar, Vienna: Böhlau, 2003.

Koch, H. W. *The Hitler Youth: Origins and Development, 1922–1945*. New York: Barnes and Noble, 1996.

Koehl, Robert L. *The Black Corps. The Structure and Power Struggles of the Nazi SS*. Madison: University of Wisconsin Press, 1983.

Koehl, Robert L. "Feudal Aspects of National Socialism." *American Political Science Review* 54, no. 4 (December 1960): 921–933.

König, Wolfgang. *Volkswagen, Volksempfänger, Volksgemeinschaft. „Volksprodukte" im Dritten Reich: Vom Scheitern einer nationalsozialistischen Konsumgesellschaft*. Paderborn: Ferdinand Schöningh, 2004.

König, Wolfgang. "Mythen um den Volksempfänger. Revisionistische Untersuchungen zur nationalsozialistischen Rundfunkpolitik." *Technikgeschichte* 70 (2003a): 73–102.

König, Wolfgang. "Der Volksempfänger und die Radioindustrie. Ein Beitrag zum Verhältnis von Wirtschaft und Politik im Nationalsozialismus." *Vierteljahreshefte für Sozial- und Wirtschaftsgeschichte* 90 (2003b): 269–289.

König, Wolfgang. "Das Scheitern einer nationalsozialistischen Konsumgesellschaft. «Volksprodukte» in Politik, Propaganda und Gesellschaft des «Dritten Reiches»." *Zeitschrift für Unternehmensgeschichte* 48. No 2. (2003c): 131–163.

Koerfer, Daniel. *Hertha unter dem Hakenkreuz. ein Berliner Fußballclub im Dritten Reich*. Göttingen: Verlag der Werkstatt, 2009.

Körner, W(olfgang) F(elix) DL1CU. *Geschichte des Amateurfunks 1909–1963*. Hamburg: FT-Verlag Rojan & Kraft, 1963.

Kompisch, Kathrin. *Täterinnen. Frauen im Nationalsozialismus*. Köln/Weimar/ Wien: Böhlau, 2008.

Koop, Volker *Himmlers letztes Aufgebot: die NS-Organisation "Werwolf"*. Cologne: Böhlau, 2008.

Korenblat, Steven D. "A School for the Republic? Cosmopolitans and Their Enemies at the Deutsche Hochschule Für Politik, 1920–1933." *Central European History* 39, No. 3 (Sep., 2006): 394–430.

Koshar, Rudy. *Social Life, Local Politics, and Nazism. Marburg, 1880–1935*. Chapel Hill/London: The University of North Carolina Press, 1986.

Krausse, Joachim. "Volksempfänger: Zur Kulturgeschichte der Monopolware." *Kunst und Medien* (1984): 81–112.

Krabiel, Klaus-Dieter. "Brechts Vortrag 'Der Rundfunk als Kommunikationsapparat'-Kontext, Anlaß, Wirkung: Zur Frage des Utopischen bei Brecht." *Wirkendes Wort: Deutsche Sprache und Literatur in Forschung und Lehre* 43, No. 2 (August 1993): 234–48.

Kropat, Wolf-Arno, ed. *Hessen in der Stunde Null 1945/1947: Politik, Wirtschaft und Bildungswesen in Dokumenten* (Wiesbaden: Selbstverlag der Historischen Kommission fur Nassau, 1979).

Kundrus, Birthe. "Nur die halbe Geschichte. Frauen im Umfeld der Wehrmacht zwischen 1939 und 1945. " In *Die Wehrmacht. Mythos und Realität,* edited by Rolf-Dieter Müller and Hans-Erich Volkmann, 719–738. Munich: Oldenbourg, 1999.

Lacey, Kate. *Listening Publics: The Politics and Experience of Listening in the Media Age*. Cambridge, Oxford, Boston: Polity Press, 2013.

Lacey, Kate. *Feminine Frequencies: Gender, German Radio, and the Public Sphere, 1923-1945*. Social History, Popular Culture, and Politics in Germany. Ann Arbor: University of Michigan Press, 1996.

Lämmer, Manfred, ed. *175 Jahre Hasenheide. Stationen der deutschen Turnbewegung: Akademiegespräch an der Führungs-und Verwaltungsakademie Berlin des Deutschen Sportbundes 13.-15. Juni 1986*. Cologne: Academia Verlag Richarz, 1988.

Latour, Conrad F. and Thilo Vogelsang. *Okkupation und Wiederaufbau. Die Tätigkeit der Militärregierung in der amerikanischen Besatzungszone Deutschlands 1944–1947*. Stuttgart 1973.

Lauster, Martina. "Walter Benjamin's Myth of the 'Flâneur.' " *The Modern Language Review* 102, no. 1 (Jan., 2007): 139–156.

Leonhard, Joachim-Felix, Ludwig, Hans-Werner and Schwarzepp, Dietrich (eds) *Medienwissenschaft. Ein Handbuch zur Entwicklung der Medien und Kommunikationsformen* Vol. 2 (Berlin and New York: Walter de Gruyter, 2001).

Lerg, Winfried B. *Rundfunkpolitik in der Weimarer Republik* Rundfunk in Deutschland 1, ed. by Hans Bausch. Munich: Deutscher Taschenbuch Verlag, 1980.

Lerg, Winfried B. *Die Entstehung des Rundfunks in Deutschland: Herkunft und Entwicklung eines publizistischen Mittels*. Frankfurt/a. M.: Josef Knecht, 1970.

Lewis, Tom. *Empire of the Air: The Men Who Made Radio*. New York: Harper Collins, 1991.

Lindquist, Rick, WW1ME. "The CIA, the Cold War, and Amateur Radio." *QST* 102, No. 2 (February 2018): 64–68.

Lösche, Peter and Franz Walter. "Zur Organisationskultur der sozialdemokratischen Arbeiterbewegung in der Weimarer Republik. Niedergang der Klassenkultur oder solidargemeinschaftlicher Höhepunkt?" *Geschichte und Gesellschaft* 15, No. 4, Innenpolitische Probleme der Weimarer Republik (1989): 511–536.

Loewy, Ernst. "Freier Äther-freies Wort? Die Rundfunkarbeit deutscher Autoren im Exil 1933-1945." *Exilforschung: Ein internationales Jahrbuch*, 2 (1984): 238–256.

Longerich, Peter. *Joseph Goebbels: Biographie*. Berlin: Siedler, 2010.

Lovell, Stephen. *Russia in the Microphone Age: A History of Soviet Radio, 1919–1970*. Oxford: Oxford University Press, 2015.

Lüdtke, Alf, Inge Marßolek and Adelheid von Saldern. *Amerikanisierung: Traum und Alptraum im Deutschland des 20. Jahrhunderts*. Transatlantische Historische Studien 6. Stuttgart: Franz Steiner Verlag, 1996.

Mai, Gunter: *Der Alliierte Kontrollrat in Deutschland 1945–1948. Alliierte Einheit - deutsche Teilung?* (Munich: 1995).

Malzahn, Manfred. *Germany, 1945–1949 A Sourcebook*. London and New York: Routledge, 1991.

Marinaro, Michael W. WN1M. "Polar Exploration." *QST* 98, No. 6 (June 2014): 63–65.

Marßolek, Inge. " 'Join in, Go Ahead and Don't Remain Silent [...]': The National Socialist Past and Reconstruction in Postwar German Broadcasting." *New German Critique* No. 95, Special Issue for David Bathrick (Spring - Summer, 2005): 122–138.

Marßolek, Inge. "Radio in Deutschland 1923–1960. Zur Sozialgeschichte eines Mediums." *Geschichte und Gesellschaft* 27, No. 2, *Kommunikationsgeschichte* (Apr. - Jun., 2001): 207–239.

Marßloek, Inge and Adelheid von Saldern, eds. *Radiozeiten: Herrschaft, Alltag, Gesellschaft (1924 – 1960*. Veröffentlichungen des Deutschen Rundfunkarchivs 25. Potsdam: Deutsches Rundfunkarchiv, 1999.

Marßolek, Inge, Adelheid von Saldern, and Daniela Münkel. *Zuhören und Gehörtwerden: zwischen Lenkung und Ablenkung*. Edition diskord, 1998.

Martin, Carly. "Building the BioMaker Space." *The Flat Hat*, October 12, 2016. Accessed May 4, 2018. (http://flathatnews.com/2016/10/12/building-the-biomakerspace/.

Mason, Tim. *Social Policy in the Third Reich: The Working Class and the 'National Community'*. Providence, Oxford: Berg, 1993.

Maubach, Franka. *Die Stellung halten: Kriegserfahrungen und Lebensgeschichten von Wehrmachthelferinnen*. Göttingen:Vandenhoeck & Ruprecht, 2009.

McLuhan, Marshall. *Understanding Media: the Extensions of Man*. New York: McGraw Hill, 1964.

Merritt, Richard L. *Democracy Imposed: U.S. Occupation Policy and the German Public, 1945–1949*. London: Yale UP, 1995.

Meyer-Hesemann, Wolfgang. "Legalität und Revolution.—Zur juristischen Verklärung der nationalsozialistischen Machtergreifung als 'Legale Revolution.' " *Recht und Unrecht im Nationalsozialismus*, edited by Peter Salje. Münster: Wissenschaftliche Verlagsgesellschaft Regensburg & Biermann, 1984, 110–137.

Michalka, Wolfgang, ed. *Der Erste Weltkrieg. Wirkung, Wahrnehmung, Analyse.* Munich: Seehamer Verlag, 1994.

Miesbeck, Peter. *Bürgertum und Nationalsozialismus in Rosenheim. Studien zur politischen Tradition.* Quellen und Darstellungen zur Geschichte der Stadt und des Landkreises Rosenheim 13. Rosenheim, Historischer Verein 1994.

Militärgeschichtliches Forschungsamt. *Deutsche Militärgeschichte 1648–1939.* Munich: Bernard & Graefe 1983.

Montez, Susan. *Radio Free Queens.* New York: George Braziller, 1994.

Müller, Rolf-Dieter, Hans Umbreit and Derry Cook-Radmore. *Das Deutsche Reich und der Zweite Weltkrieg* 5/II: *Organization und Mobilisierung des deutschen Machtbereiches. Kriegsverwaltung, Wirtschaft und personelle Ressourcen, 1942–1944/45.* Stuttgart: Deutsche Verlagsanstalt, 1999.

Nakata, Jun. *Der Grenz- und Landesschutz in der Weimarer Republik, 1918 bis 1933. Die Geheime Aufrüstung und die deutsche Gesellschaft.* Freiburg i. Br.: Rombach, 2002.

Nesper, Dr. Eugen. *Der Radio-Amateur. "Broadcasting".* Berlin: Julius Springer, 1923.

Neufeld, Thomas. "Geschichte des Amateurfunk und die rechtlichen Grundlagen" (2017). Accessed May 28, 2018. https://www.darc.de/der-club/distrikte/q/ortsverbaende/11/chronik/geschichte-des-amateurfunks/.

Nichols, Eric P. *Propagation and Radio Science.* Newington, CT: American Radio Relay League, 2015.

Niehaus, Werner. *Die Nachrichtentruppe, 1914 bis heute. Entstehung und Einsatz.* Stuttgart: Motorbuch Verlag, 1980.

Niehaus, Werner. *Die Radarschlacht 1939–1945. Die Geschichte des Hochfrequenzkrieges.* Stuttgart, Motorbuchverlag, 1977.

Nye, David E. *Technology Matters: Questions to Live With.* Cambridge, MA: M.I.T. University Press, 2007.

Okrassa, Nina. *Peter Raabe, Dirigent, Musikschriftsteller, und Präsident der Reichsmusikkammer (1872–1945).* Cologne: Böhlau, 2004.

Oral History interview with Steve Jobs. 1995, Accessed July 24, 2017. http://americanhistory.si.edu/comphist/sj1.html.

Orlow, Dietrich. *A History of Modern Germany, 1871 to the Present* (7th. ed.). Routledge, 2016.

Overmans, Rüdiger. *Deutsche militärische Verluste im Zweiten Weltkrieg*. Beiträge zur Militärgeschichte 46. Oldenbourg: DeGruyter, 2004.

Overy, Richard. *The Bombers and the Bombed: Allied Air War Over Europe 1940–1945*. London, New York: Penguin, 2014.

Paehler, Katrin. *The Third Reich's Intelligence Services: The Career of Walter Schellenberg*. Cambridge: Cambridge University Press, 2017.

Patel, Kiran Klaus. *Soldiers of Labor: Labor Service in Nazi Germany, and New Deal America, 1933–1945*. Publications of the German Historical institute. Cambridge: Cambridge University Press, 2005.

Pfeiffer, Lorenz „… unser Verein ist judenfrei"—Die Rolle der deutschen Turn-und Sportbewegung in dem politischen und gesellschaftlichen Wandlungsprozess nach dem 30. Januar 1933." *Historische Sozialforschung* 32, No. 1 (2007): 92–109.

Peiffer, Lorenz. *Sport im Nationalsozialismus: zum aktuellen Stand der sporthistorischen Forschung; eine kommentierte Bibliografie*. Göttingen: Verlag Die Werkstatt, 2004.

Pfeiffer, Lorenz. "Die Ausgrenzung und Zerschlagung des jüdischen Sportes in Deutschland am Beispiel Hannover." in *K Problematice Jinakosti. Sport, olympismus, nacionalismus a sjednoceni Evropy*, Edited by Anna Hogenová. Prag: 2001.145–154.

Peiffer, Lorenz. "Auf zur Gefolgschaft und zur Tat!" Deutsche Turnerschaft und Nationalsozialismus—zwischen Selbstgleichschaltung und Selbstbehauptung?. " *IWK Internationale wissenschaftliche Korrespondenz zur Geschichte der Deutschen Arbeiterbewegung* 35, No. 4 (1999): 530–548.

Pfeiffer, Lorenz. "Von der Gleichschaltung zur Auflösung: Das Ende der Deutschen Turnerschaft 1936." in *175 Jahre Hasenheide. Stationen der deutschen Turnbewegung: Akademiegespräch an der Führungs-und Verwaltungsakademie Berlin des Deutschen Sportbundes 13.-15. Juni 1986* edited by Manfred Lämmer, 35–45.Köln: Academia Verlag Richarz 1988).

Pfeiffer, Lorenz, ed. *Der deutsche Sport in der Zeit ds Nationalsozialismus*. Cologne: Carl-Diem-Institut, 1980.

Petzold, Hartmut. "Zur Entstehung der elektronischen Technologie in Deutschland und den USA. Der Beginn der Massenproduktion von Elektronenröhren 1912–1918." *Geschichte und Gesellschaft* 13, No. 3, *Sozialgeschichte der Technik* (1987): 340–367.

Pilgert, Henry P. *Press, Radio and Film in West Germany, 1945–1953*. Historical Division, Office of the Executive Secretary, Office of the U.S. High Commissioner for Germany, 1953.

Pinch, Trevor and Karin Bijsterveld (eds.). *Oxford Handbook of Sound Studies*. Oxford: Oxford University Press, 2011.

Plischke, Elmer. "Denazification Law and Procedure." *The American Journal of International Law* 41, No. 4 (October 1947): 807–827.

Pohle, Heinz. *Der Rundfunk als Instrument der Politik. Zur Geschichte des deutschen Rundfunks von 1923/38* (Hamburg: Hans Bredow Institut, 1955).

Poppelreuter, Tanja. *Das neue Bauen für den neuen Menschen: zur Wandlung und Wirkung des Menschenbildes in der Architektur der 1920er Jahre in Deutschland.* Hildesheim/Zürich/New York: Olms: 2007.

Porter, Cole. "Anything Goes." 1934. Lyrics from: http://www.lyricsdepot.com/cole-porter/anything-goes.html Accessed July 21, 2017.

Potter, Pamela and Celia Applegate (eds.). *Music and German National Identity.* Chicago: University of Chicago Press, 2002.

Pöttker, Horst. "Journalismus unter Goebbels: Über die Kraft der Radioreportage." *LiLi: Zeitschrift für Literaturwissenschaft und Linguistik* 28 No. 111 (September 1998): 57–76.

Praun, Albert. *Soldat in der Telegraphen- und Nachrichtentruppe.* Würzburg, Selbstverlag, 1965.

Rabinowitch, Alexander. *The Bolsheviks in Power: The First Year of Soviet Rule in Petrograd.* Bloomington: Indiana University Press, 2007.

Raboy, Marc. *Marconi: The Man Who Networked the World.* Oxford and New York: Oxford University Press, 2016.

Radio-Bauer. *Katalog Nr. 6 April 1927.* Berlin: Buchdruckerei Max Lichtwitz, 1927.

Radio-Diehr. *Nachtrag zum Illustrierten Radio-Katalog 1929.* Berlin: Walter Hilger Radio-Verlag, 1929.

Radio-Diehr. *Neuester Illustrierter Radio-Katalog 1928–1929.* Berlin: Kreditgenossenschaft des Reichsverbandes Deutscher Funkhändler, 1928.

Rehding, Alexander, "Magic Boxes and Volksempfänger: Music on the Radio in Weimar Germany" In: Nikolaus Bacht (ed.) *Music, Theatre and Politics in Germany: 1848 to the Third Reich* (Farnham: Ashgate, 2006) pp. 225–272.

Reichardt, Sven. "Die verdorbenen Burschen wollen von sich reden machen und finden auch noch ein Echo.", in: *Frankfurter Allgemeine Zeitung* 40 (September 6, 2011): N3.

Reichardt, Sven. *Faschistische Kampfbünde. Gewalt und Gemeinschaft im italienischen Squadrismus und in der deutschen SA.* Industrielle Welt/Schriftenreihe des Arbeitskreises für moderne Sozialgeschichte. Vienna, Cologne, Weimar: Böhlau Verlag, 2009.

Reichsverband Deutscher Rundfunkteilnehmer (R.D.R.) E.V. Berlin (i.A. Dr. Heinz Weiss). *Rundfunk im Aufbruch. Handbuch des deutschen Rundfunks 1934 mit Funkkalender.* (Mit einem Gleitwort von Reichsminister Pg. Dr. Goebbels). Lahr/Baden: Verlag Moritz Schauenburg K.G., 1933.

Reiz, Edgar. "Heimat-eine Deutsche Chronik." Edgar Reitz Film (ERF), Sender Freies Berlin (SFB) and Westdeutscher Rundfunk (WDR), 1984.

Rempel, Gerhard. *Hitler's Children: The Hitler Youth and the SS.* Chapel Hill, NC: University of North Carolina Press, 1989.

Renneberg, Monika and Mark Walker, eds. *Science, Technology and National Socialism*. Cambridge: Cambridge University Press, 1994.

Reuband, Karl-Heinz. "Schwarzhöreren im Dritten Reich. Verbreitung, Erscheinungsformen und Kommunikationsmuster beim Umgang mit verbotenen Sendern." *Archiv für Sozialgeschichte* 41 (2001):245–270.

Reuth, Ralf Georg. *Goebbels*. Munich/Zürich: Piper, 1990.

Rhodes, Richard. *Hedy's Folly: The Life and Breakthrough Inventions of Hedy Lamarr, The Most Beautiful Woman in the World*. New York: Doubleday, 2011.

Richards, Elaine G4LFM. *Centenary: 100 Years Working for Amateur Radio*. Bedford: Radio Society of Great Britain, 2013.

Richter, Hans Peter. ""Die Entwicklung der Rundfunk-Teilnehmerzahlen für das Deutsche Reich und die Bundesrepublik nach den Angaben der Post für den Stichtag 1.April." *Rundfunk und Fernsehen* 3 (1955): 398.

Riedel, Heide. *Lieber Rundfunk…75 Jahre Hörergeschichte(n)*. Berlin: Vistas, 1999.

Rieger, Bernhard. *Technology and the Culture of Modernity in Britain and Germany, 1890–1945* (New Studies in European History). Cambridge: Cambridge University Press, 2009.

Robinson, J. Bradford. "The jazz essays of Theodor Adorno: some thoughts on jazz reception in Weimar Germany." *Popular Music* 13, No. 1 (January 1994): 1–25.

Rohkrämer, Thomas. "Antimodernism, Reactionary Modernism and National Socialism. Technocratic Tendencies in Germany, 1890–1945." *Contemporary European History* 8, No. 1 (March 1999): 29–50.

Rohwer, Jürgen. "Der Einfluß der alliierten Funkaufklärung auf den Verlauf des Zweiten Weltkrieges." *Vierteljahrshefte für Zeitgeschichte* 27. No. 3 (July 1979): 325–369.

Rosenhaft, Eve. "Lesewut, Kinosucht, Radiotismus. Zur(geschlechter-) politischen Relevanz neuer Massenmedien in der 1920er Jahren." in *Amerikanisierung: Traum und Alptraum im Deutschland des 20. Jahrhunderts* Transatlantische Historische Studien 6, edited by Alf Lüdtke, Inge Marßolek and Adelheid von Saldern, 119–143. Stuttgart: Franz Steiner Verlag, 1996.

Ross, Corey. *Media and the Making of Modern Germany: Mass Communications, Society, and Politics from the Empire to the Third Reich*. Oxford: Oxford University Press, 2010.

Ross, Corey. "Mass Culture and Divided Audiences: Cinema and Social Change in Inter-War Germany." *Past & Present* 193 (November 2006): 157–195.

Russell, Andrew L. and Lee Vinsel. "After Innovation, Turn to Maintenance." *Technology and Culture* 59, No. 1 (January 2018):1–25

Saldern, Adelheid von. "Volk and Heimat Culture in Radio Broadcasting during the Period of Transition from Weimar to Nazi Germany." *The Journal of Modern History* 76, No. 2 (June 2004): 312–346.

Saldern, Adelheid von. "Cultural Conflicts, Popular Mass Culture, and the Question of Nazi Success: The Eilenriede Motorcycle Races, 1924-39." *German Studies Review* 15, No. 2 (May, 1992): 317–338.

Salje, Peter. *Recht und Unrecht im Nationalsozialismus*. Münster: Wissenschaftliche Verlagsgesellschaft Regensburg & Biermann, 1984.

Salomon, Ernst von. *Die Geächteten*. Reinbek bei Hamburg: Rowohlt Verlag, 1930, 1962.

Salomon, Ernst von. *Nahe Geschichte. Ein Überblick*. Berlin: Rowohlt, 1936.

Sarkowicz, Hans. "„Nur nicht langweilig werden …": Das Radio im Dienst der nationalsozialistischen Propaganda". In *Medien im Nationalsozialismus*, edited by Bernd Heidenreich and Sönke Neitzkel, 205–235. Paderborn: Ferdinand Schöningh, Wilhelm Fink, 2010.

Sauermann, Henry and Chiara Franzoni. "Crowd Science User Contribution Patterns and Their Implications." *Proceedings of the National Academy of Sciences* (PNAS) 112, no.3 (January 20, 2015): 679–684.

Schanze, Helmut. "Rundfunk, Medium und Masse. Voraussetzungen und Folge der Medialisierung nach dem 1. Weltkrieg." In *Die Idee des Radios von den Anfängen in Europa und den USA bis 1933: Jahrbuch Medien und Geschichte 2004*, edited by Edgar Lersch and Helmut Schanze, 11–27. Konstanz: UVK Verlagsgesellschaft, 2004.

Schifferdecker, Heinz, DL7AC, Horst Eligering DL9MH and Manfred May DJ1KF. "Chronik des Amateurfunks in Deutschland und im Bereich Köln-Aachen". ("von Heinz Schifferdecker, DL7AC und Horst Ellgering, DL9MH, © 1991; für das Internet aufbereitet von Manfred May, DJ1KF (neu bearbeitet von DO1KXL)"). Accessed 25. June 2015. http://www.datv-agaf.de/links/chronik.html#a7.

Schildt, Axel. *Moderne Zeiten. Freizeit, Massenmedien und "Zeitgeist" in der Bundesrepublik der 50er Jahre*. Hamburger Beiträge zur Sozial- und Zeitgeschichte 31 (Hamburg: Hans Christian Verlag, 1995).

Schiller-Lerg, Sabine. "Walter Benjamin, Radio Journalist: Theory and Practice of Weimar Radio." *Journal of Communication Inquiry* 13 No. 1 (Winter 1989): 43–50.

Schissler, Hannah. *The Miracle Years: A Cultural History of West Germany, 1949-1968*. Princeton: Princeton University Press, 2001.

Schmidt, Uta C. "Der Volksempfänger. Tabernakel moderner Massenkultur." in: *Radiozeiten. Herrschaft, Alltag, Gesellschaft* (1924–1960) edited by Inge Marßolek and Adelheid von Saldern, 136–159. Potsdam: Verlag. f. Berlin-Brandenburg, 1999.

Schmiechen-Ackermann, Detlef. "Der 'Blockwart'. Die unteren Parteifunktionäre im nationalsozialistischen Terror- und Überwachungsapparat." *Vierteljahrshefte für Zeitgeschichte* 48, No. 4 (October 2000): 575–602.

Schröder, Bernhard. "Memoiren." Accessed June 25, 2018. http://schroeder-eversburg.de/memoiren.html.

Schroder, Gerald. *NS-Pharmazie: Gleichschaltung des deutschen Apothekenwesens im Dritten Reich*. Stuttgart:Wissenschaftliche Verlagsgesellschaft, 1988.

Schulin, Ernst. "Die Urkatastrophe des zwanzigsten Jahrhunderts." In *Der Erste Weltkrieg. Wirkung, Wahrnehmung, Analyse*, edited by Wolfgang Michalka, 3–27. Munich: Seehamer Verlag, 1994.

Schulze, Hagen. *Freikorps und Republik 1918–1920*. Militärgeschichtliche Studien 8. Boppard a/R., Harald Boldt Verlag, 1969.

Schütz, Erhard. "Das 'Dritte Reich' als Mediendiktatur: Medienpolitik und Modernisierung in Deutschland 1933 bis 1945." *Monatshefte* 87, No. 2 (Summer 1995): 129–150.

Seidler, Franz W. *Blitzmädchen. Helferinnen der Wehrmacht*. Augsburg: Bechtermünz, 2003.

Senne, Christian. *Rahmen und Organisationsbedingungen für Funkamateure in der SBZ und DDR (1945–1990). Zwischen Selbstzweck und gesellschaftlichem Auftrag*. Studien zur Zeitgeschichte 70. Hamburg: Verlag Dr. Kovač, 2008.

Shearer, Stephen Michael. *Beautiful: The Life of Hedy Lamarr*. New York: St. Martin's Press, 2010.

Siemens, Daniel. *Stormtroopers: A New History of Hitler's Brownshirts*. New Haven and London: Yale University Press, 2017.

Siemens, Daniel. *Horst Wessel: Tod und Verklaerung eines Nationalsozialisten*. Munich: Siedler Verlag, 2009a.

Siemens, Daniel. "Politik ist die Fortsetzung des Krieges mit anderen Mittel: Deutungskämpfe um den Ersten Weltkrieg in der politischen Kultur der Zwischenkriegszeit." *Geschichte für heute* 2, No. 1 (2009b): 40–58.

Siewert, H.-Jörg "Zur Thematisierung des Vereinswesens in der deutschen Soziologie", in: *Historische Zeitschrift* Beihefte New Series, Vol. 9, *Vereinswesen und bürgerliche Gesellschaft in Deutschland* (1984), pp. 151–180.

Son, Gillaume de. *Zeppelin! Germany and the Airship, 1900–1939*. Baltimore, Maryland: Johns Hopkins University Press, 2005.

Spahn, Raymond J. and Leslie I. Poste. "The Germans Hail America: Some Aspects of Communication Media in Occupied Germany." *Modern Language Journal* 33, No. 6 (October 1949): 417–26.

Squier, Susan Merrill (ed.). *Communities of the Air: Introducing the Radio World*. Durham and London: Duke University Press, 2003.

Standage, Tom. *The Victorian Internet: The Remarkable Story of the Telegraph and the Nineteenth Century's On-line Pioneers*. New York: Bloomsbury, 2014.

Statistisches Bundesamt – Wiesbaden *Die deutschen Vertreibungsverluste. Bevölkerungsbilanzen für die deutschen Vertreibungsgebiete 1939/50*. Stuttgart: Verlag W. Kohlhammer, 1958.

Steiger, Karsten. *Kooperation, Konfrontation, Untergang. Das Weimarer Tarif- und Schlichtungswesen während der Weltwirtschaftskrise und seine Vorbedingungen*. Beiträge zur Unternehmensgeschichte 5. Stuttgart: Franz Steiner Verlag, 1998.

Stein, George H. *The Waffen-SS: Hitler's Elite Guard at War, 1939–1945*. Ithaca and London: Cornell University Press, 1966.

Steiner, Kilian J. L. *Ortsempfänger, Volksfernseher und Optaphon. Entwicklung der deutschen Radio- und Fernsehindustrie und das Unternehmen Loewe 1923–1962*. Essen: Klartext Verlag, 2005.

Steinhöfer, Dieter. *Hans von Tschammer und Osten. Reichssportführer im Dritten Reich*. Turn- und Sportführer im Dritten Reich 2. Berlin, Munich, Frankfurt: Bartels & Wernitz, 1973.

Stemfeld, Joshua. "Jazz Echoes: The Cultural and Sociopolitical Reception of Jazz in Weimar and Nazi Berlin, 1925–1939." Ph.D. Diss. University of California, Los Angeles, 2007.

Stephan, Gerd-Rüdiger et al. eds. *Die Parteien und Organisationen der DDR. Ein Handbuch*. Berlin: Dietz, 2002.

Stephenson, Jill. "Women's Labor Service in Nazi Germany." *Central European History* 15, No. 3 (September 1982): 241–265.

Stern, Fritz. *Einstein's German World*. Princeton: Princeton University Press, second edition 2016.

Stern, Fritz. *The Politics of Cultural Despair: A Study in the Rise of the Germanic Ideology*. Garden City, NY: Anchor, 1965.

Stiller, E. "Die Gleichschaltung eines Arbeitersportverbandes—Der Freie Segler-Verband in Zeichen der NS-Machtergreifung." *Beiträge zur Geschichte der Arbeiterbewegung* 43 (2001): 69–87.

Sturm, Peter et al (eds.). *100 Jahre Ruderklub am Wannsee. Festschrift, 13. September 1906–2006*. Berlin: Ruderklub am Wannsee, 2006.

Swaine, Michael and Paul Freiberger, *Fire in the Valley: the Birth and Death of the Personal Computer*. Raleigh, NC: Pragmatic Bookshelf, Third Edition 2014.

Syon, Gillaume de. *Zeppelin! Germany and the Airship, 1900–1939*. Baltimore, Maryland: Johns Hopkins University Press, 2005.

Szepansky, Gerda. *"Blitzmädel", "Heldenmutter", "Kriegerwitwe". Frauenleben im Zweiten Weltkrieg*. Frankfurt/M.: Fischer, 1997.

Taylor, Frederick. *The Downfall of Money: Germany's Hyperinflation and the Destruction of the Middle Class*. London: Bloomsbury, 2013.

Tessin, Georg. *Verbände und Truppen der deutschen Wehrmacht und Waffen-SS 1939–1945*. Osnabrück: Biblio Verlag, 1967.

Tomlinson, John D. *The International Control of Radio Communications* Ann Arbor: J.W. Edwards, 1945.

Trenkle, Fritz. *Die deutschen Funknachrichtenanlagen bis 1945. Heer - Marine – Luftwaffe*. Vol. 1: *Die ersten 40 Jahre*. Heidelberg: Telefunken Systemtechnik GmbH, 1989.

Trommler, Frank. "Between Normality and Resistance: Catastrophic Gradualism in Nazi Germany." *The Journal of Modern History* 64, Supplement: Resistance Against the Third Reich (December 1992): S82-S101.

(Tworek), Heidi J.S. Evans. " 'The Path to Freedom'? Transocean and Wireless Telegraphy, 1914–1922." *Historical Social Research* 35, No. 1 (2010): 209–236.

United States, Signal Corps, U.S. Army and National Bureau of Standards. *The Principles Underlying Radio Communications*. Second Edition. Radio Communications Pamphlet 40. Washington: Government Printing Office, 1922.

Van Der Bijl, H.J. *The Thermionic Vacuum Tube and its Applications*. New York: McGraw Hill, 1920.

Vieweg, Klaus. "Gleichschaltung und Führerprinzip—Zum rechtlichen Instrumentarium der Organisation des Sports im Dritten Reich." In *Recht und Unrecht im Nationalsozialismus*, edited by Peter Salje, 244–271. Münster: Wissenschaftliche Verlagsgesellschaft Regensburg & Biermann, 1984.

Vogelsang, Thilo. "Der Chef des Ausbildungswesens (Chef AW)." In: Institut für Zeitgeschichte, *Gutachten des Instituts für Zeitgeschichte*, Vol.2. Stuttgart: Deutsche Verlagsanstalt, 1966.

Vorländer, Herwart. "NS-Volkswohlfahrt und Winterhilfswerk des Deutschen Volkes." *Vierteljahrshefte für Zeitgeschichte* 34, No. 3 (1986): 341–380.

Wagenführ, Kurt ed. *Jahrbuch Welt-Rundfunk 1937/1938*. Heidelberg-Berlin: Kurt Vowinckel Verlag, 1938.

Wagner, Johannes Volker. *Hakenkreuz über Bochum. Machtergreifung und nationalsozialistischer Alltag in einer Revierstadt*. Bochum: Brockmeyer, 1983.

Wagner, Ringo, "Wehrsport zwischen Tradition und Neuanfang. Die vormilitärische Erziehung in der DDR und ihre besondere Ausprägung in der Gesellschaft für Sport und Technik (GST)." *Zeitschrift des Forschungsverbundes SED-Staat* 22 (2007): 3–17.

Wagner, Ringo. *Der Vergessene Sportverband der DDR. Die Gesellschaft für Sport und Technik in sporthistorischer Perspekektive*. Aachen: Meyer & Meyer Sport, 2006.

Weart, Spencer R. *The Rise of Nuclear Fear*. Cambridge: Harvard University Press, 1988, 2012.

Wegner, Bernd. *Hitlers Politische Soldaten: Die Waffen-SS 1933–1945. Studien zu Leitbild, Struktur und Funktion einer Nationalsozialistischen Elite*. Paderborn: Ferdinand Schöningh, 1982.

Wehler, Hans Ulrich. *Deutsche Gesellschaftsgeschichte, 1700–1990.* Munich: C.H. Beck, 2008.

Weidenfelder, Gerhard, *VDA. Verein für das Deutschtum im Ausland. Allgemeiner Deutscher Schulverein (1881—1918). Ein Beitrag zur Geschichte des deutschen Nationalismus und Imperialismus im Kaiserreich.* Europäische Hochschulschriften III, Vol. 66. Bern-Frankfurt/M.: Herbert Lang Verlag, 1976.

Wer ist Wer? vol. XI. Berlin, Ariani Verlag, 1951.

Werner, Andreas. "SA: 'Wehrverband', 'Parteitruppe', oder 'Revolutionsarmee'? Studien zur Geschichte der SA und der NSDAP 1920–1933", PhD diss. Friedrich-Alexander-Universität zu Erlangen-Nürnberg, 1964.

Westermanns Deutscher Reichsatlas. Braunschweig, Berlin, Hamburg: Georg Westermann ca. 1933.

Widdig, Bernd. *Culture and Inflation in Weimar.* Berkeley: University of California Press, 2001.

Wiegand, Rolf, DE0065, D4cxf. *DASD-Fibel. Wie werde ich Kurzwellenamateur?* Herausgegeben in Einvernehmen mit dem Rundfunkamt der Reichsjugendführung vom Deutschen Amateur-Sende- und Empfangsdienst eV. Berlin: Weidmannsche Buchhandlung, 1936.

Wildhagen, Karl Heinz, ed. *Erich Fellgiebel. Meister operativer Nachrichtenverbindungen. Ein Beitrag zur Geschichte der Nachrichtentruppe.* Emden/Hannover: Selbstverlag, 1970.

Wildt, Michael. *Volksgemeinschaft als Selbstermächtigung. Gewalt gegen Juden in der deutschen Provinz 1919 bis 1933.* Hamburg, Hamburger Edition, 2003a.

Wildt, Michael. *Generation des Unbedingten Das Führungskorps des Reichssicherheitshauptamtes.* Hamburg: Hamburger Edition, 4th. edition 2003b.

Wipplinger, Jonathan O. *The Jazz Republic: Music, Race, and American Culture in Weimar Germany.* Ann Arbor: University of Michigan Press, 2017.

Wireless Association of America. *Wireless Book of the Wireless Association of America.* Second Annual. New York: Modern Electrics Publication, 1910.

Zimmermann, Harro. "Radio-Modernisierung der Sinne: Forschungsperspektiven zwischen Literatur- und Kulturwissenschaft (am Beispiel der zwanziger Jahre)." *LiLi: Zeitschrift für Literaturwissenschaft und Linguistik* 28, No. 111 (September 1998): 9–27.

Index[1]

[1] Note: Page numbers followed by 'n' refer to notes.

B. B. Campbell, *The Radio Hobby, Private Associations, and the Challenge of Modernity in Germany*, Palgrave Studies in the History of Science and Technology, https://doi.org/10.1007/978-3-030-26534-2

J

K

N

Q

R